APPLIED PHYSICS

(For the B.Tech Students of Amity Universities, Noida, Lucknow and Jaipur)

AF633274

By

Devraj Singh

M.Sc. (Physics), A.N.S.I. (Sugar Tech.), D. Phil. (Materials Science)
Lecturer in Applied Physics, Department of Applied Sciences
Amity School of Engineering and Technology
Bijwasan, New Delhi

RAJESH B. GAUTAM

M.Sc. (Physics), M.Phil. (Nano Tech.)
Senior Lecturer in Physics
World College of Technology and Management
Farukh Nagar, Gurgaon

ANOOP KUMAR SHUKLA

M.Sc. (Physics), D.Phil. (Materials Science)
Lecturer in Physics
Amity Institute of Applied Sciences
Amity School of Engineering and Technology
Amity University, Noida

UNIVERSITY SCIENCE PRESS

(An Imprint of Laxmi Publications Pvt. Ltd.)

BANGALORE • CHENNAI • COCHIN • GUWAHATI • HYDERABAD
JALANDHAR • KOLKATA • LUCKNOW • MUMBAI • RANCHI
NEW DELHI • BOSTON, USA

Published by :

UNIVERSITY SCIENCE PRESS

(*An Imprint of Laxmi Publications Pvt. Ltd.*)
113, Golden House, Daryaganj,
New Delhi-110002

Phone : 011-43 53 25 00
Fax : 011-43 53 25 28

www.laxmipublications.com
info@laxmipublications.com

First Edition : 2010
Reprint : 2013

OFFICES

✆ **Bangalore**	080-26 75 69 30	✆ **Chennai**	044-24 34 47 26
✆ **Cochin**	0484-237 70 04, 405 13 03	✆ **Guwahati**	0361-251 36 69, 251 38 81
✆ **Hyderabad**	040-24 65 23 33	✆ **Jalandhar**	0181-222 12 72
✆ **Kolkata**	033-22 27 43 84	✆ **Lucknow**	0522-220 99 16
✆ **Mumbai**	022-24 91 54 15, 24 92 78 69	✆ **Ranchi**	0651-220 44 64

UAP-9515-295-APPLIED PHYSICS-SIN **C—6343/013/02**

Typeset at : Goswami Associates, Delhi *Printed at* : Akashdeep Enterprises, Delhi.

CONTENTS

PREFACE

- The book **Applied Physics** has been designed to cater the need of Ist year B Tech students of **Amity Universities, Noida, Lucknow and Jaipur.** Students are given a clear, practical, comprehensible and thorough coverage.
- The subject matter has been explained in a simple and lucid manner.
- The book comprises of eleven chapters, which covers wider area in Physics, such as ***Physics of Vibrations, Progressive Waves, Ultrasonics, Interference of Light, Diffraction of Light, Polarisation of Light, Electromagnetism, Special Theory of Relativity, Wave Mechanics, Atomic Physics and Solid State Physics.***
- In each chapter, the topics and their subtopics are carefully arranged and discussed from elementary level. Further the topics are supported by a large number of illustrated diagrams and numerical problems. Efforts have been made to provide an in-depth and up-to-date treatment of topics.
- Exercises are given at the end of the chapters, to test to increase of effective mental equipment and to cultivate the art of clear expression.
- We sincerely thank to scientists, whose work constituted the subject matter and the authors whose books we have freely consulted during the course of manuscript preparation.
- We are thankful to following Professors for their permanent interest to make this book:

  Prof. B.P. Singh, Sr. Director; Prof. Prem Prakash, Director;

  Dr. Shashi Chawla, Mr. Sudhanshu Tripathi, Dr. R.K. Tyagi, ASET, New Delhi

  Dr. Balvinder Shukla, Prof. A.L. Verma, Prof. S.N. Shukla, Prof. Sunita Ratan, Prof. Sujata Pandey, Dr. Pramila Shukla, Dr. Shivani A. Kumar, Dr. A.K. Gupta and Dr. Rajesh Yadav; ASET, Noida

Prof. S.K. Kor, Prof. S.L. Srivastava, Prof. Vinod Prakash, Prof. V.K. Agrawal, Prof. R.R. Yadav and Dr. Ravindar Dhar, University of Allahabad, Allahabad

Prof. S.S. Islam and Prof. D.R. Bhaskar, Jamia Milia Islamia University, Delhi

Dr. D.K. Pandey, Dr. M.B. Pandey and Dr. Alok Gupta, C.S.J.M. University, Kanpur

Dr. Krishan Lal and Dr. Janardan Singh, NPL, New Delhi

Dr. Ambuj Tripathi, Scientist 'F', IUAC, New Delhi.

- The authors are highly indebted to Dr. Ashok K Chauhan, Founder President, Amity University (RBEF) and Dr. Narender Singh, Chairman, WCTM, Gurgaon for their kind help and encouragements.
- We are thankful to our colleagues, family members and the publisher for their cooperation during preparation of the text.
- Further suggestions from teachers and students for improvements would be highly welcome.

—Authors

SYLLABUS

APPLIED PHYSICS I–FIELDS AND WAVES

Course Code: BTCPH 10101

FIRST SEMESTER

MODULE I. Oscillations and Waves

Oscillations: Introduction to S.H.M. Damped Oscillations: Differential Equation and its solution, logarithmic decrement, Quality Factor, Different conditions of damping of harmonic oscillations. Forced oscillations: Amplitude and Frequency Response, Resonance, Sharpness of Resonance.

Plane Progressive Waves: Differential Equation and Solution, Superposition of Progressive Waves stationary waves.

Ultrasonics: Generation and application of ultrasonic waves.

MODULE II. Wave Nature of Light

Interference: Coherent Sources, Conditions of interference, Interference due to division of wavefront. Fresnels biprism, interference due to division of amplitude, Newton's rings, wedge shaped film, interference due to thin films.

Diffraction: Fresnel and Fraunhofer diffraction, Fraunhofer diffraction at a single slit, double slit, N slits, Transmission grating, Rayleigh criterion and Resolving power of grating.

Polarization: Birefringence, Nicol prism, Production and analysis of plane, circularly and elliptically polarized light, Half and quarter wave plates, Optical rotation, Polarimeters (Lorentz half shade and biquartz).

MODULE III. Electromagnetics

Scalar and vector fields, gradient of a scalar field, physical significance of gradient, equipotential surface, Line, surface and volume integrals, Divergence and curl of vector field–mathematical analysis and physical significance, Electric flux, Gauss' Law: Proof and Applications, Gauss divergence and Stokes theorems. Differential form of Gauss' Law, Ampere's Law, Displacement current, Faraday's Law, Maxwell's equations in free space and isotropic media (Integral form and differential form), EM wave propagation in free space, poynting vector.

APPLIED PHYSICS II-MODERN PHYSICS

Course Code: BTCPH 10201

SECOND SEMESTER

MODULE I. Special Theory of Relativity

Michelson-Morley experiment, Importance of negative result, Inertial and non-inertial frames of reference, Einstein's postulates of special theory of Relativity, space-time coordinate system, Relativistic space time transformation (Lorentz transformation equations), Length contraction and time dilation, Addition of velocities, Variation of mass with velocity and mass-energy equivalence (Einstein's mass energy relation).

MODULE II. Wave Mechanics

Wave particle duality, De-Broglie matter waves, phase and group velocity, Heisenberg uncertainty principle, wave function and its physical interpretation, Operators, expectation values. Time dependent and time independent Schrödinger wave equation for free and bound states, square well potential (rigid wall), Step potential.

MODULE III. Atomic Physics

Vector atom model, LS and j-j coupling, Zeeman effect (normal and anomalous), Paschen-Bach effect, X-ray spectra and energy level diagram, Moseleys Law, Lasers–Einstein coefficients, conditions for light amplification, population inversion, optical pumping, three level and four level lasers. He-Ne and Ruby laser, properties and applications of lasers.

MODULE IV. Solid State Physics

Sommerfeld's free electron theory of metals, Fermi energy, Introduction to periodic potential and Kronig-Penny model (Qualitative) Band Theory of Solids, Semiconductors: Intrinsic and Extrinsic semiconductors, photoconductivity and photovoltaics, Basic aspects of Superconductivity, Meissner effect.

Chapter 1

PHYSICS OF VIBRATIONS

1.1 INTRODUCTION

(a) Periodic Motion

If an object repeats its motion on a definite path after a regular interval of time, its motion is said to be periodic motion. The definite time after which it repeats its motion, is called the periodic time and it is represented by the letter T. In periodic motion, the path of the body can be straight line, circular or curve.

Examples:

(*i*) In a wall clock, the hour arm completes one round in 12 hours and the minute arm complete one round in 1 hour. Thus the motion of the arms of the clock is periodic in which the time period of hour arm is 12 hours and that of the minute arm is 1 hour.

(*ii*) The earth completes one round on its axis in 24 hours and one round around the sun in nearly 365 days. Thus the rotational motion of earth on its own axis and the circular motion around the sun are also the periodic motions in which the time period of rotational motion about its own axis is 24 hours and that of circular motion around the sun is nearly 365 days.

(*iii*) The orbital motion of moon around the sun is also a periodic motion, the time period of which is $27\frac{1}{3}$ days.

Similarly, the oscillations of pendulum of a clock, motion of swing, vibration of strings in the musical instruments, beating of heart etc., are also the periodic motion.

At any instant t the displacement x of the particle moving periodically is a function of time t.

It can be expressed as follows

$$x = f(t)$$

where f is the function of time t.

If time period of motion is T, then obviously

$$f(t) = f(t + T)$$

It is clear that if the function f is the sine or cosine function (*i.e.*, if $x = \sin \omega t$ or $x = \cos \omega t$), the value of the displacement x changes between –1 and +1. Such function is called *a simple harmonic function*. Its time period is $T = \frac{2\pi}{\omega}$, since

$$x = \sin \omega t = \sin \omega\left(t + \frac{2\pi}{\omega}\right) \quad \text{and} \quad x = \cos \omega t = \cos \omega\left(t + \frac{2\pi}{\omega}\right)$$

On the other hand, the *tan* and *cot* functions are only the periodic functions, and not the simple harmonic functions, the time period of which is $T = \frac{\pi}{\omega}$, since

$$\tan \omega t = \tan \omega\left(t + \frac{\pi}{\omega}\right) \quad \text{and} \quad \cot \omega t = \cot \omega\left(t + \frac{\pi}{\omega}\right)$$

(*b*) Vibratory or Oscillatory Motion

If a body in periodic motion moves to and fro about a definite point on a single path, the motion of the body is said to be the vibratory or oscillatory motion. For example, the motion of simple pendulum, the motion of arms of a tuning fork are the vibratory motions. Obviously, the vibratory or oscillatory motion is definitely a periodic motion.

Mean or Equilibrium Position

The point on either side of which the body vibrates is called the mean or equilibrium position of the body.

One Vibration or One Oscillation

In a vibratory motion, one vibration is said to be complete, when the vibrating body goes from its equilibrium position to one extreme after which returns back to the equilibrium position, then goes to the extreme on other side and then returns back to the same position of equilibrium.

For example, in Figure 1.1, the vibration of the body from O to A, A to O, O to B and then from B to O, is said to be one complete vibration.

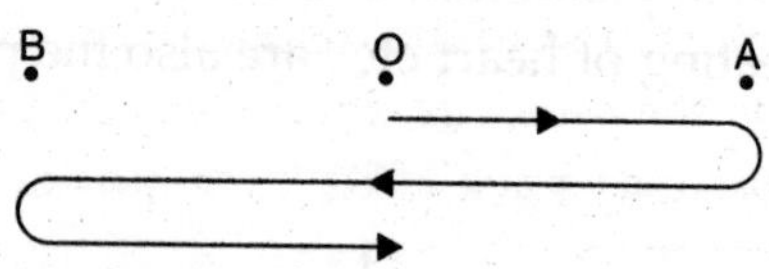

FIGURE 1.1 One complete vibration

Period of Vibration

The time taken to complete one vibration, is called the period of vibration (T) of a body. Its unit is second (s).

Frequency

The number of complete vibrations made by the body in 1 second is called the frequency of that body. It is represented by the letter f or n or ν. Its unit is s^{-1} or Hertz (Hz).

Relationship Between the Period of Vibration and Frequency

If n be the frequency of vibratory body and T is its time period, then number of oscillations in T second = 1

Therefore, number of oscillations in 1 second (*i.e.*, frequency) $n = \frac{1}{T}$

$$\text{Thus, frequency} = \frac{1}{\text{time period}} \quad \text{or} \quad n = \frac{1}{T}$$

Amplitude: In vibrating motion, the maximum displacement of the body on either side of the mean position is called the amplitude. It is represented by the letter (a). Its unit is metre (m).

1.2 SIMPLE HARMONIC MOTION

When a body moves periodically in a straight line on either side of a point, the motion of body is called the *simple harmonic motion.*

Thus, the simple harmonic motion is a special case of the periodic motion, obviously, a simple harmonic motion is definitely a periodic motion, but all the periodic motions are not the simple harmonic motions.

For example, the motion of earth around the sun is a periodic motion, but it is not the simple harmonic motion. On the other hand, the motion of a simple pendulum is simple harmonic motion as well as the periodic.

In a simple harmonic motion, the body moves periodically in a straight line on either side of its mean position such that its acceleration is proportional to the displacement of the particle and the direction of acceleration is always towards the mean position. In other words, the motion of a body under a restoring force is a simple harmonic motion.

Restoring Force

The force, which is directly proportional to the displacement of the body from its mean position and is directed towards the mean position is called the restoring force *i.e.*, the restoring force tends to bring the body back to its mean position.

1.2.1 Conditions or Characteristics of Simple Harmonic Motion

Following are the conditions (or characteristics) of simple harmonic motion:

(*i*) The motion must be in straight line on either side of a definite point (mean position).

(*ii*) The moving body must pass from its mean position repeatedly after a definite time *i.e.*, motion must be periodic.

(*iii*) The acceleration of the moving body must always be proportional to the displacement of the body from its mean position and the direction of acceleration must always be towards the mean position, *i.e.*,

Acceleration ∝ displacement and in the direction opposite to displacement.

If at any instant, the displacement of body from its mean position is x, the acceleration of the body is

$$f \propto -x$$

But by Newton's law of motion, force = mass × acceleration

$$\therefore \qquad \text{force} \propto -x$$

Thus, the force acting on the body must be proportional to the displacement of the body from its mean position and its direction must be towards the mean position (*i.e.*, the force must be the restoring force).

A system, whose motion is simple harmonic, is known as the *simple harmonic oscillator*. For example, motion of simple pendulum for small amplitude, motion of the compound pendulum, motion of the mass attached at the free end of a spring rigidly fixed at its other end, motion of torsional pendulum etc.

1.3 DIFFERENTIAL EQUATION OF A SIMPLE HARMONIC OSCILLATOR AND ITS SOLUTION

A system executing simple harmonic motion is called a simple harmonic oscillator. In simple harmonic motion, the force acting on the system at any instant, is directly proportional to the displacement from a fixed point in its path and the direction of this force is towards that fixed point. Thus, the system executes the motion under a linear restoring force. If the displacement of the system from a fixed point is x, the linear restoring force is $-Kx$, where K is a constant which is called the *force constant*. Thus no other force except the linear restoring force acts on a simple harmonic oscillator. As a result, the oscillator executes vibrations of constant amplitude and with a constant frequency. These oscillations are called the *free oscillations*.

Let a particle of mass m be executing simple harmonic oscillations. The acceleration of the particle at displacement x from a fixed point will be $\frac{d^2x}{dt^2}$. For the particle, restoring force ∝–displacement.

$$\text{or} \qquad m\frac{d^2x}{dt^2} \propto -x \qquad \text{or} \qquad m\frac{d^2x}{dt^2} = -Kx$$

where K is a constant, which is called *force constant of the particle*. Here the negative sign tells that the direction of force acting on the particle (or acceleration) is opposite to the direction of increase in displacement (Figure 1.2).

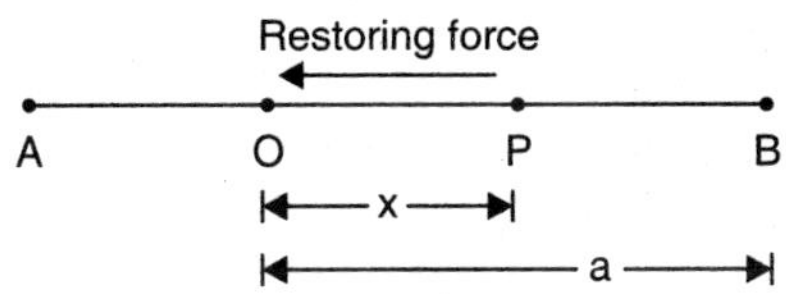

FIGURE 1.2 **Direction of restoring force**

Acceleration of the particle $\frac{d^2x}{dt^2} = \frac{-Kx}{m}$

Let $\frac{K}{m} = \omega^2$

then

Acceleration of the particle $\frac{d^2x}{dt^2} = -\omega^2 x$

or
$$\boxed{\frac{d^2x}{dt^2} + \omega^2 x = 0} \quad \text{...(1)}$$

Equation (1) is known as differential equation of simple harmonic oscillator.

1.3.1 Solution of Differential Equation of Simple Harmonic Oscillator

Now we have to find the displacement x of the particle at any instant t by solving the differential equation (1) of the simple harmonic oscillator.

In equation (1), multiplying by $2\left(\frac{dx}{dt}\right)$, we get

$$2\frac{dx}{dt}\frac{d^2x}{dt^2} + \omega^2 x \,.\, 2\frac{dx}{dt} = 0$$

or
$$\frac{d}{dt}\left(\frac{dx}{dt}\right)^2 + \omega^2 \frac{d}{dt}(x^2) = 0$$

On integrating $\left(\frac{dx}{dt}\right)^2 + \omega^2 x^2 = A$ (constant)

At the position of maximum displacement, *i.e.*, at $x = \pm a$, velocity of particle $\frac{dx}{dt} = 0$

$\therefore \quad 0 + \omega^2 a^2 = A \quad \text{or} \quad A = \omega^2 a^2$

Then $$\left(\frac{dx}{dt}\right)^2 + \omega^2x^2 = \omega^2a^2$$

Hence $$\left(\frac{dx}{dt}\right)^2 = \omega^2 (a^2 - x^2)$$

or velocity of the particle $$\frac{dx}{dt} = \omega\sqrt{(a^2 - x^2)} \quad ...(2)$$

Equation (2) tells us the velocity of particle at position x

From equation (2) $$\frac{dx}{\sqrt{(a^2 - x^2)}} = \omega\, dt$$

Again integrating, $$\sin^{-1}\frac{x}{a} = \omega t + \phi \text{ (constant)}$$

or $$\frac{x}{a} = \sin(\omega t + \phi) \quad \text{or} \quad \boxed{x = a\sin(\omega t + \phi)} \quad ...(3)$$

Here a is the amplitude of oscillations and ϕ is the initial phase of the motion of particle (whose value can be known from the initial conditions).

(*i*) If $x = 0$ at $t = 0$ (*i.e.* the particle initiates its oscillations from its mean position), then $\sin\phi = 0$ or $\phi = 0$, then the displacement equation of the particle executing simple harmonic motion at any time t will be

$$x = a\sin\omega t \quad ...(4)$$

(*ii*) If $x = 0$ at $t = 0$ (*i.e.*, the particle initiates its oscillations from its maximum displaced position), then $\sin\phi = 1$ or $\phi = \pi/2$, then the displacement equation of the particle executing simple harmonic motion at any instant t will be

$$x = a\sin(\omega t + \pi/2) = a\cos\omega t \quad ...(5)$$

1.3.2 Time Period and Frequency

It is clear from equations (4) and (5), that

$$a\sin\omega\left(t + \frac{2\pi}{\omega}\right) = a\sin\omega t \quad \text{and} \quad a\cos\omega\left(t + \frac{2\pi}{\omega}\right) = a\cos\omega t$$

It is concluded from here that the displacement of the particle at any instant $\left(t + \frac{2\pi}{\omega}\right)$ is exactly the same as it was at the instant t *i.e.*, the particle comes back to its initial position during its motion exactly after time $\frac{2\pi}{\omega}$. Hence time period of the particle executing simple harmonic motion is

$$T = \frac{2\pi}{\omega} = \frac{2\pi}{\sqrt{K/m}} \qquad \left(\text{Since } \frac{K}{m} = \omega^2\right)$$

or

$$\boxed{T = 2\pi\sqrt{\frac{m}{K}}}$$

It is clear from equation (1) that numerically, acceleration = $\omega^2 \times$ displacement.

$$\therefore \quad \omega = \sqrt{\frac{\text{acceleration}}{\text{displacement}}} \text{ and time period } T = 2\pi\sqrt{\frac{\text{displacement}}{\text{acceleration}}}$$

Since frequency $n = \frac{1}{T}$, hence frequency $n = \frac{1}{2\pi}\sqrt{\frac{K}{m}} = \frac{1}{2\pi}\sqrt{\frac{\text{acceleration}}{\text{displacement}}}$

Remember that the quantity $\omega = 2\pi n$ is known as the angular frequency of motion.

1.4 POTENTIAL AND KINETIC ENERGIES IN SIMPLE HARMONIC MOTION

We have read that a particle executing simple harmonic is acted upon by a restoring force, which tends to bring back the particle to its mean position. Hence the particle possesses the potential energy. Since the particle remains in motion, hence it also possesses the kinetic energy. The two forms of energy transform in each other during the motion, but their sum always remains constant.

Potential Energy

Let at any instant t, the displacement of particle of mass m from its mean position be x at that instant, the acceleration of particle be f. Then in simple harmonic motion, acceleration $f \propto -x$

or $$f = -\omega^2 x, \text{ where } \omega^2 \text{ is a constant}$$

$\therefore$ The restoring force acting on the particle is

$$F = \text{mass} \times \text{acceleration} = mf = -m\omega^2 x$$

Here the negative sign only gives the direction of force which is opposite to the direction of displacement. Obviously, the magnitude of force is proportional to the displacement.

Work done against the restoring force in displacing the particle from x to $(x + dx)$ is $dW = m\,\omega^2 x\, dx$.

This work done is stored in the particle in the form of potential energy. Hence total work done in displacing the particle from mean position $x = 0$ to displaced position x

$$= \text{potential energy of the particle at displacement } x$$

or $$U = W_{0 \to x} = \int_0^x m\,\omega^2 x\, dx$$

or $$U = \frac{1}{2} m\,\omega^2 x^2 \qquad ...(1)$$

Kinetic Energy

From equation (2) of article 1.3, at an instant t, when the displacement of particle is x, the velocity of particle is

$$u = \omega\sqrt{(a^2 - x^2)}$$

$\therefore$ Kinetic energy of particle $K = \frac{1}{2} mu^2$

or
$$K = \frac{1}{2} m\,\omega^2(a^2 - x^2) \qquad \text{...(2)}$$

Total Energy

Total energy of particle = Potential energy + Kinetic energy

$$= \frac{1}{2} m\,\omega^2 x^2 + \frac{1}{2} m\,\omega^2(a^2 - x^2)$$

$$= \frac{1}{2} m\,\omega^2 a^2$$

But $\omega = 2\pi n$ (where n is frequency)

$$\boxed{E = 2\pi^2 m\, n^2 a^2} \qquad \text{...(3)}$$

From the above equation (3), we conclude that

(i) The total energy of the particle executing simple harmonic motion does not depend on its displacement x and time t, *i.e.* during the motion, the total energy of the particle does not change with time (or position).

(ii) In the extreme position ($x = \pm a$), kinetic energy is zero, hence the total energy is in form of potential energy.

(iii) In mean position ($x = 0$), potential energy is zero, hence total energy is in form of kinetic energy.

(iv) As the displacement of particle from its mean position increases, its kinetic energy decreases and potential energy increases.

Similarly, as the particle approaches from the extreme position to the mean position, its potential energy decreases and kinetic energy increases, but in ideal condition when there is no loss of energy due to friction etc. in any position, the total sum of kinetic energy and potential energy remains constant, *i.e.*, the mechanical energy remains conserved.*

(v) Total energy of the particle executing simple harmonic motion is directly proportional to the square of amplitude of motion *i.e.*, $E \propto a^2$.

---

*If the frequency of oscillations of the particle excuting simple harmonic motion is n, the frequency of its kinetic energy or potential energy will be $2n$, because if the displacement is $x = a \sin \omega t$, then kinetic energy $= K = \frac{1}{2} ma^2\,\omega^2 \cos^2 \omega t = \frac{1}{4} ma^2\,\omega^2\,(1 + \cos 2\omega t)$ and potential energy $U = \frac{1}{2} ma^2\,\omega^2 \sin^2 \omega t = \frac{1}{4} ma^2\,\omega^2\,(1 - \cos 2\omega t)$.

(*vi*) For the particle executing simple harmonic motion, is directly proportional to the square of frequency of motion *i.e.*, $E \propto n^2$.

(*vii*) For a particle executing simple harmonic motion, the average kinetic energy is equal to average potential energy and it is equal to half of the total energy *i.e.*,

$$K_{av} = U_{av} = \frac{1}{2} E$$

The average kinetic energy of a particle in one period

$$K_{av} = \frac{1}{T}\int_0^T K\,dt = \frac{1}{T}\int_0^T \frac{1}{2} ma^2\omega^2 \cos^2 \omega t\, dt$$

$$= \frac{ma^2\omega^2}{2T}\int_0^T \frac{1+\cos 2\omega t}{2}\,dt$$

$$= \frac{ma^2\omega^2}{4T}\left(t + \frac{\sin 2\omega t}{2\omega}\right)_0^T$$

$$= \frac{1}{4} ma^2\omega^2 = \frac{E}{2} \quad \left(\because\ T = \frac{2\pi}{\omega}\right) \qquad ...(4)$$

The average potential energy of a particle in one time period

$$U_{av} = \frac{1}{T}\int_0^T U\,dt = \frac{1}{T}\int_0^T \frac{1}{2} ma^2\omega^2 \sin^2 \omega t\, dt$$

$$= \frac{ma^2\omega^2}{2T}\int_0^T \frac{1-\cos 2\omega t}{2}\,dt$$

$$= \frac{ma^2\omega^2}{4T}\left(t - \frac{\sin 2\omega t}{2\omega}\right)_0^T$$

$$= \frac{1}{4} ma^2\omega^2 = \frac{E}{2} \qquad ...(5)$$

It is clear from equations (4) and (5), that $K_{av} = U_{av} = \frac{E}{2}$

or
$$\boxed{K_{av} + U_{av} = E}$$

Graphical Representation of Potential Energy and Kinetic Energy

Let the displacement of particle of mass m executing simple harmonic motion at any instant t be x, then in that position of potential energy and kinetic energy of particle are given as

$$\text{Potential energy } U = \frac{1}{2} m\omega^2 x^2$$

and kinetic energy $K = \frac{1}{2} m\omega^2 (a^2 - x^2)$

where a is the amplitude of motion and ω the angular velocity

Total energy $E = U + K = \frac{1}{2} m\omega^2 a^2$

(*i*) When $x = 0$ (*i.e.*, at mean position), potential energy $U = 0$

$$\text{Kinetic energy } K = \frac{1}{2} m\omega^2 a^2 = E$$

(*ii*) When $x = \pm \frac{a}{2}$, potential energy $U = \frac{1}{2} m\omega^2 \left(\frac{a}{2}\right)^2$

$$= \frac{1}{4} \times \frac{1}{2} m\omega^2 a^2 = \frac{E}{4}$$

and kinetic energy

$$K = \frac{1}{2} m\omega^2 \left[a^2 - \left(\frac{a}{2}\right)^2\right]$$

$$= \frac{3}{4} \times \frac{1}{4} m\,\omega^2 a^2 = \frac{3}{4} E$$

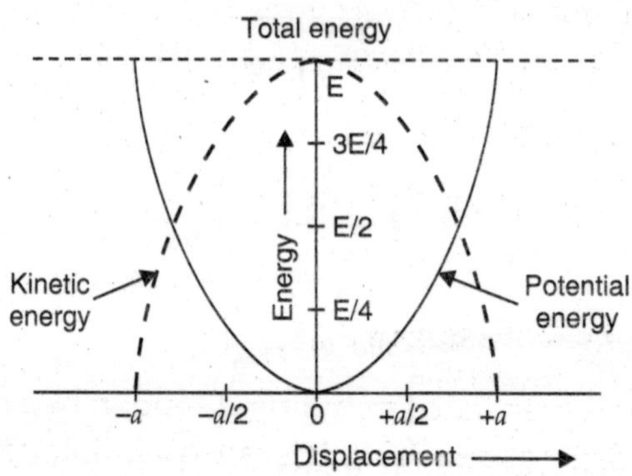

FIGURE 1.3 **Graphical representation of potential energy and kinetic energy in simple harmonic motion**

(*iii*) When $x = \pm a$, potential energy $U = \frac{1}{2} m\,\omega^2 a^2 = E$

and kinetic energy $K = 0$

Now taking the displacement x on X-axis and the energy on Y-axis, the graph plotted is shown in Figure 1.3.

The dotted line drawn parallel to X-axis represents the total energy of the particle, which remains constant during the entire motion.

1.5 EQUATION OF MOTION FROM PRINCIPLE OF ENERGY CONSERVATION

Total energy of oscillator is $\frac{1}{2} m v^2 + \frac{1}{2} m \omega^2 x^2 = E =$ constant.

Differentiating both sides with respect to t, we get

$$m v \frac{dv}{dt} + m \omega^2 x \frac{dx}{dt} = 0$$

or $$m v \frac{dv}{dt} + m \omega^2 x v = 0$$

or $$m v \left[\frac{dv}{dt} + \omega^2 x\right] = 0$$

or $$\frac{dv}{dt} + \omega^2 x = 0$$

or $$\frac{d^2x}{dt^2} + \frac{s}{m} x = 0$$

or $$\boxed{m \frac{d^2x}{dt^2} = - s x}$$

The above expression is differential equation representing SHM.

1.6 APPLICATIONS OF SIMPLE HARMONIC MOTION

1.6.1 Motion of Simple Pendulum

If a heavy point mass is suspended from a rigid support by a massless, inextensible and perfectly elastic string, it is called a simple pendulum.

In actual practice, it is impossible to obtain such a simple pendulum because there can't be a heavy point mass and a massless, inextensible and perfectly elastic string, hence it is only ideal definition. In practice a simple pendulum is obtained by suspending a metallic solid sphere by means of a light, thin but strong thread from a rigid support as shown in Figure 1.4. The sphere is called the bob. The distance of the centre of sphere O from the point of suspension S is called the effective length of the pendulum.

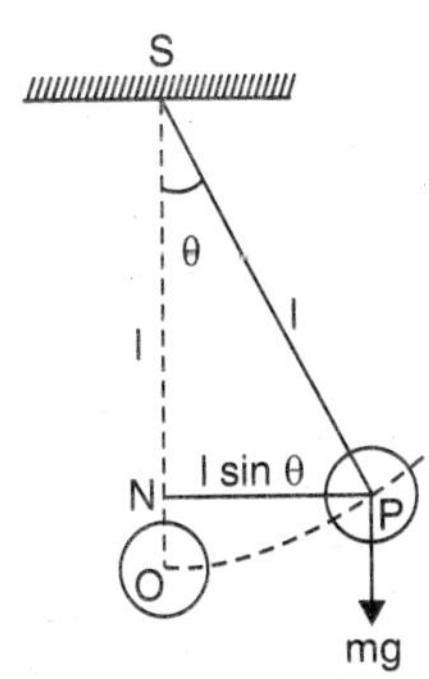

FIGURE 1.4 **Motion of a simple pendulum**

Figure 1.4 shows a simple pendulum, which consists of a solid sphere O suspended from the rigid support S by means of a thin and light thread. The equilibrium position of the simple pendulum is SO. If the bob from its equilibrium position is slightly displaced to the position P and then released, the pendulum begins to oscillates in simple harmonic motion.

Let the effective length of pendulum (*i.e.*, the distance from the point of suspension to the centre of gravity of the bob) be l and the mass of the bob be m.

Let θ be the angular displacement of the bob at any instant t at its displaced position P. Restoring torque about the point of suspension S due to the weight mg of the bob is

$$\tau = -mg\, l \sin\theta \qquad (\because \quad PN = l\sin\theta)$$

Now assuming the bob as a point mass, the moment of inertia of the bob about the point of suspension S is $I = ml^2$ and inertial torque $= I\frac{d^2\theta}{dt^2}$, where $\frac{d^2\theta}{dt^2}$ is the angular acceleration of the pendulum at displacement θ.

Hence the equation of motion of pendulum will be

$$I\frac{d^2\theta}{dt^2} = -mg\, l\sin\theta$$

or
$$ml^2\frac{d^2\theta}{dt^2} + mg\, l\sin\theta = 0 \quad \text{or} \quad \frac{d^2\theta}{dt^2} + \frac{g}{l}\sin\theta = 0$$

But
$$\sin\theta = \theta - \frac{\theta^3}{3!} + \ldots\text{. If } \theta \text{ is small, then } \sin\theta = \theta.$$

Then
$$\frac{d^2\theta}{dt^2} + \frac{g}{l}\theta = 0 \quad \text{or} \quad \frac{d^2\theta}{dt^2} + \omega^2\theta = 0$$

where
$$\omega^2 = \frac{g}{l}$$

The above equation represents the simple harmonic motion, whose solution will be $\theta = \theta_0 \sin(\omega t + \phi)$ and time period $\boxed{T = \frac{2\pi}{\omega} = 2\pi\sqrt{l/g}\,.}$

1.6.2 Vertical Spring

Take a spring mounted vertically, with the top fixed. The weight extends the spring.

The equation is now

$$F = m\frac{d^2x}{dt^2} \quad \text{and} \quad F = -sx + mg$$

$$\therefore \qquad m\frac{d^2x}{dt^2} = -sx + mg = -s\left(x - \frac{mg}{s}\right) = -s\,x'$$

where $\qquad x' = x - mg/s \quad \text{and} \quad \ddot{x}' = \ddot{x}$

So we can write $\qquad m\,\ddot{x}' = -s\,x'$

This is just the same as before, except for x' instead of x. It has the same SHM solutions $\left(\text{with the same } \omega = \sqrt{s/m}\right)$.

The difference between x and x' is mg/s, which is the amount by which the spring is extended by the weight x' is just the deviation, measuring distances from the equilibrium point for the spring stretched by gravity, rather than the natural length of the spring.

1.6.3 Superposition of Two Mutually Perpendicular Harmonic Motions of the Same Frequencies: Lissajous Figures

When a particle is acted upon simultaneously by two SHMs at right angles to each other, the resultant path traced out by the particle is called Lissajous Figures. (Figures 1.5–1.9), named after J.A. Lissajous, who made an extensive study of these motions.

The nature of the resultant path or the curve traced out depends upon; (*i*) the amplitude, (*ii*) the period (or frequencies), and (*iii*) the phase difference between the component vibrations.

Let the displacement of particle executing simple harmonic motion on *X*- and *Y*-axes are x and y respectively, where

$$x = a\sin(\omega t + \phi) \qquad \ldots(1)$$

and

$$y = b\sin\omega t \qquad \ldots(2)$$

Here a and b are respectively the amplitudes of the moving particle along *X*- and *Y*-axes. The frequency of both the displacements is same ($= \omega/2\pi$) and their phase difference is ϕ.

From equation (2) $\qquad \frac{y}{b} = \sin\omega t$

$$\cos\omega t = \sqrt{(1 - \sin^2\omega t)} = \sqrt{(1 - y^2/b^2)}$$

From equation (1) $\qquad \frac{x}{a} = \sin(\omega t + \phi)$

$$\Rightarrow \qquad \frac{x}{a} = \sin\omega t\cos\phi + \cos\omega t\sin\phi$$

Substituting the value of $\cos\omega t$ in the above equation

$$\frac{x}{a} = \frac{y}{b}\cos\phi - \sqrt{\left(1 - \frac{y^2}{b^2}\right)}\sin\phi$$

or

$$\frac{x}{a} - \frac{y}{b}\cos\phi = -\sqrt{\left(1 - \frac{y^2}{b^2}\right)}\sin\phi$$

Squaring both the sides

$$\frac{x^2}{a^2} + \frac{y^2}{b^2}\cos^2\phi - \frac{2xy}{ab}\cos\phi = \left(1 - \frac{y^2}{b^2}\right)\sin^2\phi$$

$$\Rightarrow \quad \boxed{\frac{x^2}{a^2} + \frac{y^2}{b^2} - \frac{2xy}{ab}\cos\phi = \sin^2\phi} \quad ...(3)$$

Equation (3) is a general equation of ellipse.

Let us consider a few special cases:

(*i*) When $\phi = 0°$, *i.e.*, if the phase difference between the two motions is zero, then $\sin\phi = 0, \cos\phi = 1$

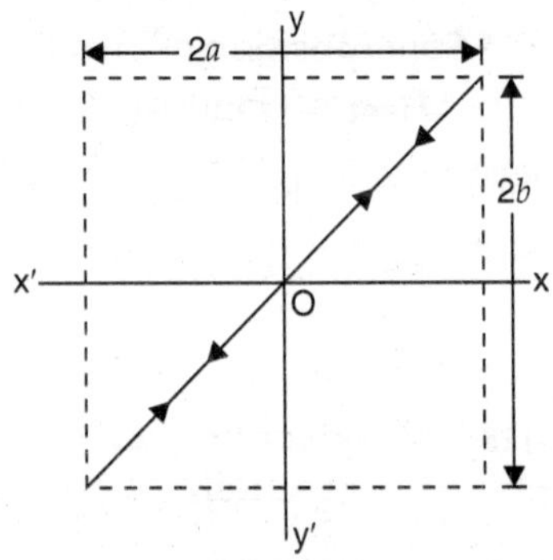

FIGURE 1.5 **When phase difference is zero**

Hence, $$\frac{x^2}{a^2} + \frac{y^2}{b^2} - \frac{2xy}{ab} = 0$$

or $$\left(\frac{x}{a} - \frac{y}{b}\right)^2 = 0$$

or $$\boxed{\pm\left(\frac{x}{a} - \frac{y}{b}\right) = 0} \quad ...(4)$$

Equation (4) represents a pair of coinciding straight line $y = \frac{b}{a}x$ passing through the origin in the first and third quadrant (Figure 1.5).

Now, if $a = b$ *i.e.*, if the amplitudes of the two motions are same, then equation of resultant curve will be $x - y = 0$ *i.e.*, the straight time will be at 45° angle from the two axes.

(*ii*) When $\phi = \pi/4$ *i.e.*, the phase difference between the two motions is $\pi/4$, then

$$\sin\frac{\pi}{4} = \frac{1}{\sqrt{2}} \text{ and } \cos\pi/4 = 1/\sqrt{2}$$

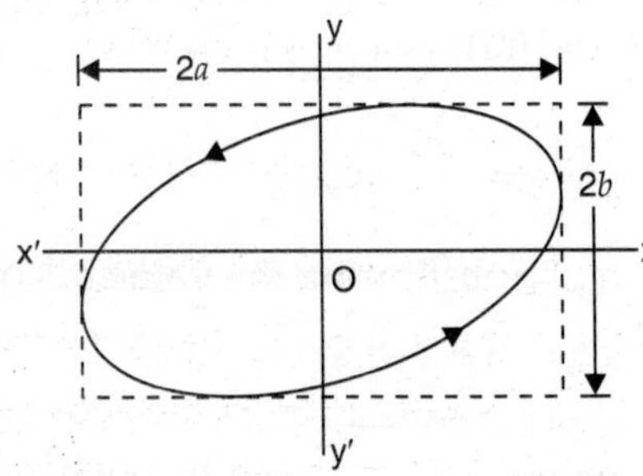

FIGURE 1.6 **When phase difference = $\frac{\pi}{4}$**

Hence, $$\frac{x^2}{a^2} + \frac{y^2}{b^2} - \frac{2xy}{ab} \times \frac{1}{\sqrt{2}} = \frac{1}{2}$$

or $$\boxed{\frac{x^2}{a^2} + \frac{y^2}{b^2} - \sqrt{2}\,\frac{xy}{ab} = \left(\frac{1}{\sqrt{2}}\right)^2} \quad ...(5)$$

This is an equation of oblique ellipse (Figure 1.6).

Now, if $a = b$ then $x^2 + y^2 - \sqrt{2}\,xy = a^2/2$.

(*iii*) When $\phi = \pi/2$, *i.e.*, the phase difference between the two motions is $\frac{\pi}{2}$, then

$\sin\phi = 1$ and $\cos\phi = 0$

Hence $$\frac{x^2}{a^2} + \frac{y^2}{b^2} = 1 \qquad ...(6)$$

This is the equation of an ellipse, whose major axis is along X-axis and minor axis is along Y-axis (Figure 1.7).

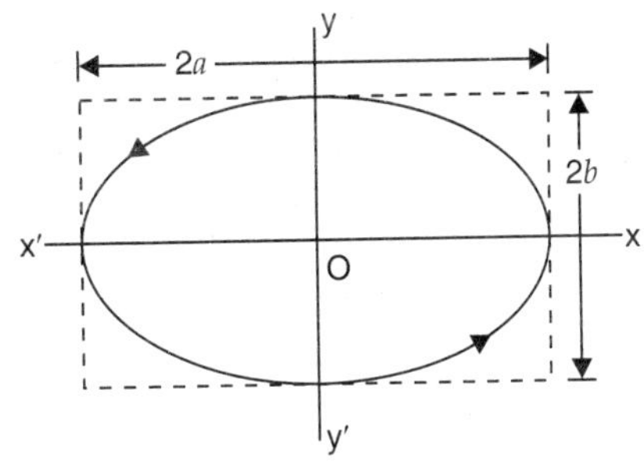

FIGURE 1.7 When $\phi = \frac{\pi}{2}$

Now if $a = b$, then the equation of resultant curve will be $x^2 + y^2 = a^2$ *i.e.*, the Lissajous figure will now be a **circle.**

Hence two mutually perpendicular simple harmonic motions of same frequency (or same time period) with a phase difference $\frac{\pi}{2}$, are equivalent to a uniform circular motion of radius equal to the amplitude of either of the SHM. In other words, we can say that a uniform circular motion can be resolved in two mutually perpendicular SHMs of same frequency and same amplitude, but having a phase difference of $\frac{\pi}{2}$.

(*iv*) When $\phi = \pi$, *i.e.*, the phase difference between two motions is π, then $\sin\phi = 0$ and $\cos\phi = -1$.

Hence $$\frac{x^2}{a^2} + \frac{y^2}{b^2} + \frac{2xy}{ab} = 0$$

or $$\left(\frac{x}{a} + \frac{y}{b}\right)^2 = 0 \quad \text{or} \quad \pm\left(\frac{x}{a} + \frac{y}{b}\right) = 0 \qquad ...(7)$$

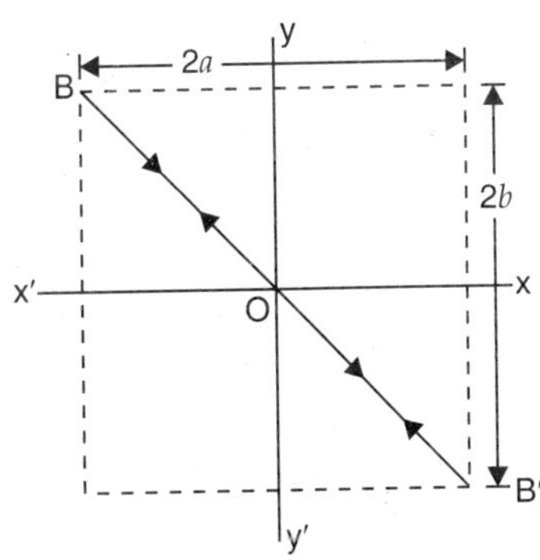

FIGURE 1.8 When $\theta = \pi$

This equation represents a pair of coinciding straight line $y = -\frac{bx}{a}$ and passing through the origin, situated in second and fourth quadrant (Figure 1.8).

Now if $a = b$, then the straight line will make an angle of 135° with the X-axis.

Similarly, from the above equation (3), the shape of the resultant curve for any particular phase difference can be obtained.

Figure 1.10 below shows the Lissajous figure obtained from the superposition of two mutually perpendicular SHMs of same frequency for the different phase differences.

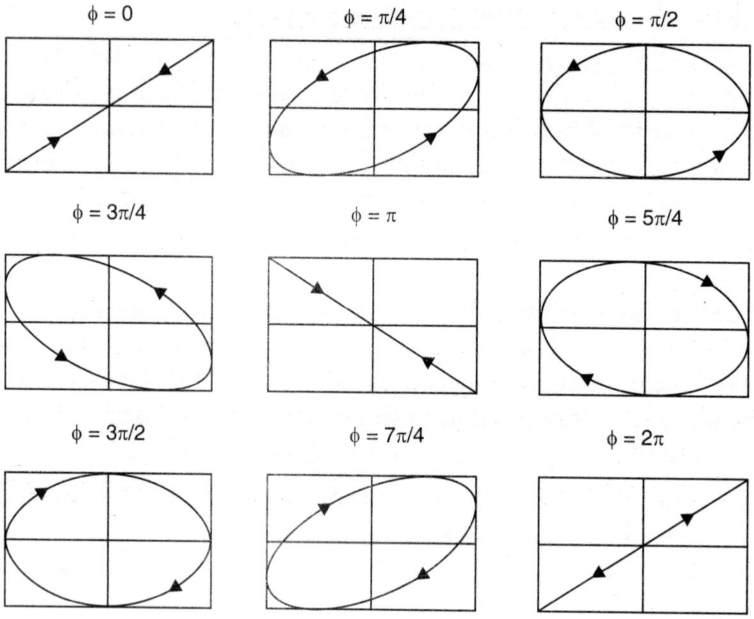

FIGURE 1.9 **Lissajous Figures**

It should be remembered that if the frequency of both the component motions are exactly equal, then the elliptical paths as given above are steady, but if there is a slight difference in the frequencies of two component motions, the relative phase difference ϕ between the two motions increases gradually due to which the elliptical shape also changes gradually, but all these ellipses remain confined inside the rectangle of sides $2a$ and $2b$. If both the motions are in same phase ($\phi = 0°$), then the ellipse coincides with the diagonal $\frac{x}{a} - \frac{y}{b} = 0$ of the rectangle. As the value of phase difference ϕ increase from 0 to $\frac{\pi}{2}$, the ellipse becomes oblique and then at the last it becomes $\frac{x^2}{a^2} + \frac{y^2}{b^2} = 1$. Now with the increase in the value of ϕ from $\frac{\pi}{2}$ to π, the ellipse coincides with the other diagonal $\frac{x}{a} + \frac{y}{b} = 0$ of the rectangle. Thereafter with the further increase in the value of ϕ from 0 to 2π, the reverse process takes place until the ellipse coincides with the first diagonal of the rectangle. All these changes are shown in Figure 1.9. The frequency of entire cycle is equal to the difference in the frequencies of two simple harmonic motions.

1.7 DAMPED HARMONIC OSCILLATOR

In a simple harmonic oscillator, no other force except the linear restoring force acts. As a result, the oscillator executes vibrations of constant amplitude and with a constant frequency. These oscillations are called the *free oscillations*. The total mechanical energy (*i.e.*, kinetic energy + potential energy) of the oscillator always remains conserved in such oscillations.

But in practice, free oscillations are not possible. This is because the medium in which the oscillator executes vibrations, exerts a frictional or viscous force on the oscillator. This force is called the *damping force*. This force acts in direction opposite to the direction of motion due to which the amplitude of vibration gradually decreases. Hence, the energy of oscillator also decreases. Such a vibrating system is called *damped harmonic oscillator*. For example, the simple pendulum executing oscillations in air or in any other medium, tuning fork, ballistic galvanometer are the damped harmonic oscillators.

Thus, a particle executing damped harmonic motion in a medium (*i.e.*, damped harmonic oscillator) is acted upon by two forces:

(*i*) Restoring force which is directly proportional to the displacement from a fixed point on its path and is in a direction opposite to the displacement, *i.e.*, $-Kx$ where K is froce constant and x is the displacement, is the displacement of the particle at any instant.

(*ii*) Damping force, which is proportional to the velocity of the particle and is in a direction opposite to motion, *i.e.*, $-r\dfrac{dx}{dt}$, where r is a positive constant, which is called the *damping coefficient*.

If m is the mass of the particle, by Newton's law, the total force acting on the particle = $m\dfrac{d^2x}{dt^2}$, where $\dfrac{d^2x}{dt^2}$ = acceleration of the particle.

Thus the equation of motion of a damped harmonic oscillator will be

$$m\frac{d^2x}{dt^2} = -r\frac{dx}{dt} - Kx \quad \text{or} \quad \frac{d^2x}{dt^2} + \frac{r}{m}\frac{dx}{dt} + \frac{K}{m}x = 0$$

or

$$\boxed{\frac{d^2x}{dt^2} + 2b\frac{dx}{dt} + n^2x = 0} \qquad ...(1)$$

where $2b = \dfrac{r}{m}$ and $n^2 = K/m$.

It must be noted that unit of damping coefficient r is kg/sec. Therefore, the unit of r/m (or $2b$) is s^{-1}. Thus the unit of m/r is same as that of the time. Hence the quantity m/r (or $1/2b$) is called the *time constant* or *the relaxation time*. It is represented by τ (Tau). Relaxation time is defined as the time in which the energy of particle reduces to $1/e$ (or 0.37) times its maximum initial energy.

Now, we are to find the solution of equation (1) for damped motion. Let the solution of equation (1) be $x = Ae^{\alpha t}$

Then, $$\frac{dx}{dt} = A\,\alpha\, e^{\alpha t} \quad \text{and} \quad \frac{d^2x}{dt^2} = A\,\alpha^2 e^{\alpha t}$$

Substituting these values in equation (1)

$$A\,\alpha^2 e^{\alpha t} + 2b\,A\,\alpha\, e^{\alpha t} + n^2 A\, e^{\alpha t} = 0$$

or $$(\alpha^2 + 2b\alpha + n^2)\, Ae^{\alpha t} = 0 \quad \text{or} \quad \alpha^2 + 2b\alpha + n^2 = 0$$

$\therefore$ $$\alpha = -b \pm \sqrt{b^2 - n^2} = -b \pm p, \quad \text{where} \quad p = \sqrt{(b^2 - n^2)}$$

Thus the two possible solutions of equation (1) are

$$x = A_1\, e^{(-b+p)t} \quad \text{and} \quad x = A_2\, e^{(-b-p)\,t}$$

The general solution of equation (1) is

$$x = A_1\, e^{(-b+p)\,t} + A_2\, e^{(-b-p)\,t}$$
$$= e^{-bt}\,[A_1\, e^{pt} + A_2\, e^{-pt}] \qquad ...(2)$$

Here, A_1 and A_2 are the constants whose values can be obtained from the initial condition of motion. There are following three cases possible:

1. When $b < n$, under damped.
2. When $b > n$, over damped.
3. When $b = n$, critically damped.

Case 1. Under-damped Case: If the damping is so small that $b << n$ or $r/2m << \sqrt{K/m}$,

then, $p = \sqrt{(b^2 - n^2)} = \sqrt{-(n^2 - b^2)} = j\sqrt{(n^2 - b^2)}$ or $p = j\omega$, where $\omega = \sqrt{(n^2 - b^2)}$.

Obviously, p is an imaginary quantity, but ω is a real quantity.

Then from equation (2) $$x = e^{-bt}\left[A_1\, e^{j\omega t} + A_2 e^{-j\omega t}\right]$$
$$= e^{-bt}\left[A_1\,(\cos\omega t + j\sin\omega t) + A_2\,(\cos\omega t - j\sin\omega t)\right]$$
$$= e^{-bt}\left[(A_1 + A_2)\cos\omega t + j\,(A_1 - A_2)\sin\omega t\right] \qquad ...(3)$$

Since x is a real quantity, therefore $(A_1 + A_2)$ and $j(A_1 - A_2)$ must also be real quantities. But A_1 and A_2 are the complex quantities, therefore

$(A_1 + A_2) = a_0 \sin\phi$ and $j(A_1 - A_2) = a_0 \cos\phi$ in above equation (3),

we get,

$$x = e^{-bt}\left[a_0 \sin\phi\cos\omega t + a_0\cos\phi\sin\omega t\right]$$

$$\boxed{x = a_0\, e^{-bt}\sin(\omega t + \phi)} \qquad ...(4)$$

This expression represents the damped harmonic motion where amplitude decreases exponentially with time. Hence, it is not a simple harmonic motion but it is an oscillatory motion, whose angular frequency is

$$\omega = \sqrt{(n^2 - b^2)} = \sqrt{\left(\frac{K}{m} - \frac{r^2}{4m^2}\right)}$$

and time period $T = \frac{2\pi}{\omega} = \frac{2\pi}{\sqrt{(n^2 - b^2)}} = \frac{2\pi}{\sqrt{\left(\frac{K}{m} - \frac{r^2}{4m^2}\right)}}$ ...(5)

The amplitude of motion is $a = a_0\, e^{-bt} = a_0\, e^{-t/2\tau}$

The displacement–time graph for the damped harmonic oscillations is shown in Figure 1.10. It is evident that the amplitude of motion decreases exponentially.

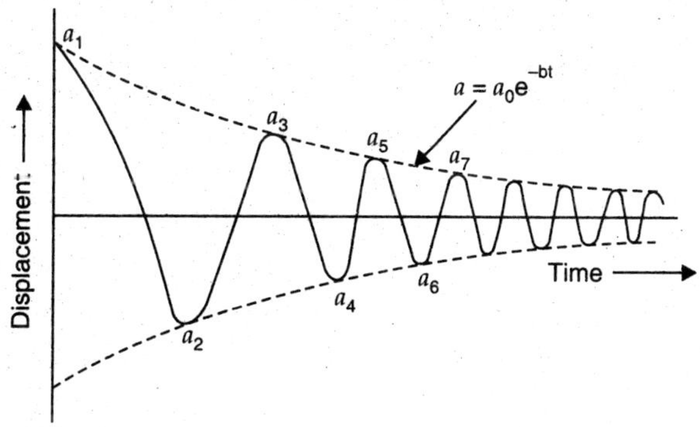

FIGURE 1.10 **Displacement-time graph for the damped harmonic oscillations**

Logarithmic Decrement: The ratio of two successive amplitudes remains constant (= d) and this ratio is called *decrement i.e.,*

$$\frac{a_1}{a_2} = \frac{a_2}{a_3} = \frac{a_3}{a_4} = \ldots = \frac{a_n}{a_{n-1}} = d \text{ (constant)} \qquad \ldots(6)$$

Since, the time interval between the two consecutive amplitudes on either side of mean position is $T/2$, therefore

$$a_n = a_0\, e^{-bt} \quad \text{and} \quad a_{n+1} = a_0 e^{-b[t+(T/2)]}$$

Hence, $d = \frac{a_n}{a_{n+1}} = \frac{a_0\, e^{-bt}}{a_0\, e^{-b(t+T/2)}} = e^{bT/2}$

or $\log_e d = \frac{bT}{2} = \lambda \text{ (say)}$ ...(7)

Here, λ is called the *logarithmic decrement*. Thus from equation (7)., $d = e^{\lambda}$.

or $\frac{a_1}{a_2} = \frac{a_2}{a_3} = \frac{a_3}{a_4} = \ldots = \frac{a_n}{a_{n+1}} = e^{\lambda}$

Hence, $\frac{a_1}{a_2} \times \frac{a_2}{a_3} \times \frac{a_3}{a_4} \times \ldots \times \frac{a_n}{a_{n+1}} = e^{n\lambda}$ or $\frac{a_n}{a_{n+1}} = e^{n\lambda}$

$$\therefore \quad \boxed{\lambda = \frac{1}{n} \log_e \frac{a_n}{a_{n+1}} = \frac{2.3026}{n} \log_{10} \frac{a_n}{a_{n+1}}} \quad \ldots(8)$$

Case 2. Over-damped Case: If the damping is too large that $b > n$, then $p = \sqrt{(b^2 - n^2)}$ is a real quantity and $p < b$.

Then from equation (4) $\boxed{x = A_1 e^{-(b-p)t} + A_2 e^{-(b+p)t}}$

Since, $p < b$, hence both the quantities on RHS of above equation decrease exponentially with time and there will no oscillations (Figure 1.11).

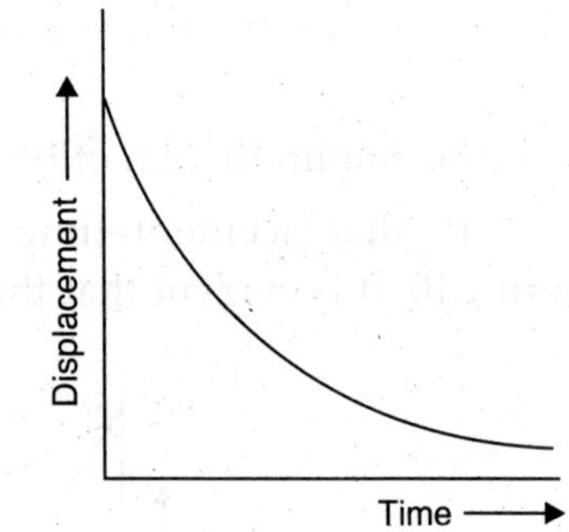

FIGURE 1.11 **Displacement-time graph of the overdamped oscillator**

Such a motion is called the *dead beat* or *a periodic* motion. This type of motion is used in the dead beat galvanometer.

Case 3. Critically Damped Case: If $b = n$, then $p = 0$

From equation (2) $x = (A_1 + A_2) e^{-bt} = C e^{-bt}$,

where $A_1 + A_2 = C$ (a constant)

Since, there is only one constant in above equation, therefore it cannot be the solution of second order differential equation (1)

Now if we consider that $\sqrt{(b^2 - n^2)} = p$ is a very small quantity, then

$$x = e^{-bt}[A_1 e^{pt} + A_2 e^{-pt}]$$
$$= e^{-bt}[A_1(1 + pt + \ldots) + A_2(1 - pt + \ldots)]$$

Since, p is very small, therefore neglecting the terms p^2 and p^3, ...

$$x = e^{-bt}[(A_1 + A_2) + p(A_1 - A_2)t] = e^{-bt}[P + Qt]$$

where, $A_1 + A_2 = P$ and $p[A_1 + A_2] = Q$

If initially at $t = 0$, the displacement of particle is $x = x_0$ and velocity $v = dx/dt = v_0$, then at $t = 0$, $x_0 = P$ and from $\frac{dx}{dt} = Qe^{-bt} - be^{-bt}(P + Qt)$ at $t = 0$, $v_0 = Q - bP$.

Hence, $Q = v_0 + bx_0$

Then, $\boxed{x = [x_0 + (v_0 + bx_0)t]e^{-bt}}$ ...(9)

In the above expression, due to the coefficient $[x_0 + (v_0 + bx_0)t]$, first the value of x increases with increase in t, but later on the exponential term e^{-bt} becomes more pronounced due to which the value of x falls to zero in a small time interval and no oscillations occur.

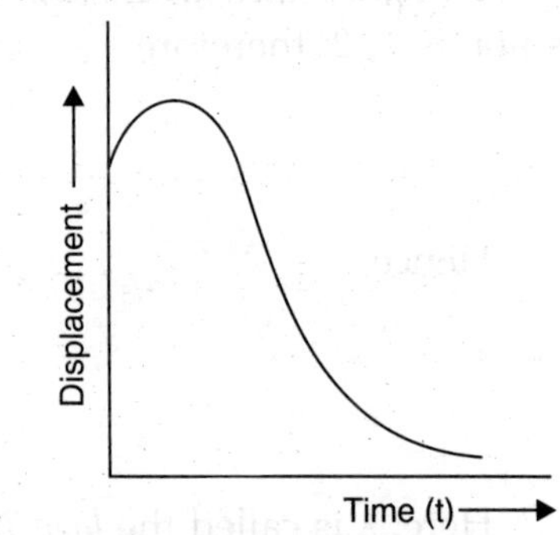

FIGURE 1.12 **Displacement-time graph for critically damped oscillator**

1.7.1 Power Dissipation in Damped Harmonic Oscillator

We have read that in damped oscillations, the mechanical energy of the particle continuously decreases due to the damping forces. As a result, the amplitude of motion also decreases with time. The displacement of particle executing damped harmonic motion at any instant t is

$$x = a_0 e^{-bt} \sin(\omega t + \phi) \text{ from equation (4)}$$

$\therefore$ Velocity of the particle $= \dfrac{dx}{dt} = a_0 e^{-bt} [(-b)\sin(\omega t + \phi) + \omega \cos(\omega t + \phi)]$ ...(10)

If m is the mass of the particle, the kinetic energy of the particle is

$$K = \frac{1}{2} m \left(\frac{dx}{dt}\right)^2$$

$$= \frac{1}{2} m a_0^2 e^{-2bt} [b^2 \sin^2(\omega t + \phi) + \omega^2 \cos^2(\omega t + \phi) - 2b\omega \sin(\omega t + \phi)\cos(\omega t + \phi)]$$

and potential energy $U = \int_0^x m\omega^2 x \, dx = \dfrac{m\omega^2 x^2}{2}$ ...(11)

$$= \frac{1}{2} m \omega^2 a_0^2 e^{-2bt} \sin^2(\omega t + \phi) \quad ...(12)$$

Average total energy of oscillator in a periodic time

= Average kinetic energy + Average potential energy

Now, to evaluate the average kinetic energy and the average potential energy in one time period, we can assume that the amplitude of oscillation remains nearly unchanged *i.e.*, e^{-2bt} remains nearly constant. Since the average value of $\sin^2(\omega t + \phi)$ and $\cos^2(\omega t + \phi)$ in $\frac{1}{2}$ for one time period and average value of $\sin(\omega t + \phi)\cos(\omega t + \phi)$ is zero over one time period, therefore in a period of time.

Average kinetic energy $= \dfrac{1}{2} m a_0^2 e^{-2bt} \times \left(\dfrac{b^2 + \omega^2}{2}\right) = \dfrac{1}{4} m a_0^2 \omega^2 e^{-2bt}$ ...(13)

and the average potential energy $= \dfrac{1}{2} m a_0^2 \omega^2 e^{-2bt} \times \dfrac{1}{2} = \dfrac{1}{4} m a_0^2 \omega^2 e^{-2bt}$ ...(14)

Hence, the average total energy in one periodic time

$$E_{av} = \frac{1}{4} m a_0^2 \omega^2 e^{-2bt} + \frac{1}{4} m a_0^2 \omega^2 e^{-2bt}$$

$$= \frac{1}{2} m a_0^2 \omega^2 e^{-2bt} = \frac{1}{2} m a_0^2 \omega^2 e^{-t/\tau} \quad (\because \tau = 1/2b)$$

or $$E_{av} = E_0\, e^{-2bt} \quad ...(15)$$

where, $$E_0 = \frac{1}{2} m a_0^2\, \omega^2$$

It is clear from equation (15) that the average energy of system in damped oscillations decreases exponentially with time as shown in Figure 1.13.

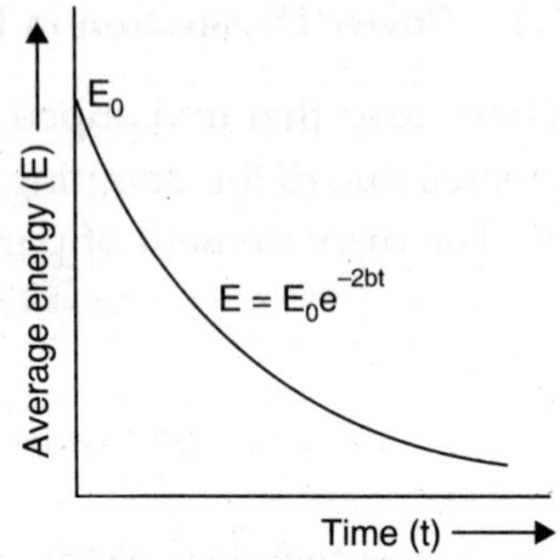

FIGURE 1.13 Variation of average energy with time in damped oscillator

The rate of loss of energy or the average power loss at any instant

$$P = \frac{-dE_{av}}{dt} = \frac{1}{2}\frac{m a_0^2\, \omega^2}{\tau} e^{-t/\tau} = \frac{E}{\tau} = 2b < E_{av} > \quad ...(16)$$

Clearly, the rate of loss of energy at any instant depends on the damping of the system (more the value of b, more is the rate of loss in energy).

This loss in energy generally appears in the form of heat energy of the oscillating system.

1.7.2 Relaxation Time of Damped Harmonic Oscillator

If $t = \tau$ in equation $a = a_0\, e^{-bt} = a_0\, e^{-t/2\tau}$ then $a = a_0\, e^{-1/2} = 0.6065\, a_0$.

Hence, relaxation time in damped simple harmonic oscillator is that time in which its amplitude decreases to 0.6065 time its initial value.

In other way, from equation (15)

$$E_{av} = E_0\, e^{-2bt} = E_0\, e^{-t/\tau}, \text{ hence at } t = \tau,$$

$$\boxed{E_{av} = E_0\, e^{-1} = 0.37\, E_0}$$

Hence, the relaxation time in damped simple harmonic oscillator is that time in which its total energy reduces to 0.37 times its initial value.

1.7.3 Quality Factor of Damped Harmonic Oscillator

The quality factor of an oscillator expresses its efficiency. It is represented by the letter Q. Since there is no energy loss in free oscillations of an ideal oscillator, its efficiency is maximum.

Definition: *The quality factor of an oscillator is defined as the product of 2π with the ratio of the average energy stored in the oscillator at any instant to the loss in energy in one periodic time.*

i.e., Quality factor $$Q = 2\pi \times \frac{\text{energy stored at any instant}}{\text{energy lost in one second}} \quad ...(17)$$

or $$Q = \frac{2\pi \times E_{av}}{PT} = \frac{E_{av}\, \omega}{P} \qquad \left[\because \frac{2\pi}{T} = \omega\right]$$

But from equation (16) $\quad P = E_{av}/\tau \quad$ or $\quad E_{av}/P = \tau$

$$\boxed{Q = \omega\tau} \quad ...(18)$$

Thus, quality factor is a dimensionless quantity. More the relaxation time of an oscillator, more is its quality factor.

1.8 FORCED OR DRIVEN HARMONIC OSCILLATOR

When an external periodic force is applied on a system, the force imports a periodic impulse to the system so that the loss in energy in doing work against the dissipative forces is recovered. As a result, the system is continuously oscillates. In the initial stages, the system tends to execute oscillations with the natural frequency (or frequency of free vibrations), while the impressed periodic force tries to impose its own frequency on it. Therefore, the free vibrations of the body soon die out and ultimately, the system starts oscillating with a constant amplitude and with the frequency equal to that of the impressed force. This is called the *steady state* of an oscillator. These vibrations are called the *forced vibrations*. The force impressed on the system is called the *driver* and the system which executes *forced vibrations*, is called the *forced* or *driven harmonic oscillator*.

Thus a particle executing the forced harmonic oscillations is acted upon by the following three forces:

(*i*) A linear restoring force ($= -Kx$), which is directly proportional to the displacement from a fixed point and is in a direction opposite to the displacement.

(*ii*) A damping force $\left(= -r\dfrac{dx}{dt}\right)$, which is directly proportional to the velocity and is in a direction opposite to the motion.

(*iii*) An external periodic force ($F = F_0 \sin \omega t$), where F_0 is the amplitude of the impressed force and ω is the angular frequency of the impressed force.

If m be the mass of particle executing forced oscillations and $\dfrac{d^2x}{dt^2}$ is its acceleration at any instant, then by Newton's law.

Total force acting on the particle $= m\dfrac{d^2x}{dt^2}$

The equation of motion of forced harmonic oscillations will be

$$m\frac{d^2x}{dt^2} = -Kx - r\frac{dx}{dt} + F$$

or
$$m\frac{d^2x}{dt^2} + r\frac{dx}{dt} + Kx = F_0 \sin \omega t$$

or
$$\frac{d^2x}{dt^2} + \frac{r}{m}\frac{dx}{dt} + \frac{K}{m}x = \frac{F_0}{m}\sin \omega t$$

or $$\boxed{\frac{d^2x}{dt^2} + 2b\frac{dx}{dt} + n^2x = f \sin \omega t} \qquad ...(1)$$

where, $r/m = 2b$, $K/m = n^2$ and $F_0/m = f$

Obviously from equation (1), the differential equation for the free oscillations of the system will be $\frac{d^2x}{dt^2} + n^2x = 0$, where $n = \sqrt{\frac{K}{m}}$ = angular frequency of free oscillations.

Transient and Steady States

We have read that on applying an external periodic force, initially the system tends to oscillate with an angular frequency $\omega_1 = \sqrt{(n^2 - b^2)}$ due to damping, but the driving force of angular frequency ω acting on the system forces it to oscillate with its own frequency. In this way, the actual motion of system is obtained by the superposition of two oscillations, whose angular frequencies are $\omega_1 = \sqrt{(n^2 - b^2)}$ and $\omega_2 = \omega$ respectively. Hence if $\omega \neq n$, the solution of the equation (1) can be written as follows

$$x = x_1 + x_2$$

where, x_1 is the solution of equation (1), when the external force is zero. Then from equation (1)

$$x_1 = a\, e^{-bt} \sin (nt + \phi) \qquad ...(2)$$

and $$\frac{d^2x_2}{dt^2} + 2b\frac{dx_2}{dt} + n^2x_2 = f \sin \omega t \qquad ...(3)$$

Remember that in the complete solution $x = x_1 + x_2$, x_1 is known as the *complementary function* and x_2 is known as the *particular integral.*

The complementary function represented by equation (2) decreases exponentially with time and after some time, this term vanishes, hence it is also known as the transient solution of the forced harmonic oscillator. In this way, the system in transient state, oscillates with a frequency different from the natural frequency or the frequency of the driving force.

After a long time, when $t >> \tau$, the natural oscillations of the system vanishes due to damping and then the system oscillates with the frequency of the driving force. This state of the system is known as the *steady state.*

Let the solution of equation (3) in the steady state be $x_2 = A \sin (\omega t - \theta)$

[Since in the steady state, the amplitude of forced oscillation is constant (= A, say) and the frequency is equal to the frequency of impressed force *i.e.*, $\frac{\omega}{2\pi}$.]

Here θ is the phase difference between the displacement and the impressed force.

Then, $$\frac{dx_2}{dt} = A\,\omega \cos (\omega t - \theta) \quad \text{and} \quad \frac{d^2x_2}{dt^2} = -A\,\omega^2 \sin (\omega t - \theta)$$

Substituting these values in equation (3)

$$- A\,\omega^2 \sin(\omega t - \theta) + 2bA\omega \cos(\omega t - \theta) + n^2 A \sin(\omega t - \theta) = f \sin(\omega t - \theta + \theta)$$

or
$$A(n^2 - \omega^2) \sin(\omega t - \theta) + 2bA\omega \cos(\omega t - \theta) = f\,[\sin(\omega t - \theta) \cos\theta + \cos(\omega t - \theta) \sin\theta]$$

Since this equation is valid for all value of t, therefore by equating the coefficients of sin $(\omega t - \theta)$ and cos $(\omega t - \theta)$ separately, we get

$$A\,(n^2 - \omega^2) = f \cos\theta \text{ and } 2bA\omega = f \sin\theta$$

Squaring and adding the above equations

$$A^2[(n^2 - \omega^2)^2 + 4b^2\omega^2] = f^2[\cos^2\theta + \sin^2\theta]$$

or
$$A = \frac{f}{[4b^2\omega^2 + (n^2 - \omega^2)^2]^{1/2}} \qquad ...(4)$$

and
$$\tan\theta = \frac{f \sin\theta}{f \cos\theta} = \frac{2bA\omega}{A(n^2 - \omega^2)} = \frac{2b\omega}{n^2 - \omega^2}$$

or
$$\theta = \tan^{-1}\left(\frac{2b\omega}{n^2 - \omega^2}\right) \qquad ...(5)$$

Putting these values of A and θ in $x_2 = A \sin(\omega t - \theta)$, the solution of equation (3) is

$$x_2 = \frac{f}{[4b^2\omega^2 + (n^2 - \omega^2)^2]^{1/2}} \sin\left[\omega t - \tan^{-1}\left(\frac{2b\omega}{n^2 - \omega^2}\right)\right] \qquad ....(6)$$

Substituting the values of x_1 and x_2; the complete solution of differential equation (1) of forced oscillator is given as

$$\boxed{x = a_0\, e^{-bt} \sin(nt + \phi) + \frac{f}{[4b^2\omega^2 + (n^2 - \omega^2)^2]^{1/2}} \sin\left[\omega t - \tan^{-1}\frac{2b\omega}{n^2 - \omega^2}\right]} \qquad ...(7)$$

In the above equation, the first term on the RHS which is transient part, decreases with time and finally its role vanishes. The time upto which it plays its role, depends on the amount of damping. More damping, more rapidly this term decreases to zero.

When the damping is zero (*i.e.*, when $b = 0$), in steady state $x = \dfrac{f}{(n^2 - \omega^2)} \sin \omega t$ and $\theta = 0°$ (*i.e.*, the driving force and displacement will be in same phase). It is concluded that the phase difference between the displacement and driving force of a forced oscillator is due to damping. It is also clear that when $\omega = n$, the amplitude of oscillations become infinite. This condition is known as **Resonance**.

Figure 1.14 represents the variation of the transient displacement x_1, steady displacement x_2 and their sum, *i.e.*, total displacement x with time t of a forced oscillator for low damping.

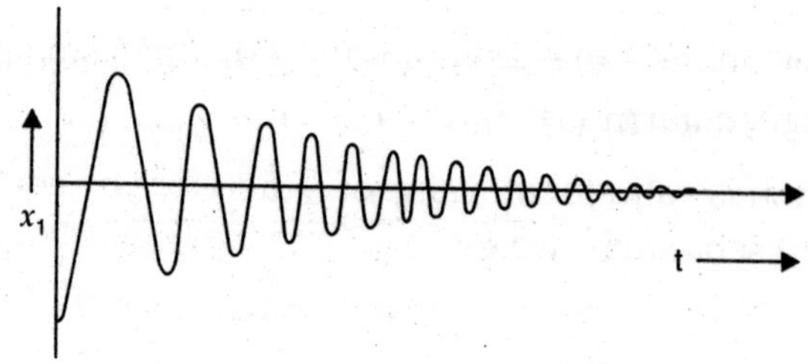

(a) Variation of transient displacement x_1 with time

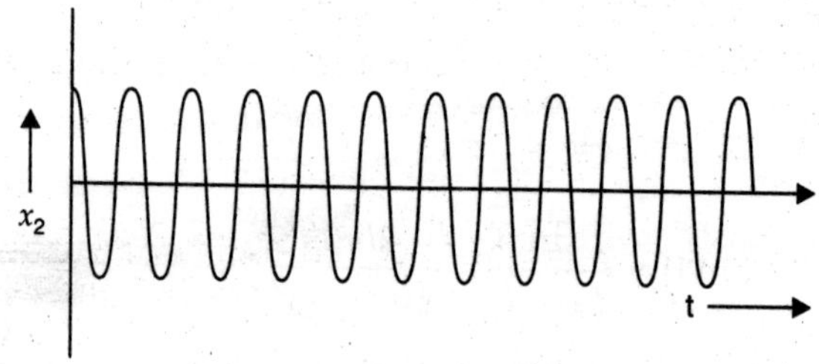

(b) Variation of steady displacement x_2 with time

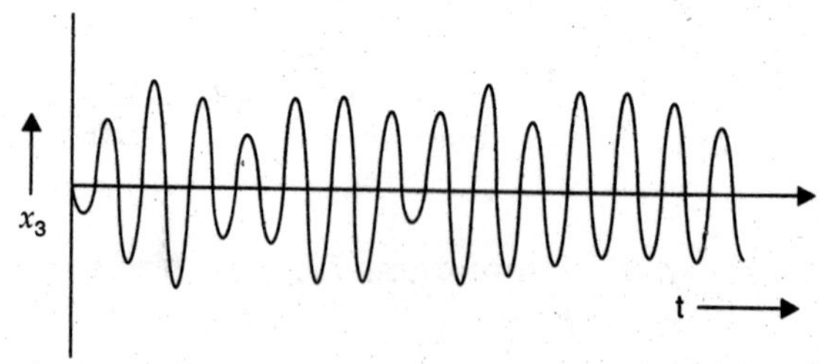

(c) Variation of total displacement x with time

FIGURE 1.14 **Variation of displacement of a forced oscillator with time**

Above equation (7) gives the displacement of the forced oscillator at any instant t. It is clear that the amplitude of the driven oscillator in steady state does not depend on time t (*i.e.*, it remains constant with time), but it depends on the frequency ω of the external periodic force.

Now we will study the following three cases:

(*i*) When $\omega << n$, *i.e.*, the frequency of the driving force is much less.

(*ii*) When $\omega = n$, *i.e.*, the state of resonance.

(*iii*) When $\omega > n$, *i.e.*, the frequency of the driving force is much high.

Case (*i*). When $\omega << n$, *i.e.*, the driving frequency is less than the natural frequency of the driven.

For low damping (when $b \to 0$), from equation (4) and equation (5)

$$A = \frac{f}{n^2} = \frac{F_0/m}{K/m} = \frac{F_0}{K} \text{ and } \tan\theta = 0 \quad \text{or} \quad \theta = 0° \qquad ...(8)$$

Thus in this case, the displacement is in phase with the driving force and the amplitude of oscillations does not depend on the mass and damping, but only depends on the force constant.

Case (*ii*). When $\omega = n$, *i.e.*, the frequency of driver is equal to the natural frequency of the driven. Then from equation (4)

$$A = \frac{f}{2b\omega} = \frac{f\tau}{\omega} = \frac{f\tau}{n} = \frac{F_0/m}{(r/m)\,n} = \frac{F_0}{nr} \qquad ...(9)$$

$$\left[\text{Since, } \tau = \frac{1}{2b} \text{ and } \omega = n\right]$$

This is called the *state of resonance*. Thus at resonance, the amplitude of oscillation depends on the damping coefficient r. Low the damping, more is the amplitude. If damping is zero (*i.e.*, $r = 0$ or $\tau = \infty$), $A_{max} = \infty$ (infinite)

From, equation (5) $\tan\theta = \infty$ or $\theta = \pi/2$ ...(10)

i.e., at resonance, the displacement of the oscillator lags behind the driving force in phase by $\pi/2$.

Remember that the amplitude represented by equation (9) is not maximum. The reason behind it is as follows:

From, $A = \dfrac{f}{\sqrt{\left[4b^2\omega^2 + (n^2 - \omega^2)^2\right]}}$, it is clear that for A to be maximum, the value of the term $\sqrt{(4b^2\omega^2 + (n^2 - \omega^2))}$ must be minimum. *i.e.*,

$$\frac{d}{d\omega}\left[\sqrt{(4b^2\omega^2 + (n^2 - \omega^2)^2}\right] = 0$$

or $$\frac{d}{d\omega}\left[4b^2\omega^2 + (n^2 - \omega^2)^2\right] = 0$$

or $$8b^2\omega + 2(n^2 - \omega^2)(-2\omega) = 0$$

or $$\omega^2 = n^2 - 2b^2 = \omega_r^2 \text{ (Say)}$$

Then, $$\boxed{A_{max} = \frac{f}{\sqrt{(4b^2n^2 - 4b^4)}} = \frac{f}{2b\sqrt{(n^2 - b^2)}}} \qquad ...(11)$$

Thus, at a particular frequency of the driver, the amplitude of oscillator becomes maximum. This phenomenon is called the *amplitude resonance* and this particular frequency is called the *resonance frequency*. The resonant frequency of forced harmonic oscillator

$$f_r = \frac{\omega_r}{2\pi} = \frac{1}{2\pi}\sqrt{(n^2 - 2b^2)} \qquad ...(12)$$

If damping is zero (*i.e.*, $b = 0$), than $\omega_r = n$ (*i.e.*, the resonant angular frequency of the oscillator is equal to the natural angular frequency of the driven) and maximum amplitude A_{max} = infinite. Figure 1.15 shows the resonant amplitude of different cases.

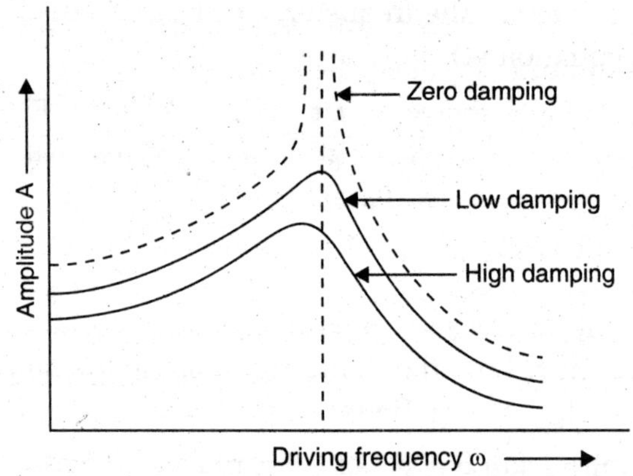

FIGURE 1.15 **Variation of amplitude in steady state with the driving frequency**

Case (*iii*). When $\omega > n$, *i.e.*, the frequency of driver in more than the natural frequency of driven. In case if damping is very low (*i.e.*, $b \to 0$),

$$\text{Amplitude } A = \frac{f}{\omega^2} \text{ (nearly)} \quad \text{and} \quad \tan\theta = -\frac{2b}{\omega} \qquad ...(13)$$

As the frequency of the impressed force increases, the amplitude of oscillation decreases and $\tan\theta \to 0$ or $\theta \to \pi$ (*i.e.*, the displacement tends to be in opposite phase with the impressed force).

Thus, it is clear that

(*i*) In steady state, amplitude A of oscillation changes with the frequency of the impressed force and at $\omega_0 = \sqrt{(n^2 - 2b^2)}$, its value becomes maximum $\left(= \frac{f}{2b\sqrt{(n^2 - b^2)}}\right)$. Again on increasing the frequency (when $\omega > \omega_r$), the value of A decreases inversely as the square of frequency $\left(i.e., \; A \propto \frac{1}{\omega^2}\right)$.

(*ii*) The displacement lags behind the impressed force by angle θ in phase. The value of angle θ increases from 0 to π with increase in frequency. At $\omega = 0$, $\theta = 0$, *at* $\omega \neq n$, $\theta = \pi/2$ and at high frequency $\theta \to \pi$.

Figure 1.15 shows the variation of amplitude A of a forced harmonic oscillator with angular frequency ω of impressed force. As the angular frequency ω of the impressed force increases, the amplitude of oscillations first increases and at a particular value of ω (which is equal to the natural frequency of oscillator at zero damping or is slightly less than it), the equal to natural frequency of oscillator at zero damping or is slightly less than it), the amplitude becomes maximum, than with further increase in the angular frequency of impressed force, the amplitude of oscillation slowly decreases. As low the damping, more is the resonant amplitude.

When damping is zero, the amplitude is infinite at resonance. The frequency corresponding to maximum value of amplitude is called the *resonant frequency*. It is clear from Figure 1.15, that when the damping is high, the resonant frequency is slightly smaller than the natural frequency of oscillator, but when the damping is low, the resonant frequency is nearly equal to the free (or natural) frequency of oscillator.

Figure 1.16 shows the variation of phase difference θ with the angular frequency ω of the impressed periodic force. As the angular frequency of the impressed force increases from zero, the phase difference θ between the displacement and the impressed force increases from 0 to π. In the initial stages, when the frequency of impressed force is less than the free frequency of the oscillator (*i.e.*, $\omega << n$), the displacement is in phase with the driving force. As the angular frequency of the driving force increase, the phase difference between the displacement and the driving force increases. When $\omega = n$ (*i.e.*, at resonance), the displacement lags behind the driving force in phase by $\pi/2$. Now with further increase in frequency of the driving force, the phase difference increases and when $\omega >> n$, the displacement lags behind the driving force in phase by π.

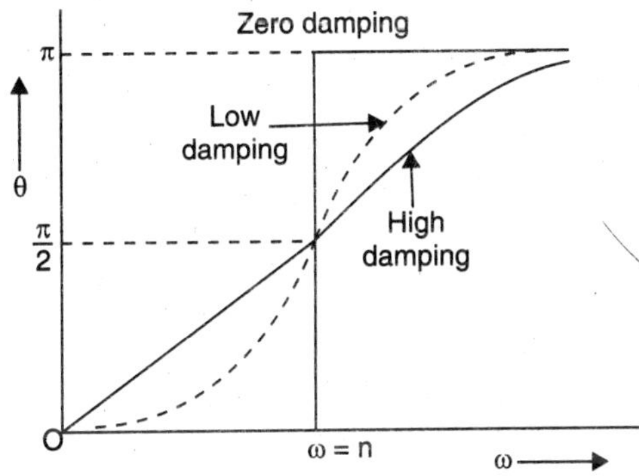

FIGURE 1.16 **Variation of phase difference with frequency in steady state**

1.8.1 Power Absorption by a Forced Oscillator

We know that there is some loss in energy in doing work against damping by each oscillating system but in a forced oscillator, the oscillations are maintained by providing energy by external impressed force. Hence it becomes essential to know that at what rate of the energy should be supplied to the system to maintain the oscillations in the steady state, *i.e.*, what is the absorption of average energy by the oscillating system? By definition at any instant t, power P = force $F \times$ velocity v.

But the solution in steady state

$$x = \frac{f}{\left[(n^2 - \omega^2)^2 + 4b^2\omega^2\right]^{1/2}} \sin(\omega t - \theta)$$

$$= v_0 \sin(\omega t - \theta + \pi/2) = v_0 \sin(\omega t - \phi)$$

where $\phi = \theta - \frac{\pi}{2}$ and $v_0 = \frac{f\omega}{\sqrt{(n^2-\omega^2)^2 + 4b^2\omega^2}}$

and $$\theta = \tan^{-1}\left(\frac{2b\omega}{n^2-\omega^2}\right)$$

Then at any instant t, power $P = F_0 \sin \omega t \times v_0 \sin(\omega t - \phi)$ or average power absorbed

$$P_{av} = \frac{1}{T}\int_0^T F\,v\,dt$$

$$= \frac{1}{T}\int_0^T F_0 \sin \omega t\; v_0 \sin(\omega t - \phi)\,dt$$

$$= \frac{F_0 v_0}{T}\int_0^T (\sin^2 \omega t \cos\phi - \sin\omega t \cos\omega t \sin\phi)\,dt$$

$$= \frac{F_0 v_0}{T}\left[\cos\phi\int_0^T\left(\frac{1-\cos 2\omega t}{2}\right)dt - \sin\phi\int_0^T \frac{\sin 2\omega t}{2}\,dt\right]$$

$$= \frac{F_0 v_0}{T}\left[\frac{1}{2}\cos\phi\left(t - \frac{\sin 2\omega t}{2\omega}\right)_0^T - \frac{1}{2}\sin\phi\left(\frac{-\cos 2\omega t}{2\omega}\right)_0^T\right]$$

$$= \frac{1}{2} F_0 v_0 \cos\phi$$

$$= \frac{b F_0^2}{m}\left[\frac{\omega^2}{(n^2-\omega^2)^2 + 4b^2\omega^2}\right] \quad ...(14)$$

In the condition of resonance

$$(v_0)_{max} = \frac{f}{2b} = \frac{F_0/m}{r/m} = \frac{F_0}{r} \quad \text{and} \quad \phi = 0° \text{ or } \cos\phi = 1$$

$$\therefore \quad \boxed{(P_{av})_{max} = \frac{1}{2}F_0 \times \frac{F_0}{r} = \frac{F_0^2}{2r}} \quad ...(15)$$

Thus, in the condition of resonance, the average power absorbed by the forced oscillator is maximum because in the condition of resonance, velocity and impressed force are in the same phase.

The value of maximum absorbed average power depends on the amplitude of the impressed force. Figure 1.17 represents the variation of absorbed average power with frequency.

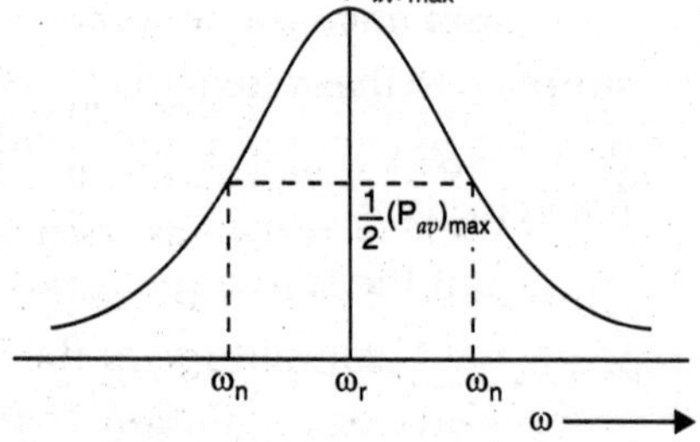

FIGURE 1.17 **Variation of average power with frequency**

1.8.2 Quality Factor Q of the Forced Harmonic Oscillator

We have read that the quality factor

$$Q = 2\pi \times \frac{\text{average energy stored in one cycle}}{\text{average energy loss in one cycle}}$$

Now, average kinetic energy $K_{av} = \frac{1}{2} mv^2$

$$= \frac{1}{2} m\,\omega^2 A^2 \times [\text{average value of } \cos^2(\omega t - \phi)]$$

$$= \frac{1}{4} m\,\omega^2 A^2$$

$\therefore$ The average value of cos 2 $(\omega t - \phi)$ is $\frac{1}{2}$.

and average potential energy $U_{av} = \frac{1}{2} K\,x^{-2}$

$$U_{av} = \frac{1}{2} m\,\omega^2 A^2 \times [\text{the average value of } \sin^2(\omega t - \theta)]$$

$$= \frac{1}{4} m\,\omega^2 A^2$$

since, the average value of $\sin^2(\omega t - \phi) = \frac{1}{2}$

$\therefore$ Total average energy in one cycle is $E_{av} = K_{av} + U_{av} = \frac{1}{2} m\,\omega^2 A^2$ ...(16)

Now, average energy loss per second = average value of $r\,v^2$

$$= m\,b\,\omega^2 A^2 \qquad \text{...(17)}$$

Hence, quality factor $Q = 2\pi \times \dfrac{\frac{1}{2} m\,\omega^2 A^2}{T \times m\,b\,\omega^2 A^2} = \frac{1}{2} \times \frac{\omega}{b} = \omega\tau$

$$\boxed{Q = \omega\tau} \qquad \text{...(18)}$$

1.8.3 Resonance and Sharpness of Resonance

We have read above that when the frequency of impressed force becomes equal to the natural frequency of the system, the amplitude of oscillation of forced oscillator becomes maximum. This condition is known as *resonance*. At resonance, the displacement of oscillator lags behind the impressed force in phase by $\pi/2$ and the maximum value of amplitude of oscillation depends on the amplitude of the impressed force and damping. Lesser the damping, more is the amplitude of oscillation at resonance. In ideal conditions when damping is zero, the amplitude at resonance infinite. Besides this, the average power absorbed by the system in the condition of resonance is maximum.

From equation (11), resonant amplitude $A_{max} = \dfrac{f}{2b\sqrt{n^2 - b^2}}$ and

From equation (12), resonant angular frequency $\omega_r = \sqrt{(n^2 - 2b^2)}$.

Sharpness of Resonance: Sharpness of resonance is measured by the rate at which the amplitude falls with the frequency of the driving force on either side of the resonant frequency. If the amplitude falls rapidly on small change in frequency of the driving force from its resonant value, the resonance is said to be *sharp* and if the amplitude falls slowly, the resonance is said to be *'flat'*. It is clear from Figure 1.15 that if damping is low, the resonance is sharp and if damping is high, the resonance is flat. The resonance of air column with a tuning fork is flat, while the resonance of sonometer wire with a tuning fork is sharp.

1.8.4 Half Power Point and Half Width of the Resonance Curve

If the amplitude of oscillation of a forced oscillator falls to $\dfrac{1}{\sqrt{2}}$ of the maximum amplitude at resonance at angular frequency ω_h (Figure 1.18) or the energy of oscillator falls to half of its maximum energy at resonance (Figure 1.17), then $|\omega_h - \omega_r|$ is called the half width of resonance curve. The amplitude falls to $\dfrac{1}{\sqrt{2}}$ of the resonant amplitude on either side of resonance, are called the *half power points*. Thus, the *half width of resonance curve* is $\Delta\omega = |\omega_h - \omega_r|$

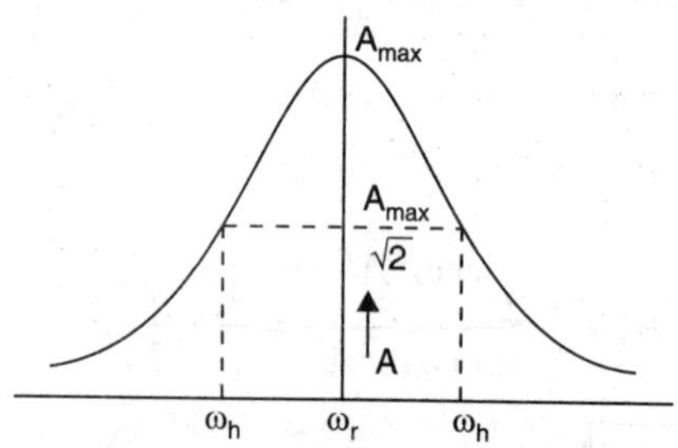

FIGURE 1.18 **Variation of amplitude with frequency**

From equation (4), the amplitude of a forced oscillator is

$$A = \frac{f}{\sqrt{4b^2\omega^2 + (n^2 - \omega^2)^2}}$$

$$= \frac{f}{\sqrt{[(n^2 - 2b^2 - \omega^2)^2 + 4b^2n^2 - 4b^4]}}$$

At resonance [From equation (11)]

$$A_{max} = \frac{f}{2b\sqrt{(n^2 - b^2)}} = \frac{f}{\sqrt{(4b^2n^2 - 4b^4)}}$$

At $\omega = \omega_h$, since $A = \frac{1}{\sqrt{2}} A_{max.}$

$$\therefore \quad \frac{f}{\sqrt{[(n^2 - 2b^2 - \omega_h^2)^2 + 4b^2n^2 - 4b^4]}} = \frac{1}{\sqrt{2}} \times \frac{f}{\sqrt{(4b^2n^2 - 4b^4)}}$$

or $(n^2 - 2b^2 - \omega_h^2)^2 + 4b^2n^2 - 4b^4 = 2(4b^2n^2 - 4b^4)$

But $\omega_r^2 = n^2 - 2b^2 \quad \therefore \quad (\omega_r^2 - \omega_h^2)^2 = (4b^2n^2 - 4b^4)$

or $(\omega_r^2 - \omega_h^2)^2 = (4b^2\omega_r^2 + 4b^4)$

If damping is low, we can neglect the term $4b^4$. Then

$$(\omega_r^2 - \omega_h^2) = \pm 2b\omega_r$$

or $\omega_h^2 = \omega_r^2 \pm (2b\omega_r)$

or $\omega_h = \omega_r \left(1 \pm \frac{2b}{\omega_r}\right)^{1/2}$ or $\omega_h = \omega_r \left(1 \pm \frac{b}{\omega_r}\right) = \omega_r \pm b$

Hence, half width of resonance curve

$$\boxed{\Delta\omega = |\ \omega_h - \omega_r\ | = b = \frac{1}{2\tau}} \qquad \text{...(19)}$$

where τ is the relaxation time

Now, quality factor $Q = 2\pi \times \frac{\text{Average energy stored}}{\text{Energy loss in one period}}$

$$= 2\pi \times \frac{E_{av}}{PT} = \frac{E_{av}\,\omega}{P} \qquad \left[\because \frac{2\pi}{T} = \omega\right]$$

But $P = E_{av}/\tau$ or $E_{av}/P = \tau \quad \therefore \quad \boxed{Q = \omega\tau}$

From equation (17)

Quality factor, $Q = \frac{\omega}{2\Delta\omega}$

$$= \frac{\text{frequency when average power is maximum}}{\text{total width of half power point}}$$

Thus, for the sharp resonance, the value of Q must be high, *i.e.*, the half width $\Delta\omega$ of the resonance curve must be small.

1.8.5 Velocity Resonance

From equation of the displacement of a forced oscillator is

$$x = \frac{f}{\sqrt{[(n^2 - \omega^2)^2 + 4b^2\omega^2]}} \sin(\omega t - \theta)$$

and velocity of the oscillator at any instant

$$v = \frac{dx}{dt} = \frac{f\omega}{\sqrt{(n^2 - \omega^2)^2 + 4b^2\omega^2}} \cos(\omega t - \theta)$$

or

$$v = v_0 \sin\left(\omega t - \theta + \frac{\pi}{2}\right) = v_0 \sin(\omega t - \phi)$$

where $\phi = \theta - \frac{\pi}{2}$

Hence amplitude of velocity

$$v_0 = \frac{f\omega}{\sqrt{(n^2 - \omega^2)^2 + 4b^2\omega^2}} = \frac{f}{\sqrt{\left(\omega - \frac{n^2}{\omega}\right)^2 + 4b^2}} \qquad ...(20)$$

and

$$\theta = \tan^{-1}\left(\frac{2b\omega}{n^2 - \omega^2}\right)$$

It is clear from the above expression (20) that as the frequency ω of the driving force changes, the amplitude of velocity also changes. Figure 1.19 shows the variation of velocity amplitude v_0 with the frequency ω of the driving force.

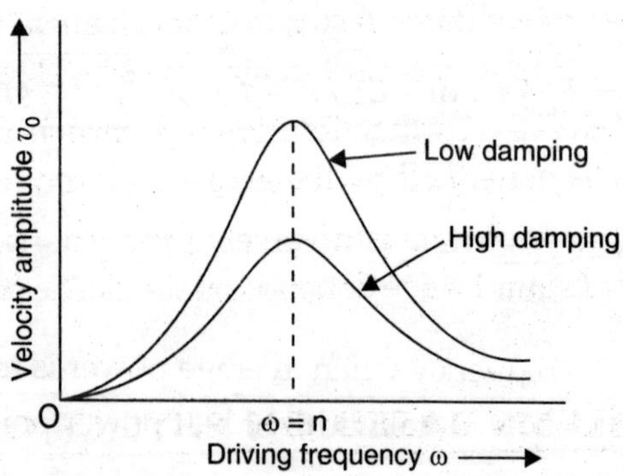

FIGURE 1.19 **Variation of velocity amplitude with frequency**

When $\omega = 0$, $\quad v_0 = 0$

When $\omega = \frac{n^2}{\omega}$ or $\omega = n$, $\quad v_0 = \frac{f}{2b}$ = maximum

This is called the condition of *velocity resonance*. When the value of ω exceeds n, the value of v_0 decreases. Thus, when the frequency of the driving force is equal to the natural frequency of the oscillator, the velocity of oscillator is maximum. Hence the kinetic energy of the oscillator is also maximum. At resonance, the amplitude of velocity depends on the damping (lower the damping, high is the velocity amplitude).

At velocity resonance, $\theta = \tan^{-1} (\infty) = \pi/2$

$\therefore$ Phase difference between the velocity and the driving force is

$$\phi = \theta - \frac{\pi}{2} = \frac{\pi}{2} - \frac{\pi}{2} = 0$$

i.e., at resonance, the velocity of oscillator is in phase with the driving force.

1.9 DIFFERENCE BETWEEN MECHANICAL WAVE AND ELECTROMAGNETIC WAVE

1. A mechanical wave needs a medium for its propagation. An electromagnetic wave can travel through vacuum and it has the highest speed as it moves through the vacuum. Sound wave is a mechanical wave and light is an electromagnetic wave.
2. A mechanical wave can be both transverse and longitudianl but an electromagnetic wave is a transverse wave.
3. The particles of the medium vibrate, when a mechanical wave passes through it. In an electromagnetic wave the oscillating parameters are the electric and magnetic field. At any point on the line of propagation the field strength varies periodically with time.
4. Electromagnetic waves move much faster than mechanical waves. Speed of light in vacuum is 3×10^8 m/s. Speed of sound in air is nearly 332 m/s.
5. Mechanical wave is generated by the vibration of a particle in any material medium. Electromagnetic wave is generated by the accelerated motion of a charged particle.
6. Speed of an electromagnetic wave is inversely proportional to the refractive index of the medium. Speed of sound wave depends on the elastic properties of the medium.

1.10 NUMERICAL PROBLEMS BASED ON PHYSICS OF VIBRATIONS

Problem 1. *A particle of mass 0.1 kg is situated in the potential field $V = 5x^2 + 10$ Joule kg^{-1}. Write down the differential equation of motion of the particle and its frequency of oscillations.*

Solution: Potential energy of the particle U = mass × potential = $0.1 \times (5x^2 + 10)$ Joule

Force $F = -\dfrac{dU}{dx} = 0.1\,(10x) = -x$

But force = mass × acceleration $\therefore\quad F = 0.1\dfrac{d^2x}{dt^2}$

Hence equation of motion of the particle will be

$$0.1\frac{d^2x}{dt^2} = -x \quad\text{or}\quad \frac{d^2x}{dt^2} + 10x = 0 \quad\text{or}\quad \frac{d^2x}{dt^2} + \omega^2 x = 0$$

The above equation represents the SHM of the particle whose frequency of oscillation is

$$n\,\frac{\omega}{2\pi} = \frac{1}{2\pi}\sqrt{10} = 0.5\text{ Hz}$$

Problem 2. *A particle of mass 10 g is moving along the X-axis under the force $4 \times 10^{-2}\,x$ acting towards the origin. If the initial position of rest of the particle is x = 0.1 m, write down the differential equation of motion of particle and thus calculate (i) the position and velocity of the particle at any instant, (ii) the frequency of oscillation and amplitude of oscillation.*

Solution: Given m = 10 g = 0.01 kg, force $F = -4 \times 10^{-2}\,x$

Hence differential equation of the motion of particle will be

$$m\frac{d^2x}{dt^2} = F \quad\text{or}\quad 0.01\frac{d^2x}{dt^2} = -4\times 10^{-2}\,x$$

or
$$\frac{d^2x}{dt^2} + 4x = 0 \quad\text{or}\quad \frac{d^2x}{dt^2} + \omega^2 x = 0 \quad\text{where}\quad \omega^2 = 4 \quad\text{or}\quad \omega = 2$$

(*i*) Let position of particle at any instant t be $x = a\sin(\omega t + \phi)$, where $\omega = 2$

Given, at $t = 0$, $x = 0.1$ metre and $\dfrac{dx}{dt} = 0$

$$\therefore\qquad 0.1 = a\sin\phi \quad\text{and}\quad 0 = a\omega\cos\phi \qquad \left[\because\ \frac{dx}{dt} = a\omega\cos(\omega t + \phi)\right]$$

i.e.,
$$\cos\phi = 0 \quad\text{or}\quad \phi = \pi/2 \quad\text{and}\quad a = 0.1\text{ metre}$$

Hence
$$x = 0.1\sin\left(2t + \frac{\pi}{2}\right) = 0.1\cos 2t$$

and velocity of the particle at any instant t is

$$u = \frac{dx}{dt} = -0.2\sin 2t$$

(*ii*) Amplitude of oscillations of the particle a = 0.1 metre and frequency of oscillation

$$n = \frac{\omega}{2\pi} = \frac{2}{2\pi} = 0.32\text{ Hz.}$$

Problem 3. *A particle executing simple harmonic motion completes 1200 oscillations in 1 minute and while passing through its mean position, its velocity is 3.14 m/s.*

(*i*) *What is the maximum displacement of the particle from its mean position?*

(*ii*) *If at an instant t = 0, displacement is zero, obtain the equation for displacement of the particle.*

Hint:

(*i*) $\omega = 2\pi n = 2\pi \times 20 = 40\pi$ radian per second

$$U_{max} = \omega a \quad \text{or} \quad 3.14 = 40\pi a \quad \Rightarrow \quad a = \frac{3.14}{40\pi} = \frac{1}{40} = 0.025\ m$$

(*ii*) $t = 0$, $x = a \sin \omega t = 0.025 \sin 40\pi t$.

Problem 4. *The amplitude of a particle of mass 0.1 kg executing simple harmonic motion is 0.1 m and its kinetic energy in mean position is 8×10^{-3} J. If the initial phase difference is 45°, write down the equation of motion of the particle.*

Hint: $a = 0.1\ m$, kinetic energy at mean position $= \frac{1}{2} ma^2\omega^2 = 8 \times 10^{-3}$

$$\therefore \qquad \omega = \frac{1}{0.1}\sqrt{\frac{2 \times 8 \times 10^{-3}}{0.1}} = 4$$

Hence the equation of motion is

$$\boxed{x = 0.1 \sin (4t + \pi/4).}$$

Problem 5. *A particle executes SHM of amplitude 25 cm and time period 3s. What minimum time will the particle take to move between the two points at a separation of 12.5 cm on either side of its mean position.*

Hint: If the particle at $t = 0$ is at mean position, its displacement equation will be $x = 25 \sin \frac{2\pi t}{3}$. If it takes time t_1 to move a distance $x = 12.5$ cm to one side of its mean position,

then $12.5 = 25 \sin \frac{2\pi t_1}{3} \quad \therefore \quad \sin \frac{2\pi t_1}{3} = \frac{1}{2} = \sin \frac{\pi}{6} \quad \text{or} \quad t_1 = \frac{1}{4}$ s.

The same will be the time to move 12.5 cm to the other side of its mean position, hence total time $t = t_1 + t_2$

$$= \frac{1}{4} + \frac{1}{4} = 0.5 \text{ s.}$$

Problem 6. *The amplitudes of two identical simple pendulums are 2 cm and 6 cm respectively. Compare their energies of oscillations.*

Hint: Energy of oscillations $E \propto a^2 \quad \therefore \quad \frac{E_1}{E_2} = \left(\frac{a_1}{a_2}\right)^2 = \left(\frac{2}{6}\right)^2 = \frac{1}{9}.$

Problem 7. *The differential equation of an oscillating system is expressed as*

$$\frac{d^2x}{dt^2} + 2b\frac{dx}{dt} + n^2x = 0$$

If b << n, find the time in which: (i) the amplitude of oscillation falls to $\frac{1}{e}$ *of its initial value, (ii) the energy falls to* $\frac{1}{e}$ *of its initial value.*

Solution: Since $b << n$, hence the given equation represents the differential equation of damped harmonic oscillator whose amplitude is $a = a_0\, e^{-bt}$

(*i*) When the amplitude falls to $1/e$ times its initial amplitude a_0, then

$$a_0\, e^{-bt} = \frac{1}{e}\, a_0 \quad \text{or} \quad e^{-bt} = e^{-1}$$

or

$$-\,bt = 1 \quad \text{or} \quad t = (1/b)\text{ s}$$

(*ii*) Since energy is proportional to the square of the amplitude $E = E_0\, e^{-2bt}$

When the energy falls of $1/e$ times the initial energy E_0, then

$$E_0\, e^{-2bt} = \frac{1}{e}\, E_0 \quad \text{or} \quad e^{-2bt} = e^{-1}$$

$$-\,2bt = -\,1 \quad \text{or} \quad t = (1/2b)\text{ s.}$$

Problem 8. *A particle executing damped oscillations acquires the first amplitude equal to 50 cm, starting from rest and it falls to 0.5 cm after 100 oscillations. If the time period of oscillations is 2.3 s, find: (i) the relaxation time and (ii) the damping constant.*

Solution: At any instant t, the amplitude of damped oscillations is

$$a = a_0\, e^{-t/2\tau} = a_0\, e^{-bt}$$

where τ is the relaxation time and b is the damping constant.

Given: $T = 2.3$ s, first amplitude at $t = T/4$ is $a_1 = 50$ cm $= 0.5$ cm, 201th amplitude after 100 oscillations at $t = 100\,T + (T/4)$ is $a_{201} = 0.5$ cm $= 0.005$ m.

Thus $$a_1 = a_0\, e^{-bT/4} \quad \text{and} \quad a_{201} = a_0\, e^{-b\,(100T\, +\, (T/4))}$$

$$\therefore \quad \frac{a_{201}}{a_1} = e^{-b \times 100\,T} \quad \text{or} \quad \frac{0.005}{0.5} = e^{-b \times 100 \times 2.3}$$

or $$10^{-2} = e^{-230\,b} \quad \text{or} \quad -\,2\log_e 10 = -\,230\,b$$

or $$b = \frac{2 \times 2.3026 \times \log_{10} 10}{230} = 0.02\text{ s}^{-1}$$

and relaxation time $$\tau = \frac{1}{2b} = \frac{1}{2 \times 0.02} = 25\text{ s.}$$

Problem 9. *The amplitude of oscillation of a forced oscillator is expressed as*

$$A = \frac{f}{\sqrt{(n^2 - \omega^2)^2 + (\omega^2/\tau^2)}}$$

Find the value of A/A_{max}, if quality factor $Q = 50$ and $\omega/n = 0.99$.

Solution: Quality factor $Q = \omega_r$, $Q = n\tau$ (at resonance)

$$\therefore \qquad \tau = Q/n = 50/n$$

Hence

$$A = \frac{f}{n^2\sqrt{\left(1 - \frac{\omega^2}{n^2}\right)^2 + \frac{1}{2500}\frac{\omega^2}{n^2}}} = \frac{f}{n^2\sqrt{\left\{1-(0.99)^2\right\}^2 + \frac{(0.99)^2}{2500}}}$$

$$= \frac{f}{0.028\, n^2}$$

At resonance (when damping is low) $A_{max} = \frac{f\tau}{n} = \frac{50 f}{n^2}$

$$\therefore \qquad \frac{A}{A_{max}} = \frac{f/0.028\, n^2}{50\, f/n^2} = \frac{1}{0.028 \times 50} = \frac{1}{1.4} = 0.714.$$

Problem 10. *The amplitude of oscillation of a forced harmonic oscillator is initially 0.02 mm, when the driving frequency is very low. It acquires a maximum value equal to 5×10^{-3} m at the frequency of driving force equal to 100 Hz. Calculate (i) the quality factor of the oscillator Q, (ii) the relaxation time τ and (iii) the half width of the resonance curve.*

Solution: (*i*) Quality factor,

$$Q = \frac{\text{Amplitude of displacement at resonance}}{\text{Amplitude of displacement at zero or very low driving frequency}}$$

$$= \frac{5\, mm}{0.02\, mm} = 250$$

(*ii*) Since quality factor $Q = \omega_r\tau$, where $\omega_r = 2\pi \times 100 = 200\pi$ rad/s (at resonance)

$$\text{Relaxation time } \tau = \frac{Q}{\omega_r} = \frac{250}{200\,\pi} = 0.4 \text{ second.}$$

(*iii*) Half width of the resonance curve

$$\Delta\omega = \frac{\omega_r}{2Q} = \frac{200\,\pi}{2 \times 250} = 0.4\,\pi$$

$$= 1.256 \text{ rad/s.}$$

EXERCISES

1. Write short note on Lissajous figures. *(Amity Univ., Noida, Jan 2007)*
2. What are Lissajous figures? Discuss the formation of these figures, when the periods of vibration of two simple harmonic motions are in the ratio of 1 : 1. *(Amity Univ., Noida, Dec. 2005)*
3. What fraction of total energy of a simple harmonic oscillator is kinetic, when its displacement from the mean position is equal to half the amplitude. *(Amity Univ., Noida, Dec. 2005)*
4. What is quality factor? *(Amity Univ., Noida, Dec. 2005)*
5. What is the difference between free, damped and forced oscillations? Give the theory of forced oscillations. Discuss the condition of resonance. *(Amity Univ., Dec. 2005)*
6. What do you mean by resonance? Discuss the sharpness of resonance. *(Amity Univ., Jan. 2007)*
7. What is sample harmonic oscillator? Establish the differential equation for it and solve it to deduce the expression for velocity, displacement and time period.
8. Show that the solution of equation $\frac{d^2x}{dt^2} + \omega^2 x = 0$ is $x = a \sin(\omega t + \phi)$, where x_0 is the initial displacement, for the initial velocity v_0 amplitude $a = \sqrt{x_0^2 + (v_0^2/\omega^2)}$ and initial phase angle $\phi = \tan^{-1}\left(\frac{x_0 \omega}{v_0}\right)$.
9. Deduce an expression for potential energy, kinetic energy and total energy for simple harmonic oscillator and prove that the average kinetic energy is equal to the average potential energy and it is equal to half the total energy.
10. What is a simple pendulum? Show that the oscillations of a simple pendulum for small amplitude are simple harmonic. Establish the expression for the time period of its oscillations.
11. The displacements of two mutually perpendicular simple harmonic oscillations of same fequency are $x = a \sin(\omega t + \phi)$ and $y = b \sin \omega t$, obtain the expression for the resultant oscillation obtained due to their superposition. What will be the resultant path if $\phi = 0°, \frac{\pi}{2}$ and π?
12. What is meant by a damped harmonic oscillator? Write the differential equaiton for it and find its solution.
13. Explain under damped case and obtain expressions for average total energy and average power loss by it.
14. What do you understand by the relaxation time, damping constant and quality factor of a damped harmonic oscillator? Obtain their expressions.
15. If the damping constant $b << n$ (where n is the angular frequency of the free oscillations). Show that (*i*) average total energy $E_{av} = \frac{1}{2} m \omega^2 a_0^2 e^{-2bt}$ and (*ii*) power dissipation $P = E_{av}/\tau$.
16. Write down the differential equation for a forced harmonic oscillator and explain the significance of each term in it. Obtain solution of this equation and discuss it. Write the condition of resonance and explain the sharpness of resonance.

17. Write down the expression for the amplitude of a forced oscillator. Explain the resonance and half power points on the amplitude-frequency curve. Show that the quality factor $Q = \frac{\omega_r}{2\Delta\omega}$, where the symbols have their usual meanings.
18. Write short notes on
    (*i*) Damped harmonic oscillator
    (*ii*) Simple harmonic oscillator
    (*iii*) Forced harmonic oscillator
    (*iv*) Quality factor
    (*v*) Resonance and its sharpness.
19. What is meant by the quality factor of a damped harmonic oscillator? Derive an expression for the same.

SUGGESTION FOR FURTHER READING

1. Frank S. Granford Jr, Berkeley Physics Course Vol. 3 : Waves, Tata McGraw-Hill Publishing Company, New York, 2009.
2. A.P. French, Vibrations and Waves, CBS Publishers, New Delhi, 2003.
3. H.J. Pain, The Physics of Vibrations and Waves, Wiley, New York, 1968.
4. Devraj Singh, Fundamentals of Engineering Physics,Vol. 1, Dhanpat Rai and Co. (P) Ltd., New Delhi, 2009.
5. Devraj Singh, Engineering Physics, Vol. 1, Dhanpat Rai and Co. (P) Ltd., New Delhi, 2009.
6. Robert D. Finch, Introduction to Acoustics, Prentice-Hall of India Pvt. Ltd., New Delhi, 2008.
7. R.E.D. Bishop, Vibrations, Cambridge University Press, New York, 1965.

Chapter 2 PROGRESSIVE WAVES

2.1 INTRODUCTION

When a particle moves through space, it carries kinetic energy itself wherever the particle goes, the energy goes with it. The energy is associated with the particle and is transported from one region of the space to the other together with the particle just like we ride a car and are taken from Amity Noida to Amity Bijwasan with the car.

There is another way to transport energy from one part of space to the other without any bulk motion of material together with it. Sound is transmitted in air in this manner. When you say "Hello" to your friend, no material particle is ejected from your lips and falls on your friend's ear. You create some disturbance in the part of the air particles either by pushing them ahead or putting them back. The density of the air in this part temporarily increases or decreases. These disturbed particles exert force on the next layer of air, transferring the disturbance to that layer of air. In this way, the disturbance proceeds in air and finally the air near the ear of the listener get disturbed. The disturbance produced in the air near the speaker travels in air, the air itself does not move. The air that is near the speaker at the time of uttering a word remain all the time near the speaker even when the message reaches the listener. This type of motion of energy is called a *wave motion*.

Progressive Wave

A wave that travels from one point of the medium to another is called a progressive wave. A progressive wave may be transverse or longitudinal.

Plane Progressive Harmonic Wave

If during the propagation of a wave through a medium, the particles of the medium vibrate simple harmonically about their mean positions, then the wave is said to be plane progressive harmonic wave. In a harmonic progressive wave of given frequency, all particles have same amplitude but the phase of oscillation changes from one particle to the next.

2.2 PROPERTIES OF LONGITUDINAL PROGRESSIVE WAVES

- The particles of the medium vibrate simple harmonically along the direction of propagation of the wave.
- All the particles have the same amplitude, frequency and time period.
- There is a gradual phase difference between the successive particles.
- All the particles vibrating in phase will be at a distance equal to $n\lambda$. Here $n = 1, 2, 3, \ldots$ etc. It means the minimum distance between two particles vibrating in phase is equal to the wavelength.
- The velocity of the particles is maximum at their mean position and it is zero at their extreme positions.
- When the particle moves in the same direction as the propagation of the wave, it is in a region of compression.
- When the particle moves in the direction opposite to the direction of propagation of the wave, it is in a region of rarefaction.
- When the particle is at the mean position, it is region of maximum compression and rarefaction.
- When the particle is at the extreme position, the medium around the particles has its normal density, with compression on one side and rarefaction on the other.
- Due to the repeated periodic motion of the particles compressions and rarefactions are produced continuously. These compressions and rarefactions travel forward along the wave and transfer energy in the direction of propagation of the wave.

2.3 EXPRESSION FOR PLANE PROGRESSIVE HARMONIC WAVE

Let us consider a wave originating at O (Figure 2.1) is travelling along $+X$-axis. If y be the simple harmonic displacement of a particle at position P, distant x from O at O, then y is expressed as

$$y = a \sin (\omega t - \phi), \qquad ...(1)$$

where a is the amplitude, ω is the angular velocity and ϕ the phase lag of the particle. Since the successive particles to the right of point O receive and repeat the harmonic motion after a definite interval of time resulting in a phase lag. This phase lag goes on increasing with increasing distance from O.

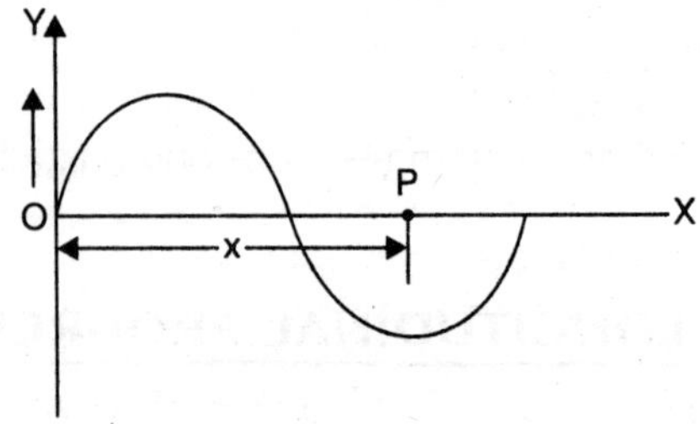

FIGURE 2.1

If λ be the wavelength, then we know a distance λ corresponds to a phase difference of 2π. Thus, a path difference of x corresponds to a phase difference $\phi = \frac{2\pi x}{\lambda}$.

Substituting $\omega = \frac{2\pi}{T}$ and $\phi = \frac{2\pi x}{\lambda}$ in equation (1), we get

$$y = a \sin \left(\frac{2\pi t}{T} - \frac{2\pi x}{\lambda} \right) \qquad ...(2)$$

If v is the wave or phase velocity, then $v = \lambda \nu = \frac{\lambda}{T}$

$$\therefore \qquad T = \frac{\lambda}{v}$$

Substituting this value of T in the above equation, we have

$$y = a \sin \left(\frac{2\pi vt}{\lambda} - \frac{2\pi x}{\lambda} \right)$$

$$\Rightarrow \qquad y = a \sin \frac{2\pi}{\lambda} (vt - x) \qquad ...(3)$$

Equation (3) can also be expressed in terms of propagation constant k. We know $\frac{2\pi}{\lambda} = k$,

therefore,

$$y = a \sin (\omega t - kx)$$

$$\Rightarrow \qquad y = a \sin \frac{2\pi}{T}\left(t - \frac{x}{v}\right) \qquad (\because \ \lambda = vT)$$

$$\Rightarrow \qquad y = a \sin \omega\left(t - \frac{x}{v}\right) \qquad \text{...(4)}$$

When the wave is travelling towards the left, *i.e.* $-x$-axis, the wave equation expressed by equation (3), can be written as

$$y = a \sin \frac{2\pi}{\lambda}\left[vt - (-x)\right] = a \sin \frac{2\pi}{\lambda}(vt + x) \qquad \text{...(5)}$$

$$\Rightarrow \qquad \boxed{y = a \sin (\omega t + kx)} \qquad \text{...(6)}$$

The quantity $(\omega t \pm kx)$ represents the phase of wave motion. We have assumed that at $t = 0, y = 0$ *i.e.*, motion starts at O. If this is not the case and the particle has some initial phase θ_0, the wave equation is represented as

$$\boxed{y = a \sin (\omega t - kx + \theta_0)} \qquad \text{...(7)}$$

2.4 PARTICLE VELOCITY AND WAVE VELOCITY

The equation for a simple harmonic wave is given by

$$y = a \sin \frac{2\pi}{\lambda}(vt - x) \qquad \text{...(1)}$$

Here v is the velocity of wave and y is the displacement of the particle.

The velocity of the particle $\quad u = \dfrac{dy}{dt}$

So differentiating equation (1) with respect to time t

$$u = \frac{dy}{dt} = \frac{2\pi a v}{\lambda} \cos \frac{2\pi}{\lambda}(vt - x) \qquad \text{...(2)}$$

The maximum value of the particle velocity is

$$u_{max} = \frac{2\pi a v}{\lambda} \qquad \text{...(3)}$$

$\therefore$ Maximum particle velocity $= \dfrac{2\pi a}{\lambda} \times$ Wave velocity

To find the particle acceleration, differentiate equation (2) with respect to time

$$f = \frac{d^2y}{dt^2} = -\frac{4\pi^2 a v^2}{\lambda^2} \sin \frac{2\pi}{\lambda}(vt - x) \qquad \text{...(4)}$$

$$f = -\frac{4\pi^2 v^2}{\lambda^2}\left[a \sin \frac{2\pi}{\lambda}(vt - x)\right]$$

$$\therefore \qquad f = -\left(\frac{4\pi^2 v^2}{\lambda^2}\right) y$$

The acceleration is maximum when $y = a$

$$\therefore \qquad f_{max} = -\left(\frac{4\pi^2 v^2}{\lambda^2}\right) a \qquad ...(5)$$

The negative sign shows that the acceleration of the particle is directed towards its mean position.

Differentiating equation (1) with respect to x

$$\frac{dy}{dx} = -\frac{2\pi a}{\lambda} \cos \frac{2\pi}{\lambda}(vt - x) \qquad ...(6)$$

$\frac{dy}{dx}$ represents the slope of the displacement curve

From equations (2) and (6)

$$\boxed{u = \frac{dy}{dt} = -v\left(\frac{dy}{dx}\right)}$$

$\therefore$ (Particle velocity at any instant)

= (wave velocity) (slope of the displacement curve at that instant)

2.5 DIFFERENTIAL EQUATION OF WAVE MOTION

The general equation of a simple harmonic wave is

$$y = a \sin \frac{2\pi}{\lambda}(vt - x) \qquad ...(1)$$

Differentiating equation (1) with respect to time

$$\frac{dy}{dt} = \frac{2\pi a v}{\lambda} \cos (vt - x) \qquad ...(2)$$

Again differentiating equation (2) with respect to time

$$\frac{d^2y}{dt^2} = -\frac{4\pi^2 a v^2}{\lambda^2} \sin (vt - x) \qquad ...(3)$$

To find the value of compression, differentiate equation (1) with respect to x

$$\frac{dy}{dx} = -\frac{2\pi a}{\lambda} \cos \frac{2\pi}{\lambda} (vt - x) \qquad ...(4)$$

To find the rate of change of compression with respect to distance, differentiate equation (4) with respect to x

$$\frac{d^2y}{dx^2} = -\frac{4\pi^2 a}{\lambda^2} \sin \frac{2\pi}{\lambda} (vt - x) \qquad ...(5)$$

From equations (2) and (4)

$$\frac{dy}{dt} = -v \frac{dy}{dx} \qquad ...(6)$$

From equations (3) and (5)

$$\boxed{\frac{d^2y}{dt^2} = v^2 \frac{d^2y}{dx^2}} \qquad ...(7)$$

Equation (7) represents the differential equation of wave motion.

The general differential equation of wave motion can be written as

$$\frac{d^2y}{dt^2} = K \frac{d^2y}{dx^2} \qquad ...(8)$$

Here $K = v^2$

or $v = \sqrt{K}$

Thus, knowing the value of K, the value of the wave velocity can be calculated.

2.6 VARIATION OF VELOCITY AND PRESSURE IN A PROGRESSIVE WAVE

A plane progressive wave is represented by the equation

$$y = a \sin \frac{2\pi}{\lambda} (vt - x) \qquad ...(1)$$

and its graphical representation of variation of displacement y is shown in Figure 2.1

The particle velocity is given by

$$u = \frac{dy}{dt} = \frac{2\pi a v}{\lambda} \cos \frac{2\pi}{\lambda} (vt - x) \qquad ...(2)$$

and its variation is shown in Figure 2.2.

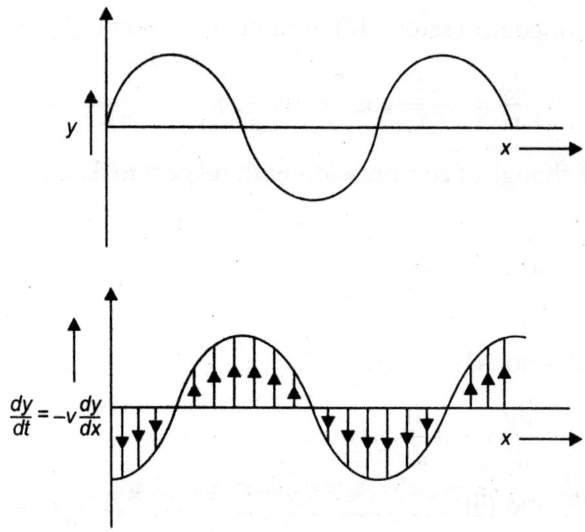

FIGURE 2.2

The volume strain in the medium is $\frac{dy}{dx}$, given by

$$\frac{dy}{dx} = -\frac{2\pi a}{\lambda}\cos\frac{2\pi}{\lambda}(vt - x) \qquad ...(3)$$

The modulus of elasticity (K) of the medium is defined as

$$K = \frac{\text{change in pressure}}{\text{volume strain}} = \frac{-dP}{(dy/dx)}$$

$$\Rightarrow \qquad dP = -K\frac{dy}{dx}$$

$$\Rightarrow \qquad dP = K\left(-\frac{dy}{dx}\right) \qquad ...(4)$$

In a region where $\frac{dy}{dx}$ is –ve, so that dP +ve *i.e.*, it is a *region of compression*. If $\frac{dy}{dx}$ is +ve, then dP is –ve *i.e.*, a *region of rarefaction*.

Using equations (3) and (4); we have

$$\boxed{dP = \frac{2\pi Ka}{\lambda}\cos\frac{2\pi}{\lambda}(vt - x)} \qquad ...(5)$$

The variation of dP is shown in Figure 2.3, where P_0 is the normal pressure of the medium when wave is not propagating.

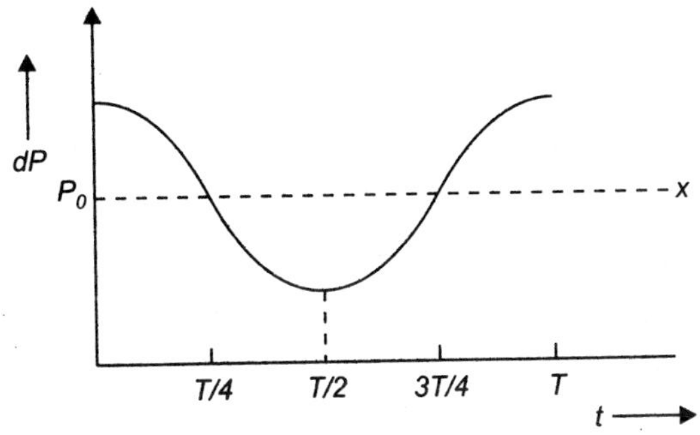

FIGURE 2.3

2.7 ENERGY OF A PLANE PROGRESSIVE WAVE

In case of a progressive wave, new waves are continuously formed at the head of the wave. This means that there is continuous transfer of energy in the direction of propagation of the wave. Thus energy is supplied from the source. The energy transferred per second also corresponds to the energy possessed by the particles in a length v, where v is the velocity of the wave. The energy of the particles in partly kinetic and partly potential. The kinetic energy is due to the velocity of the vibrating particles. For a particle executing simple harmonic motion, the velocity is maximum at the mean position and it is zero at the extreme positions. Consequently the kinetic energy of the particle at the mean position is maximum and zero at the extreme positions. Similarly the particles also possess potential energy due to their displacements from their mean positions. At the extreme positions the potential energy of a particle is maximum and at the mean position the potential energy is minimum.

In longitudinal wave motion, compressions and rarefactions are formed. The energy distribution is not uniform over the wave. At points of no velocity there is no compression and the particle do not possess energy at these points. At points of maximum velocity, there is compression and the particle possess maximum energy. However in the case of progressive wave motion, *there is no transfer of the medium in the direction of propagation of the wave, but there is always transfer of energy in the direction of propagation of the wave.*

Analytical Treatment

The equation of a simple harmonic wave is

$$y = a \sin \frac{2\pi}{\lambda}(vt - x) \qquad \text{...(1)}$$

The particle velocity u at any instant be obtained by differentiating equation (1) with respect to time

$$u = \frac{dy}{dt} = \frac{2\pi a v}{\lambda} \cos \frac{2\pi}{\lambda}(vt - x) \qquad \text{...(2)}$$

The acceleration of the particle at that instant

$$f = \frac{du}{dt} \quad \therefore \quad f = \frac{d^2y}{dt^2}$$

Differentiating equation (2) with respect to time

$$\boxed{f = \frac{d^2y}{dt^2} = -\frac{4\pi^2 a v^2}{\lambda^2} \sin \frac{2\pi}{\lambda}(vt - x)} \quad ...(3)$$

The negative sign shows that acceleration is directed towards the mean position.

Potential Energy

To move the particle from its mean position to a distance y, work has to be done against acceleration.

Workdone for a displacement $dy = F\,dy$

Let ρ be the density of the medium

Workdone per unit volume for a displacement dy

$$= \rho \left[\frac{4\pi^2 a v^2}{\lambda^2} \sin \frac{2\pi}{\lambda}(vt - x)\right] dy$$

Total work done for displacement y

$$= \int_0^y \rho \left(\frac{4\pi^2 a v^2}{\lambda^2} \sin \frac{2\pi}{\lambda}(vt - x)\right) dy$$

But $$y = a \sin \frac{2\pi}{\lambda}(vt - x)$$

$$\therefore \quad \text{Potential energy per unit volume} = \left[\frac{4\pi^2 \rho v^2}{\lambda^2}\right] \int_0^y y\,dy$$

$$= \frac{4\pi^2 \rho v^2 y^2}{2\lambda^2}$$

$$\text{P.E.} = \frac{2\pi^2 \rho v^2 y^2}{\lambda^2}$$

$$\text{P.E.} = \frac{2\pi^2 \rho v^2 a^2}{\lambda^2} \sin^2\left[\frac{2\pi}{\lambda}(vt - x)\right] \quad ...(4)$$

Kinetic energy per unit volume $= \frac{1}{2}\rho u^2$

$$= \frac{1}{2}\rho\left[\frac{2\pi a v}{\lambda}\cos\frac{2\pi}{\lambda}(vt - x)\right]^2$$

$$= \frac{2\pi^2 \rho v^2 a^2}{\lambda^2}\cos^2\left[\frac{2\pi}{\lambda}(vt - x)\right] \quad ...(5)$$

Total energy per unit volume

$$E = \text{Kinetic energy} + \text{Potential energy}$$

$$= \frac{2\pi^2 \rho v^2 a^2}{\lambda^2}\left[\sin^2\frac{2\pi}{\lambda}(vt - x) + \cos^2\frac{2\pi}{\lambda}(vt - x)\right]$$

$$\Rightarrow \quad E = \frac{2\pi^2 \rho v^2 a^2}{\lambda^2} \quad ...(6)$$

But $\quad v^2 = n\lambda$

$$\therefore \quad \boxed{E = 2\pi^2 \rho n^2 a^2} \quad ...(7)$$

The average kinetic energy per unit volume and the average potential energy per unit volume are equal and each is equal to half the total energy per unit volume

Average kinetic energy per unit volume $= \pi^2 \rho n^2 a^2$ ...(8)

Average potential energy per unit volume $= \pi^2 \rho n^2 a^2$ ...(9)

Suppose that the area of cross-section of a parallel beam of radiation is 1 unit and the velocity of wave is v

$$\text{volume} = 1 \times v = v$$

$\therefore$ Energy transfer per unit area per second $= E \times v$

$$= 2\pi^2 \rho n^2 a^2 v \quad ...(10)$$

Energy transfer per second also called current per unit area of cross-section.

It is to be remembered that the potential and kinetic energies of every particle will change with time, but the average kinetic energy per unit volume and the average potential energy per unit volume remains constant. The current density per unit area of cross-section and the total energy per unit volume also remains constant.

2.8 SUPERPOSITION OF PROGRESSIVE WAVES: STATIONARY WAVE

When two identical progressive waves travelling in a medium with same velocity but in opposite directions along the same straight line are superposed, then give rise to a system of alternative rarefaction and compression that cannot move in any direction of medium. This resultant wave is called stationary or standing wave.

They are called stationary because there is no flow of energy along the waves. There are certain points, half a wavelength apart, which are permanently at rest are known as *nodes* and there are some other points midway between the nodes where the displacement is maximum, known as *antinodes*.

Analytical Treatment of Stationary Waves and Their Properties

Consider two plane progressive waves-one travelling along positive X-axis and another along negative X-axis with amplitude a, wave velocity v and wavelength λ. So, the superposing progressive waves are

$$y_1 = a \sin \frac{2\pi}{\lambda}(vt - x) \quad \text{and} \quad y_2 = a \sin \frac{2\pi}{\lambda}(vt + x)$$

By the principle of superposition, the resultant displacement of a particle at x at time t will be

$$Y = y_1 + y_2$$

$$= a \sin \frac{2\pi}{\lambda}(vt - x) + a \sin \frac{2\pi}{\lambda}(vt + x)$$

$$= 2a \cos \frac{2\pi}{\lambda} vt \cos \frac{2\pi}{\lambda} x$$

$$\Rightarrow \qquad \boxed{Y = A \cos \frac{2\pi vt}{\lambda}} \qquad \text{...(1)}$$

where $$A = 2a \cos \frac{2\pi}{\lambda} x$$

This equation represents a simple harmonic motion of same wavelength of the superposed wave but not of same amplitude. Moreover, the amplitude, $A = 2a \cos \frac{2\pi}{\lambda} x$, is not a constant. For different values of x, A will have different values. Equation of motion given by equation (1) is not a progressive motion since its phase does not contain any term like $(vt - x)$ or $(vt + x)$. So, equation (1) represents a stationary wave.

Position of Nodes

The nodes will be obtained when

$$\cos \frac{2\pi}{\lambda} x = 0 \quad \text{or} \quad \frac{2\pi}{\lambda} x = (2n \pm 1)\frac{\pi}{2}$$

or $$\boxed{x = (2n \pm 1)\frac{\lambda}{4}}$$

where $n = 0, 1, 2, 3, 4, \ldots$

So nodes will be obtained at $x = \pm \frac{\lambda}{4}, \pm \frac{3\lambda}{4}, \pm \frac{5\lambda}{4}, \ldots$ the particles at these points will be at rest. The distance between any two successive nodes is $\frac{\lambda}{2}$.

Position of Antinodes

Antinodes are the points of maximum vibration, so the antinodes will be obtained when

$$\cos \frac{2\pi}{\lambda} x = 1 \quad \text{or} \quad \frac{2\pi}{\lambda} x = \pm n\pi$$

or

$$\boxed{x = \pm 2n \frac{\lambda}{4}}$$

So nodes will be obtained at $x = 0, \pm \frac{\lambda}{2}, \pm \lambda, \pm \frac{3\lambda}{2}, \ldots$ at these points the amplitude of vibration of the particles will be $\pm 2a$. The distance between any to successive antinodes is $\frac{\lambda}{2}$.

The distance between one node and next antinode is $\frac{\lambda}{4}$.

Characteristics of Stationary Waves

1. Stationary waves are produced when two identical waves travelling along the same straight line but in opposite direction are superposed.
2. Crests and troughs do not progress through the medium but simply appear or disappear at the same place alternatively.
3. All the particles, except those at the nodes, follow simple harmonic motion. The amplitude of the oscillation is zero at nodes and maximum at antinodes. The distance between two successive antinode and node is equal to half of wavelength.
4. The particle between two successive nodes are in the same phase of vibration while the particles on opposite sides of a node are in opposite phase of vibration.
5. Stationary waves can be produced both by longitudinal waves and transverse waves.
6. All the particles pass through their mean positions or reach their outermost positions simultaneously twice in a periodic time.
7. There is no advancement of the wave and no flow of energy in any direction.

2.9 DISTINCTION BETWEEN PROGRESSIVE AND STATIONARY WAVES

S. No.	*Progressive Wave*	*Stationary Wave*
1.	This is an advancing wave, which moves in the medium continuously with finite velocity.	There is no advancement of the wave in any direction.
2.	These waves are produced due to continuous periodic vibrations of the particles of a portion of a medium.	These waves are produced due to the superposition of two identical progressive waves moving along the same line but in opposite direction in a medium.
3.	Each particle of the medium executes identical periodic motion about the mean position of rest.	The particles between two successive nodes are in the same phase of vibration, while the particles on opposite sides of a nodes are in opposite phase of vibration.
4.	The phase of vibration varies continuously from point to point.	The particles between two successive nodes are in the same phase of vibration, while the particles on opposite sides of a node are in opposite phase of vibration.
5.	All the particles do not pass through their mean positions or reach their outer most positions simultaneously.	All the particles pass through their mean positions or reach their outer most positions simultaneously twice in a periodic time.
6.	Wave curve advances without change of shape.	Wave curve changes its shape from time to time-it shrinks to straight line twice in each period.
7.	The amplitude of each particles is same.	The amplitude varies continuously from point to point-it is maximum at the antinodes and minimum at nodes.

2.10 NUMERICALS BASED ON PROGRESSIVE WAVES

Problem 1. *When a simple harmonic wave is propagated through a medium, the displacement of a particle (in cm) at any instant of time is given by*

$$y = 10 \sin \frac{2\pi}{100} (3600\, t - 20)$$

Calculate, the amplitude of vibrating particle, wave velocity, wavelength, frequency and time period.

Hint: $$y = a \sin \frac{2\pi}{\lambda} (vt - x)$$

$$n = \frac{v}{\lambda}, T = \frac{1}{n}$$

[**Ans.** $a = 10$ cm, $\lambda = 100$ cm, $v = 3600$ cm/sec, $n = 360$ Hz, $T = \frac{1}{360}$ second.]

Problem 2. *A source of sound has a frequency of 512 Hz and an amplitude of 0.25 cm. What is flow of energy across a square cm per second, if the velocity of sound in air is 340 m/s and the density of air is 0.00129 g/cc?*

Hint: Total energy per unit volume = $2\pi^2\rho n^2a^2$

Energy transferred across a sq. cm per second = $2\pi^2\rho n^2a^2v$.

[**Ans.** 1.417×10^7 erg/cm^2 8]

Problem 3. *The equation of stationary wave formed by superposition of two identical progressive waves travelling in opposite direction is given by, $y = A \cos kx \sin \omega t$, where $k = \frac{2\pi}{\lambda}$, $A = 1.0$ mm, $k = 1.57$ cm^{-1} and $\omega = 78.5$ rad/sec. Calculate (a) velocity of superposing wave, (b) amplitude of vibration of the wave at $x = 2.33$ cm.*

Solution: The equation of the superposing waves are

$$y_1 = \frac{A}{2} \sin(\omega t - kx) \quad \text{and} \quad y_2 = \frac{A}{2} \sin(\omega t + kx)$$

Then (*a*) velocity $\quad v = \frac{\omega}{k} = \boxed{50 \text{ m/sec,}}$

(*b*) Amplitude $\quad a = A \cos kx = \boxed{0.866 \text{ mm.}}$

Problem 4. *A simple harmonic wave of amplitude 8 units transverses a line of particles in the direction of the positive X-axis. At any given instant of time, for a particle at a distance of 10 cm from the origin, the displacement is +6 units and for a particle at a distance of 25 cm from the origin, the displacement is + 4 units. Calculate the wavelength.*

Solution: $\quad y = a \sin \frac{2\pi}{\lambda}(vt - x)$

$$\frac{y}{a} = \sin 2\pi\left(\frac{t}{T} - \frac{x}{X}\right) \frac{6}{8} = \sin 2\pi\left(\frac{t}{T} - \frac{10}{\lambda}\right) \quad \text{and} \quad \frac{4}{8} = \sin 2\pi\left(\frac{t}{T} - \frac{25}{\lambda}\right)$$

$$\boxed{\lambda = 290.8 \text{ cm}}$$

Problem 5. *Show that the equation $y = 4 \cos 2\pi\left(\frac{t}{0.02} - \frac{x}{400}\right)$ cm represents a progressive wave. Calculate its amplitude, wavelength, wave velocity and frequency.*

Solution: $y = 4 \cos 2\pi\left(\frac{t}{0.02} - \frac{x}{400}\right)$ compare with $y = a \cos 2\pi\left(\frac{t}{T} - \frac{x}{X}\right)$ $a = 4, \lambda = 400$ cm,

$v = 200$ m/sec, $n = \frac{v}{\lambda} = \frac{20000}{400} = 50$ Hz

At $(t + 1)$ sec, displacement at distance $(x + 20000)$ cm is

$$y_1 = 4 \cos \frac{2\pi}{400} [20000 (t + 1) - (x + 20000)]$$

$$= 4 \cos \frac{2\pi}{400} (20000\, t - x)$$

$\therefore$ $$y = y_1$$

$\therefore$ After 1 second at a distance 20000 cm the displacement is same. Hence the given equation represents a progressive wave.

EXERCISES

1. What are progressive waves? Describe briefly the properties of longitudinal progressive waves.
2. Derive the relation
$$\frac{d^2y}{dt^2} = v^2 \frac{d^2y}{dx^2}.$$
3. Distinguish between particle velocity and wave velocity and obtain the relation between the two.
4. Show that for a plane progressive wave, on the average, half the energy is kinetic and half potential. Hence derive the expression for intensity of sound in terms of pressure amplitude.
5. Show that $u = -v \frac{dy}{dx}$, where u is the particle velocity and v is the wave velocity.
6. Show that the energy of a plane progressive wave is given by
$$E = 2\pi^2 \rho n^2 a^2$$
7. Derive the equation of wave motion in the form of
$$y = a \sin \frac{2\pi}{\lambda} (vt - x)$$
8. Deduce the equation of simple harmonic wave travelling in the positive x direction in the form of
$$y = a \sin \frac{2\pi}{\lambda} (vt - x)$$
and explain how energy is distributed in such a progressive wave.
9. Differentiate between
(*a*) wave velocity and particle velocity
(*b*) progressive and stationary waves.
10. What are nodes and antinodes? Calculate the average kinetic and potential energy for a plane progressive wave.

SUGGESTION FOR FURTHER READING

1. W.C. Elmore and M.A. Heald, The Physics of Waves, McGraw-Hill, New York, 1969.
2. A.P. French, Vibrations and Waves, CBS Publishers, New Delhi, 2003.
3. Lain G Main, Vibrations and Waves in Physics, Cambridge University Press, 1993.
4. C.A. Coulson, Waves, Oliver and Boyd, Edinburgh, 1941.

Chapter 3

ULTRASONICS

3.1 INTRODUCTION

The human ear can detect sound waves of frequency lying in the range 20 to 20,000 Hz. Sound waves having frequencies greater than 20,000 Hz are known as ultrasonics. Ultrasonic waves, because of high frequencies, have a *large energy* as $E = h\nu$ and therefore a high penetrating power. Due to this property, they find wide applications in processes such as flaw detection, drilling and soldering etc. Recently, they have also been employed for the detection of various diseases in the human body.

Ultrasonic waves are sound waves in nature and can therefore undergo processes such as reflection, refraction and absorption. It may be pointed out that ultrasonic waves travels with the same speed as sound waves but have a greater frequency, or smaller wavelength than them.

Some other properties of ultrasonic waves are as:

1. The speed of propagation of ultrasonic waves increases with increase in frequency.
2. The wavelength of the waves is very small and the waves exhibit negligible diffraction effect.
3. They can travel over long distances as a highly directional beam and without appreciable loss of energy.
4. They produce cavitation effects in liquids.

3.2 GENERATION/PRODUCTION OF ULTRASONIC WAVES

Ultrasonics are generated by means of following:

1. Galton Whistle
2. Magnetostriction Generator
3. Piezoelectric Generator.

1. Galton Whistle

Galton whistle works on the principle of organ pipe. It consists of a closed end air column A whose length can be adjusted with the help of a movable piston. The piston P can be moved to the desired position with the help of a screw S_1. The open end of the pipe A is fitted with a lip L. The position of the pipe C can be adjusted with the help of the screw S_2. The gap between the ends of A and C can be adjusted with the help of the screw S_2 (Figure 3.1)

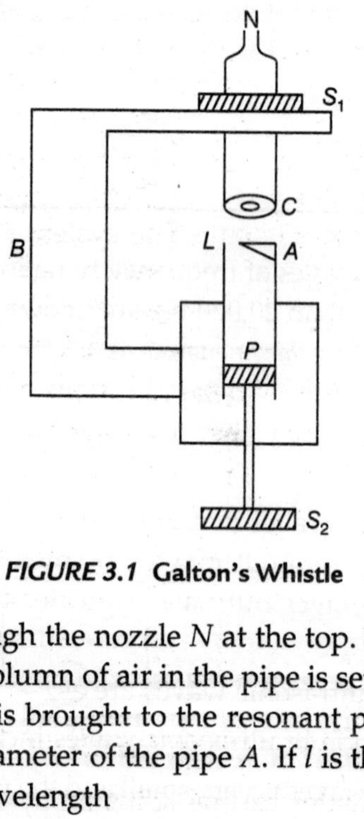

FIGURE 3.1 **Galton's Whistle**

An air blast is blown through the nozzle N at the top. The blast of air coming out of C strikes against the lip L and the column of air in the pipe is set into vibration. By adjusting the length of the air column in A, it is brought to the resonant position. The resonant frequency will depend on the length and diameter of the pipe A. If l is the length of the air column in A, x the end correction, then the wavelength

$$\lambda = 4\,(l + x)$$

The frequency of sound is

$$n = \frac{v}{\lambda} = \frac{v}{4(l + x)}$$

with the help of this whistle, frequencies of the order of 30,000 Hz can be produced. The micrometer screw S_1 can also be calibrated to give directly the frequency of the sound.

2. Magnetostriction Generator

It is found that the length of a bar of a ferromagnetic material such a iron or nickel changes when the bar is subjected to strong magnetic field parallel to its length. This phenomenon is known as *magnetostriction*. However, if the bar is subjected to an alternating magnetic field, it expands and contracts allternately. Due to this linear contraction and expansion, longitudinal waves are produced in the medium surrounding the bar. If the rod is clamped in the centre, the frequency of vibration n is given by

$$n = \frac{1}{2L}\sqrt{\frac{Y}{\rho}}$$

where L is the length of rod, Y is its Young's modulus and r is the density of the material of the rod.

Figure 3.2 shows the electric circuit used for the generation of ultrasonic waves using magnetostriction. The coils L_1 and L_2 are wraped round the ferromagnetic rod AB; One is connected in the grid circuit and the other to the plate circuit of a triode valve. The rod is clamped in the middle. It is magnetised by the plate current flowing in the coil L_1. A change in turn changes its length due to the magnetostriction effect. The change in the length of the rod alters the magnetic field across the coil L_2 due to converse magnetostriction effect. The varying field, so produced across L_2 changes its flux causing an induced emf across this coil, which changes the potential difference across the grid circuit. These vibrations are amplified by the triode valve and passed on the plate circuit. The system thus provides a feedback for the triode valve as an oscillator.

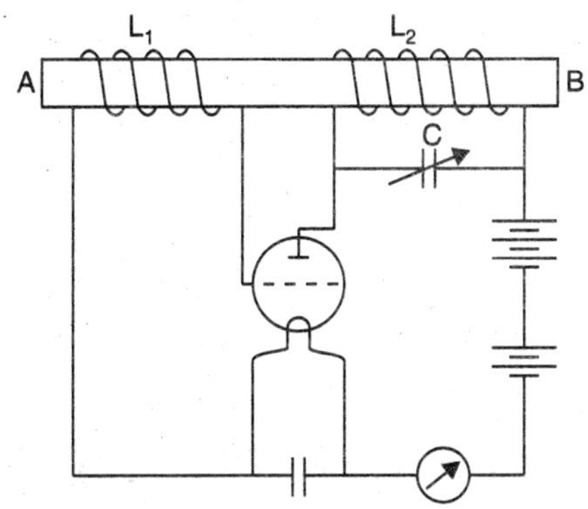

FIGURE 3.2 **Generation of Ultrasonic waves using the effect of magnetostriction**

The frequency of the oscillator can be adjusted by changing the capacitance of the condenser C. A magnetostriction generator produces ultrasonic waves of comparatively low frequency, upto 200 kHz.

3. Piezoelectric Generator

For generating ultrasonic waves of high frequency (about 50 MHz) a generator using the piezoelectric effect is employed. It is found that when crystals of some materials such as quartz, tourmaline, rocksalt etc. are subjected to a mechanical pressure in a certain direction, each charges of opposite sign develop as their faces, normal to the direction of the applied pressure. This phenomenon is known as the *piezoelectric effect*.

Referring to Figure 3.3, which is a cross-sectional view of a quartz crystal, if pressure is applied along the axis X_2-X_2 electrical charges appear on the faces *ab* and *a′b′*. Conversely if two opposite faces of a crystal are subjected to a potential difference (in order to provide charge on them), a tensile pressure appears on the crystal. This pressure alters the length of the crystal. This pressure alters the length of the crystal in that direction. If the applied potential is alternating the crystal, the crystal begins to oscillate with a frequency which lies in the ultrasonic range.

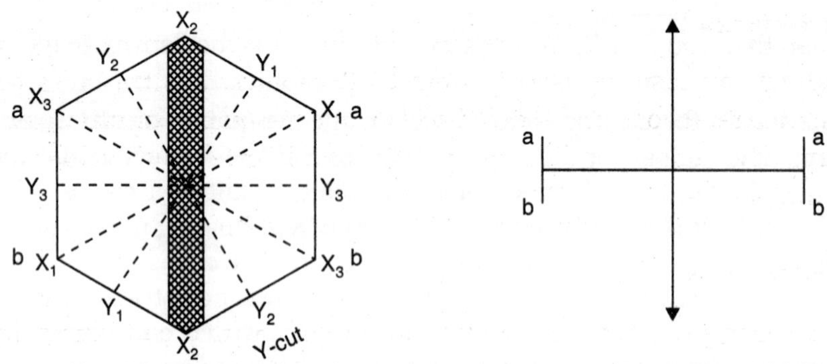

FIGURE 3.3 **Cross-sectional view of a quartz crystal. If pressure is applied along the axis X_2–X_2, electric charge appears on the faces *ab* and *a′b′***

Figure 3.4 shows circuit arrangements that can be used to generate ultrasonic waves by using the piezoelectric effect. A thin slice of quartz crystal *R* is placed between two metal plates *A* and *B* to form a parallel plate capacitor, with the quartz crystal as dielectric. The plates are connected to the terminals of a coil which is inductively coupled to the oscillating circuit of a triode valve. An alternating potential difference developed across the condenser plates due to the electrical circuits. The quartz slab is thus subjected to an alternating electric field which produces alternate contraction and expansion of the slab in the perpendicular direction leading to the oscillation of the crystal.

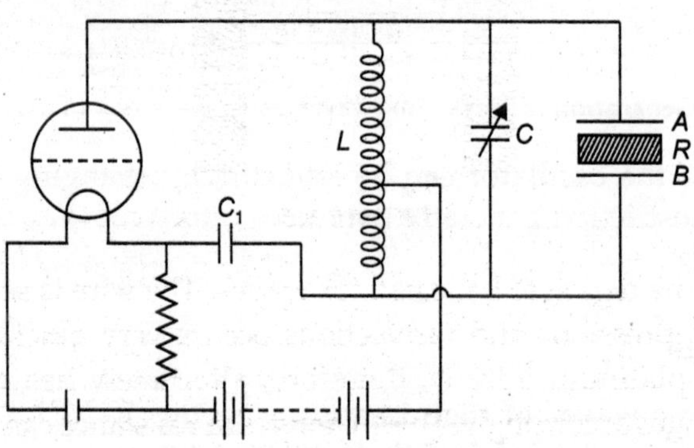

FIGURE 3.4 **Circuit arrangement used to generate ultrasonic waves using piezoelectric effect**

The variable condenser C is adjusted so that the frequency of the oscillatory circuit is equal to the natural frequency of one of the modes of vibration of the crystal. This produces resonant mechanical vibrations in the crystal due to the linear expansion and contraction. If one of the faces of the crystal is placed in contact with some medium in which elastic waves can be propagated, ultrasonic waves are generated.

3.3 DETECTION OF ULTRASONIC WAVES

1. Piezoelectric Detector

Piezoelectric effect can also be used to detect ultrasonics. If ultrasonics comprising of compressions and rarefactions are allowed to fall upon a quartz crystal a certain potential difference is developed across the faces which after amplification by a value amplifier can be used to detect ultrasonics.

2. Kundt's Tube Method

Kundt's tube is a long glass tube supported horizontally with a air column in it when the ultrasonic waves are passed the Kundt's tube, the lycopodium powder sprinkled in the tube collects in the form of heaps at the nodal points and is blown off at the antinodal points. This method is used provided that the wavelength is not very small.

3. Thermal Detector

Thermal detectors can also be used to detect ultrasonics. A thermal detector is nothing but a fine platinum wire fixed at two ends as shown in Figure 3.5.

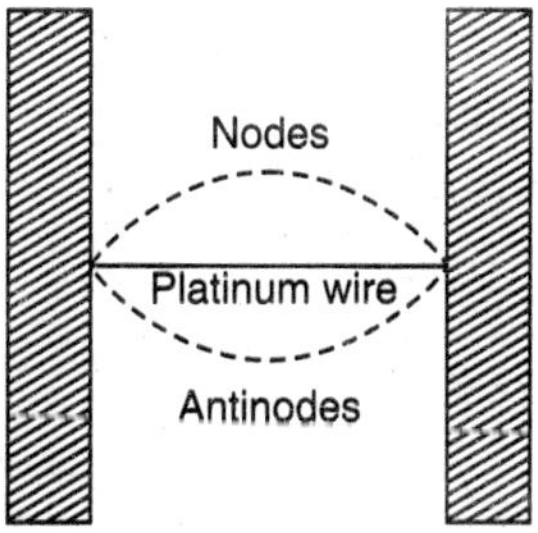

FIGURE 3.5

It is placed in the region to be tested for waves. The wire is set into rapid vibrations. At a node compressions and rarefactions occurs very quickly and cause adiabatic changes. The platinum wire is, therefore, alternately heated and cooled and its resistance changes accordingly. These changes in resistance can be detected by suitable means. No such changes occur at an antinode.

4. Flame Method

A narrow sensitive flame is moved along the medium. At the position of the antinode, the flame is steady. At the position of the node the flame flickers because there is change in pressure. In this way the positions of nodes and antinodes can be found out in a medium. The average distance between two adjacent nodes is equal to half the wavelength. If the value of frequency of ultrasonic wave is known, the velocity of the ultrasonic wave through the medium can be calculated.

5. Acoustic Diffraction Method

This method is based on the fact that ultrasonic waves which consist of alternate compressions and rarefactions changes the density of the medium through which they pass.

This leads to a periodic variation of refractive index of the liquid, such a liquid column is subjected to ultrasonic waves constitutes an *acoustical grating*. If monochromatic light is passed through the waves the liquid causes the diffraction of light.

Figure 3.6 shows the experimental arrangement, standing ultrasonic waves are produced in a liquid contained in a glass tube. The density and so the refractive index of the liquid is maximum at the nodal point and minimum at antinodal points. Hence the nodal area acts as opaque region, while antinodal area acts as transparent region for light. The liquid column thus resembles the rules grating.

The grating period d equal to $\lambda/2$ and is given by

$$\boxed{d \sin \theta = m\lambda}$$

λ = wavelength of monochromatic light beam

m = order of minima.

An acoustic diffraction grating produced by a liquid column subjected to ultrasonic waves.

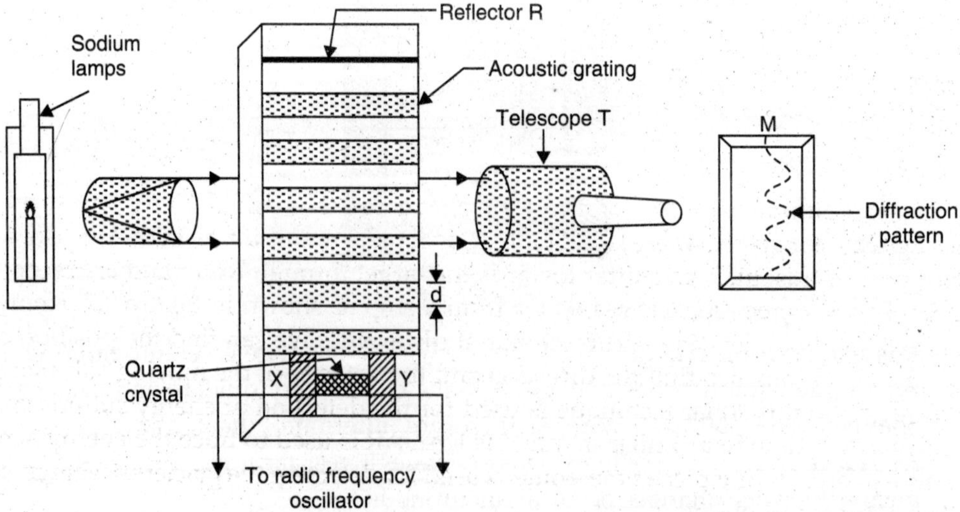

FIGURE 3.6 **Experimental set up of acoustic grating**

3.4 ENGINEERING APPLICATIONS OF ULTRASONIC WAVES

1. Depth Sounding

The depth of sea or depth of water below a ship can be found out by using the echo sounding principle. Ultrasonic waves generated by a crystal transducer in the range of 40–50 kHz are directed towards the bottom of the sea. The time lag between the sending of ultrasonic signals and return back as an echo to the ship is recorded. Knowing the velocity and with measurement of the time lag, the depth of sea water below the ship can be calculated.

$$\text{Depth of sea} = \frac{\text{velocity of sound in sea} \times \text{time}}{2}$$

The velocity v of sound in sea water at t°C is given by

$$\boxed{v = v_0 + 1.14\,S + 4.21\,t - 0.037\,t^2}$$

where v_0 = velocity of sound in sea water at 0°C = 1510 m/sec

S = The salinity (gm/litre)

t = temperature of sea water in °C.

2. Signalling

Since the frequency of the ultrasonics can be made very high so that wavelength is reduced to a small value. Therefore ultrasonics can be transmitted in the form of a narrow beam of small amplitude and large energy. The ship captain uses ultrasonics to steer its path in mist when light fails within few metres.

3. Sonar Exploration

Sonar stands for sound navigation and ranging.

SO stands for sound

N stands for Navigation

A stands for And

R stands for Ranging

Sonar is similar to RADAR in working. In radar we use microwaves, which are electromagnetic in nature, whereas in SONAR, we make use of ultrasonics. Ultrasonic waves are sent by the ultrasonic transmitter towards the target through water and are received back by the receiver lowered into the sea water from a ship as shown in Figure 3.7. Knowing the time interval between sending and receiving of ultrasonics we can find the position of target or sub-marine. We can also find the direction and the velocity of the moving sub-marine with this apparatus. So this sonar technique is used for the detection of energy sub-marine inside water during war time in a similar manner as RADAR is used to trace the enemy aeroplanes in the atmosphere. During piece time, sonar is used for detecting the glaciers, icebergs, searocks and other submerged objects.

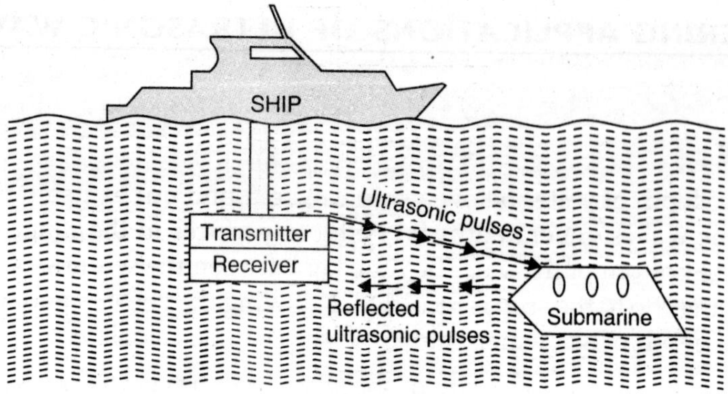

FIGURE 3.7

4. Drilling and Etching

Drilling and etching, which is very difficult or impossible with the usual machine process, can be conveniently done by ultrasonics. This is particularly useful for square drills and when the dimensions of the required drill are of tight tolerance. A transducer producing ultrasonics having a core of soft material is fixed with a bit of hard material and set over the plate to be processed. Slurry of water and carborandum powder are passed over the plate to be drilled just below the bit and ultrasonics applied. Cavities when break send carborandum particles with very high velocity drilling a hole (circular or square) in the plate depending upon the section of the hard material bit. (Figure 3.8).

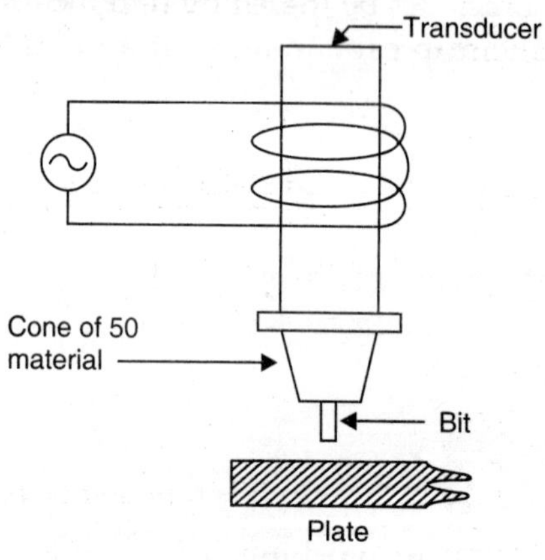

FIGURE 3.8

5. Cold Welding

In ordinary process, the temperature of the plates to be welded is raised which might bring certain undesirable change in the physical properties, like brittleness, magnetic etc. So by ultrasonics, two pieces to be welded are placed together and ultrasonic source is applied. The particles of two pieces vibrate violently resulting in the diffusion and binding forces are developed so that welding is completed. It is clear from Figure 3.9.

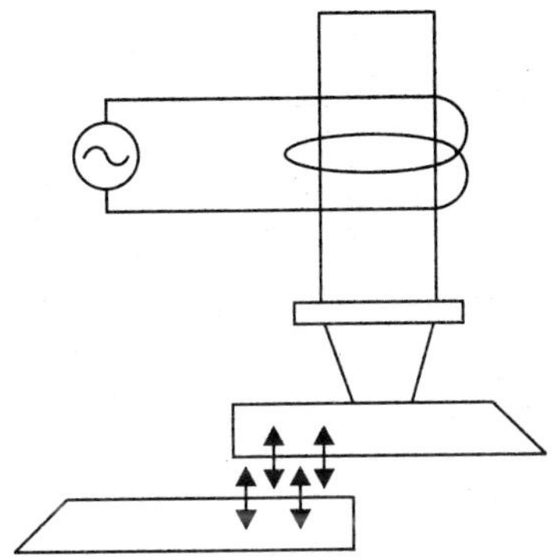

FIGURE 3.9

6. Flaw Detection

Ultrasonic waves can be used to detect cracks (flaws) or discontinuity in metal structures. In this case, an emitter and detector of ultrasonic waves are used. Ultrasonic waves from the emitter are directed towards the metal. The reflected beam is detected by the detector. If there is a flaw or discontinuity there will be rise in energy received by the detector, if the emitter and detector are on the same side. If the emitter and the detector are on the opposite side of the metal, there will be fall in energy at the regions of flaws or discontinuity.

7. Formation of Alloys

The alloys of different compositions are prepared by using ultrasonic waves. These alloys are homogenous in nature and well mixed.

3.5 NUMERICALS BASED ON ULTRASONICS

Problem 1. *Calculate the fundamental frequency of a quartz crystal of 2 mm thickness. The density of the crystal is 2650 kg/m³ and Young's modulus is 7.9 × 10¹⁰ N/m².*

Hint: $$n = \frac{1}{2L}\sqrt{\frac{Y}{\rho}}$$ [**Ans.** 1.365 MHz]

Problem 2. *An ultrasonic beam of 1 cm wavelength sent by the ship returns from sea bed after 2 seconds. If velocity of ultrasonic beam in sea water is 1510 m/sec at 0°C, its salinity at 30°C is 2 gm/lit., calculate the depth of sea bed at 30°C.*

Hint: $$v_t = v_0 + 1.14s + 4.21t - 0.037\, t^2$$

and Depth of sea bed at 30°C $= \dfrac{V_{30} \times t}{2}$ [**Ans.** 1637.20 m]

Problem 3. *The difference between the transmission time and receiving time is 0.44 s. Calculate the depth of sea bed in this case if velocity of ultrasonic wave is 1440 m/sec.*

Hint: $$\frac{2d}{t} = v$$ [**Ans.** d = 316.8 m]

Problem 4. *A quartz crystal of thickness 0.001 m is vibrating at resonance. Calculate the fundamental frequency. Given Y for quartz* 7.9×10^{10} N/m² *and* ρ *for quartz* = 2650 kg/m³.

Hint: $v = \sqrt{\frac{Y}{\rho}}$, thickness $= \frac{\lambda}{2}$, $n = \frac{v}{\lambda}$ **[Ans.** $n = 2.73$ MHz]

Problem 5. *A piezoelectric X-cut quartz plate has a thickness of 1.5 mm. If the velocity of propagation of longitudinal sound waves along the X-direction is 5760 m/s, calculate the fundamental frequency of the crystal.*

Hint: Thickness $= \frac{\lambda}{2}$, $n = \frac{v}{\lambda}$ **[Ans.** $n = 1.92$ MHz]

EXERCISES

1. What are ultrasonics? Describe their production.
2. How are ultrasonics detected? Explain the principle and working of a thermal detector.
3. Discuss briefly the applications of ultrasonics in
   (*i*) Depth sounding and (*ii*) Signalling.
4. Discuss briefly applications of ultrasonics.
5. What are ultrasonic waves? Describe a method of producing ultrasonic waves using piezoelectric crystals. Mention some of the uses of ultrasonic waves.
6. Write short notes on
   (*i*) Galton's Whistle (*ii*) Magnetostriction Oscillator
   (*iii*) Piezoelectric Oscillator (*iv*) Acoustic Grating
   (*v*) Ultrasonics and its uses (*vi*) Depth Sounding.
7. Write an essay on ultrasonic waves.
8. Give one method for the production of ultrasonic waves.
9. Specify different methods for the production of ultrasonic waves and describe one of them in detail.
10. Discuss the various applications of ultrasonic waves.

SUGGESTION FOR FURTHER READING

1. L.E. Kinsler and A.R. Frey, Fundamentals of Acoustics, Wiley India Ltd., New Delhi, 2002.
2. Devraj Singh, Engineering Physics, Vol. 2 Dhanpat Rai and Co. (P) Ltd., New Delhi, 2009.
3. J. David N. Cheeke, Fundamental and Applications of Ultrasonic Waves, CRC Press, New York, 2002.
4. D.R. Raichel, The Science and Applications of Acoustics, Springer Verlag, New York, 2000.
5. A.P. Cracknell, Ultrasonics, Wykeham Publications, London, 1980.
6. W.P. Mason, Piezoelectric Crystals and Their Applications to Ultrasonics, D. Von Nostrand, Princeton, 1950.
7. J-Blittz, Elements of Acoustics, Butterworth, London 1964.
8. Devraj Singh, Fundamentals of Engineering Physics, Vol. 1, Dhanpat Rai and Co. (P) Ltd., New Delhi, 2009.
9. P. Vigoureux, Ultrasonics, Chapman and Hall, London, 1970.

Chapter 4

INTERFERENCE OF LIGHT

4.1 INTRODUCTION

With a single source of light the distribution of light energy in the surrounding space is uniform but when there are two such sources, under certain condition the distribution is no longer uniform. There are certain regions where on account of superposition of waves enhanced light intensity is observed while at certain other places the two set of waves destroy each other so darkened is produced. This modification of the intensity of light distribution obtained by the superposition of two or more waves under certain condition is called *phenomenon of interference.*

The two set of waves coming from the two sources cross each other than they pass through any point. The presence of one wave does not affect the flow of other waves but total amplitude at any point will be the algebraic sum of the amplitudes of parent waves. At certain point the wave superimpose in such a way that resultant intensity is greater than the sum of intensities due to individual waves. The interference is called *constructive interference*. While at certain other points the resultant intensity is less than the sum of the intensities due to individual waves—the phenomenon is called *destructive interference.* It must be clearly understood that there is no destruction of energy anywhere. Energy is merely redistributed so that energy missing at one point appears at another point. The interference phenomenon was demonstrated experimentally by Thomas Young for the first time which established that light travel in the form of wave.

4.2 PRINCIPLE OF SUPERPOSITION

According to this principle the resultant displacement of a particle of medium acted upon by two or more waves simultaneously is equal to the algebraic sum of the displacement of the particle due to individual waves.

Thus if two waves are

$$y_1 = a_1 \sin \omega t \text{ and } y_2 = a_2 \sin (\omega t + \delta)$$

which have same frequency but differing in phase and amplitude, then from the principle of superposition the resultant displacement y will be

$$y = y_1 + y_2 = a_1 \sin \omega t + a_2 \sin (\omega t + \delta)$$

$$= (a_1 + a_2 \cos \delta) \sin \omega t + a_2 \cos \omega t \sin \delta$$

Here we assume $\quad a_1 + a_2 \cos \delta = A \cos \phi$

and $\quad a_2 \sin \delta = A \sin \phi$

$$y = A \cos \phi \sin \omega t + A \sin \phi \cos \omega t$$

$$\Rightarrow \quad \boxed{y = A \sin (\omega t + \phi)} \quad \text{...(1)}$$

$$A^2 = a_1^2 + a_2^2 + 2a_1a_2 \cos \delta \quad \text{...(2)}$$

$$\tan \phi = \frac{a_2 \sin \delta}{a_1 + a_2 \cos \delta} \quad \text{...(3)}$$

which gives the amplitude and phase of resultant wave.

The intensity at any point is given by

$$I = A^2 = a_1^2 + a_2^2 + 2a_1a_2 \cos \delta \quad \text{...(4)}$$

Condition for maximum and minimum intensities

Intensity will be maximum of points where

$$\cos \delta = +1 \quad \text{or} \quad \delta = 2n\pi \qquad [n = 0, 1, 2, 3, \text{......}]$$

i.e., intensity will be maximum when phase difference is even multiple of π or path difference (Δ) is even multiple of $\lambda/2$.

$$\boxed{\begin{aligned} &\delta = 2n\pi, \quad n = 0, 1, 2, \text{......} \\ &\Delta = 2n\lambda/2 \\ &I_{max} = (a_1 + a_2)^2 \end{aligned}} \quad \text{...(5)}$$

The intensity will be minimum, when

$$\boxed{\begin{aligned} &\delta = (2n + 1)\pi \quad n = 0, 1, 2, \text{......} \\ &\Delta = (2n + 1)\lambda/2 \\ &I_{min} = (a_1 \sim a_2)^2. \end{aligned}} \quad \text{...(6)}$$

4.3 COHERENT SOURCES

It is found that it is not possible to show interference due to two independent sources of light because a large number of difficulties are involved. Two sources may emit light wave of different amplitude and wavelength and the phase difference between the two may change with time. The fundamental requirement to get will defined interference pattern is that the phase difference between the two waves should be constant. The two sources are said to be *coherent* if they emit light waves of same frequency nearly the same amplitude and are always in phase with each other. In actual practice it is not possible to have two independent sources which are coherent. But for experimental purpose two virtual sources formed from a single source can act as *coherent sources*. Methods for producing coherent sources is divided in following two parts have been divide (*i*) interference of light take place between waves from two sources formed due to single source (division of wavefront) (*ii*) interference takes place between the waves from the real source and virtual source (division of amplitude).

4.4 ANALYTICAL THEORY OF INTERFERENCE

Let S_1 and S_2 are two coherent sources separated by a distance d and made from a monochromatic source S. Let a screen be placed at a distance D from the coherent source as shown in Figure 4.1 point O. On the screen is equidistance from S_1 and S_2.

∴ As such path difference at O will be zero and the intensity at O will be maximum.

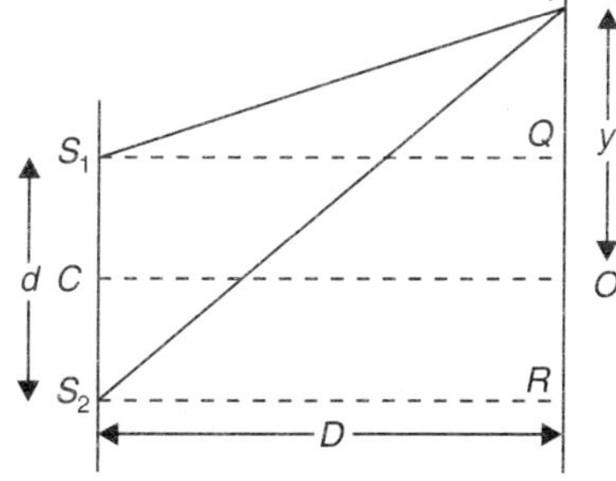

FIGURE 4.1

Consider a point P at a distance y from O. The wave reaches the point P from S_1 and S_2.

From Figure 4.1, $PQ = (y - d/2)$ and $PR = (y + d/2)$ ...(1)

The path $$S_1P = \sqrt{\{D^2 + (y - d/2)^2\}} = D + \frac{(y - d/2)^2}{2D} \quad ...(2)$$

[Using binomial expansion]

$$S_2P = \sqrt{\{D^2 + (y + d/2)^2\}} = D + \frac{(y + d/2)^2}{2D} \quad ...(3)$$

[Using binomial expansion]

Path difference

$$(S_2P - S_1P) = \left[D + \frac{(y + d/2)^2}{2D}\right] - \left[D + \frac{(y - d/2)}{2D}\right]^2$$

$$\Delta = \frac{yd}{D} \qquad ...(4)$$

Corresponding phase difference will be

$$\Delta\phi = \frac{2\pi}{\lambda}\Delta = \frac{2\pi}{\lambda}\left(\frac{yd}{D}\right) \qquad ...(5)$$

Position of bright fringes:

Path difference = even multiple of $(\lambda/2)$

So, $$\frac{yd}{D} = 2n\lambda/2 = n\lambda,$$

or $$y_n = \frac{n\lambda D}{d} \qquad ...(6)$$

which gives position of different maximum at

$$y_0 = 0,\ y_1 = \frac{\lambda D}{d},\ y_2 = \frac{2\lambda D}{d},\ y_3 = \frac{3\lambda D}{d},\ ......$$

$$\Delta y = y_1 - y_0 = y_2 - y_1$$

$$= ..... = y_n - y_{n-1} = \frac{\lambda D}{d} \qquad ...(7)$$

Position of dark fringes:

Path difference odd multiple of $(\lambda/2)$

$$\frac{yd}{D} = (2n + 1)\ \lambda/2$$

$$y_n = \frac{(2n+1)\,\lambda D}{2d} \qquad ...(8)$$

which gives the position of different minima

$$y_0 = \frac{\lambda D}{2d},\ y_1 = \frac{3\lambda D}{2d},\ y_3 = \frac{5\lambda D}{2d}$$

$$\Delta y = y_1 - y_0 = y_2 - y_1 = y_3 - y_2 = ... = y_n - y_{n-1} = \frac{\lambda D}{d} \qquad ...(9)$$

Fringe width: The distance between any two consecutive bright or dark fringes is called fringe width and is denoted by β.

$$(\beta)_{\text{bright}} = (\Delta y)_{\min} = \frac{\lambda D}{d}$$

$$(\beta)_{\text{dark}} = (\Delta y)_{\max} = \frac{\lambda D}{d}$$

$$\beta = (\beta)_{\text{bright}} = (\beta)_{\text{dark}} = \frac{\lambda D}{d} \quad ...(10)$$

It is found that

$$\beta \propto \lambda; \beta \propto D \quad \text{and} \quad \beta \propto \frac{1}{d}$$

Angular fringe width θ_β: It is given by

$$\theta_\beta = \theta_{n+1} - \theta_n = \frac{y_{n+1}}{D} - \frac{y_n}{D} = \frac{\beta}{D}$$

$$= \frac{\lambda}{d} \quad ...(11)$$

1. At the centre we get bright fringe.
2. Then alternately dark, bright, dark ...... fringe on both sides of the central maximum.
3. Fringe width of dark and bright fringe are same.
4. Width of fringe depends upon wavelength, separation of sources and source to screen separation.

If white light is used each wavelength will give its own fringe pattern upto a short distance from O the fringes of different colour would stand side by side but beyond that a general illumination is observed. The central fringe with white light is white as for all colours the central fringe is formed at O and they together produce white light.

4.5 CONDITION FOR SUSTAINED INTERFERENCE

1. The two sources must be coherent.
2. The sources should emit light of same frequency.
3. The amplitudes of two wave must be equal otherwise a clear cut distinction between maximum $(a_1 + a_2)$ and minimum $(a_1 - a_2)$ is not possible.
4. The two waves must propagate along the same line.
5. The two sources must be narrow otherwise a single source may itself be equivalent to a large number of fine sources which will give their own pattern.
6. The separation between sources should be small so that wide fringes are obtained.
7. Separation between source and screen should be large to get wide fringes.
8. If the interfering waves are polarised they must be in the same state of polarisation.

4.6 EFFECT OF INTRODUCING A THIN TRANSPARENT PLATE IN THE PATH OF ONE WAY

Let S_1 and S_2 be two monochromatic coherent sources of wavelength λ. Let a thin plate of thickness t and refractive index μ is introduced in the path of one ray from source S_1 as shown in Figure 4.2. Now, this ray travel partially in the material if c and v are the velocities of light in air and medium respectively then path difference between two rays.

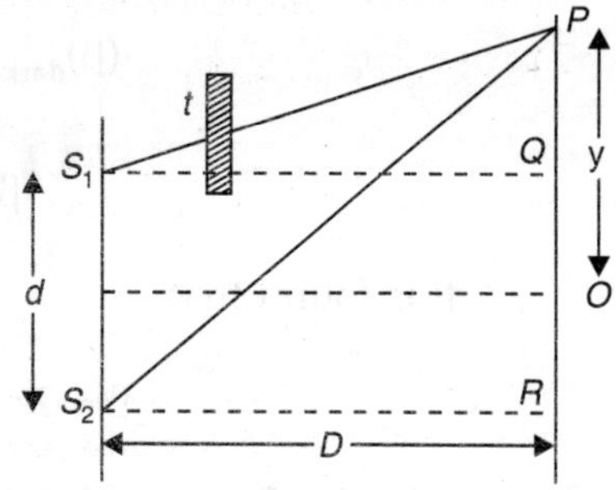

FIGURE 4.2

$$(S_2P - S_1P) = [(S_2P)_{air} - \{(S_1P - t)_{air} + (t)_{plate}\}] \quad ...(1)$$

$$= [(S_2P - S_1P)_{air} + (t_{air} - t_{plate})]$$

we know that distance t travelled by light through plate is equivalent to a distance μt travelled in air.

$$\text{Path difference} = [(S_2P - S_1P) + t - \mu t]_{air}$$

$$= [S_2P - S_1P - (\mu - 1)\, t]$$

$$= \left[\frac{yd}{D} - (\mu - 1)\, t\right] \quad ...(2)$$

For maxima

$$\text{Path difference} = \left[\frac{yd}{D} - (\mu - 1)\, t\right] = n\lambda$$

or

$$y = \frac{D}{d}[n\lambda + (\mu - 1)\, t] \quad ...(3)$$

Without plate (*i.e.* $t = 0$) the position of maximum is given by

$$y_0 = \frac{nD\lambda}{d} \quad ...(4)$$

The plate causes a shift of maximum and minimum by an account.

$$\delta' = (y - y_0) = \frac{D}{d}(\mu - 1)\, t$$

We know that $\beta = \dfrac{\lambda D}{d} \Rightarrow \dfrac{D}{d} = \dfrac{\beta}{\lambda}$.

$$\boxed{\delta' = \frac{\beta}{\lambda}(\mu - 1)\, t} \quad ...(5)$$

where β is fringe width

This shift is +ve if $\mu > 1$ and the pattern will move by an amount δ' toward the side on which the plate is placed keeping the nature of pattern same.

If the central fringe moves through a distance formely occupied by the nth bright fringe

$$\delta' = n\beta = \frac{\beta}{\lambda}(\mu - 1)\,t$$

$$\boxed{n = \frac{(\mu - 1)\,t}{\lambda}} \qquad ...(6)$$

The gives the number of bands through which the system has shifted.

4.7 TWO CLASSES OF INTERFERENCE

1. **Division of wavefront:** In this class the wavefront originate from source *S* is divided into two parts by using any optical system (like Young Double slit experiment, Fresnel biprism, Fresnel mirror Lloyd's mirror *etc.*). The two wavefront after travelling certain distance again brought together to give interference pattern. Here, point sources are required.
2. **Division of amplitude:** Here the amplitude is divided into two or more parts either by partial reflection and which after travelling through different path reach to a point and produce interference. Here broad source is required. Newton's Rings, Michelson interferometer and Fabry Perot interferometer are the examples of this class.

4.8 FRESNEL BIPRISM

Used a biprism to get interference pattern by division of wavefront method. The biprism consist of two active angled prisms with their bases in contact. Here two sources S_1 and S_2 are the virtual image of the fine slit *S* as shown in Figure 4.3. The experimental arrangement consist of a slit *S* the biprism *ABC* and the microscope *M*. All are mounted on an optical bench. These are adjusted at the same height and can move and rotate as required. The light emerging from the slit fall on the biprism. The edge *A* of the biprism divides the incident wave front into two parts. One is through upper half *AB* of biprism and appears to coming from virtual source S_1. Other is from lower half *AC* of biprism and appears to coming virtual S_2. The interference fringes are seen in the overlapping region *XY* and can be seen by eyepiece.

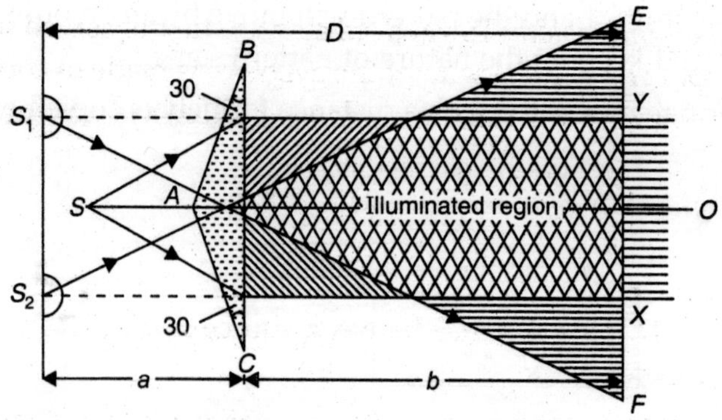

FIGURE 4.3 **Fresnel's biprism**

Adjustment of Biprism: To obtain well defined interference fringes following adjustments of apparatus are required.

1. The optical bench should be levelled.
2. Slit, biprism and eyepiece should be adjusted to the same height.
3. Eyepiece should be adjusted to make fine image at crosswires.
4. The slit should be narrow and vertical.
5. Biprism is positioned in such away, may be by giving lateral movement that overlapping region should be in field of view of eyepiece.
6. The fringes so obtained should be made clear by adjusting biprism slit and eyepiece.
7. Lateral shift of the fringes should be removed with the help of biprism and eyepiece. The reason for lateral shift is that the line joining the slit and biprism is not parallel to the length of optical bench.
8. The bench should not be disturbed once these adjustment have been done.

Theory: On the screen at O we get a central maximum then alternatively dark, bright ... fringes on both sides.

$$(y_n)_{\text{bright}} = \frac{n\lambda D}{d} \quad ...(1)$$

$$(y_n)_{\text{dark}} = \frac{(2n+1)\lambda D}{2d} \quad ...(2)$$

and fringe width is

$$\beta = \frac{\lambda D}{d} \quad ...(3)$$

(*a*) **Determination of wavelength:** Biprism can be used to determine the wavelength of given monochromatic light using the expression.

$$\lambda = \beta\frac{d}{D} \quad ...(4)$$

(*i*) *Measurement of fringe width:* To get β, fringes are first observed in the field of view of the microscope. The vertical wire of the eyepiece is made to coincide with one of the fringes and screw of micrometer is moved sideways and number of fringes are counted.

$$\beta = \frac{\text{Distance moved}}{\text{number of fringes passed}} \qquad ...(5)$$

(*ii*) *Measurement of D:* This distance between source and eyepiece is directly measured on the optical bench scale.

(*iii*) *Determination of D:* For this we make the use of displacement method. A convex lens is placed between biprism and the eyepiece in such a way that for two positions of lens the image of virtual sources S_1 and S_2 are seen in eyepiece. Then

$$d = \sqrt{d_1 d_2} \qquad ...(6)$$

where d_1 and d_2 are the distance between S_1 and S_2 for two position of lens.

(*b*) **Determination of thickness of a thin film:** We know that the interference pattern shifts one side when a thin transparent film is put in the path of one ray. This shift is

$$\delta' = \frac{\beta}{\lambda}(\mu - 1)\, t \qquad ...(7)$$

Thus, measuring fringe width β and shift δ' using biprism the t can be calculated if refractive index μ and wavelength λ are known.

4.9 INTERFERENCE IN THIN FILMS

The film of transparent material like a drop of oil spread on the surface of water, show brilliant colours when exposed to an extended source of light. This phenomenon can be explained on interference basis. Here interference takes place between rays reflected from the upper and from the lower surface of the film.

Case I: Thin film of uniform thickness:

(*a*) **Reflection Pattern:** Let us consider a thin film of thikckness t, refractive index μ and a ray AB of monochromatic light of λ is falling on it at an angle i. This ray is partly reflected and partly refracted as shown in Figure 4.4.

Path difference

$$= (BD + DE)_{\text{film}} - (BH)_{\text{air}}$$
$$= [\mu\,(BD + DE) - BH]_{\text{air}}$$
$$= [\mu\,(BD + DG + GE) - BH]$$
$$= \mu\,(BD + DG) \qquad ...(1)$$

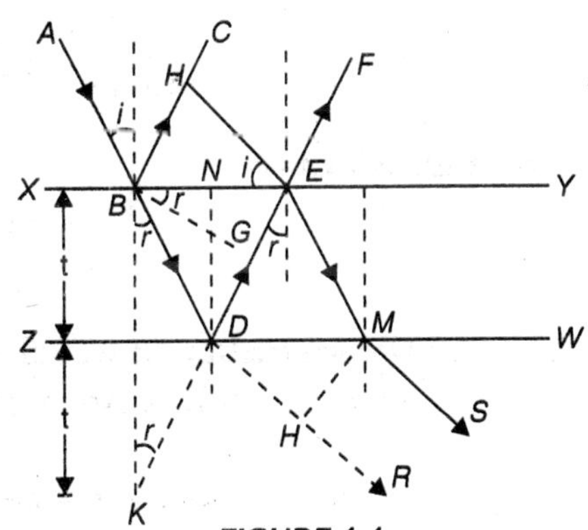

FIGURE 4.4

as $$\mu\, GE = BH$$
$$BH = EB \sin i$$
$$GE = EB \sin r$$
$$\mu\, GE = EB = \sin i$$
$$\mu = \frac{\sin i}{\sin r}$$

Extend BL and ED to meet at K we have

$$\angle LKD = \angle NDE = r \qquad ...(2)$$

Triangles $\angle BLD$ and $\angle LKD$ are congruents

$$BD = DK$$

Then path difference $= \mu\,(BD + DG) = \mu\,(KD + DG)$ ...(3)

$$= \mu\, KG = \mu\, KB \cos r$$

$$\Delta = 2\,\mu t \cos r \qquad ...(4)$$

Now, here since reflection from denser medium is taking place by one ray so additional path difference of $\lambda/2$ is taken into account. Thus the total path difference is

$$\boxed{\Delta_T = 2\,\mu t \cos r - \lambda/2} \qquad ...(5)$$

Condition for maxima

$$\Delta_T = (2\,\mu t \cos r - \lambda/2) = n\lambda$$
$$2\,\mu t \cos r = (2n + 1)\,\lambda/2 \quad n = 0, 1, 2, 3, ... \qquad ...(6)$$

Condition for minima

$$\Delta_T = (2\,\mu t \cos r - \lambda/2) = (2n + 1)\,\lambda/2$$
$$2\,\mu t \cos r = (n + 1)\,\lambda \qquad ...(7)$$

n is the integer so $(n + 1)$ is also integer and can be taken as n

$$\boxed{2\,\mu t \cos r = n\lambda} \quad n = 0, 1, 2 \qquad ...(8)$$

when $\cos r$ is kept constant and thickness increases gradually, the path difference becomes $\lambda/2$, λ, $3\lambda/2$, 2λ, $5\lambda/2$, ... and as a result the film will appear dark ($t = 0$), bright, dark and so on. On the other hand if t is constant and r is varied we again get a series of maxima and minima.

(*b*) **Transmitted Pattern:** Here the path difference two rays DEMS and PR will be

$$\Delta = \mu\,(DE + EM) - DH$$
$$= 2\,\mu t \cos r \qquad ...(9)$$

In this case there will be no additional path difference so the total path difference

$$\boxed{\Delta_T = 2\,\mu t \cos r} \qquad ...(10)$$

Condition for maxima

$$\Delta = 2\,\mu t \cos r = n\lambda \quad n = 0, 1, 2, 3, ... \qquad ...(11)$$

Condition for minima

$$\Delta = 2\,\mu t \cos r = (2n+1)\,\lambda/2 \quad n = 0, 1, 2, 3, \ldots\ldots \qquad \ldots(12)$$

We find that the conditions for maxima and minima are found in case of transmitted pattern are opposite to those found in case of reflected pattern. Under the same conditions of the film looks dark in reflected pattern it will look bright in transmitted pattern.

Colours of thin films: When white light is incident on a thin film only few wavelengths will satisfy the condition of maxima and therefore corresponding colours will seen in the pattern. For other wavelengths condition of minima is satisfied, and so corresponding colours will be missing in the pattern. The colouration of film vary with t and r. Therefore if one vary either t or r a different set of colours will be observed.

Since the condition for maxima and minima are opposite in case of refleced and transmitted pattern. So the colours found in two patterns will be complimentary to each other.

Necessity of broad source: When point source is used only a small portion of the film can be seen through eye and as a result the whole interference pattern cannot be seen. But when a broad source is used rays of light are incident at different angles and reflected parallel beam reach the eye and whole beam and complete pattern is visible.

Case II: The film is of varying thickness (Wedge shaped thin film): Consider the wedge shaped film as shown in Figure 4.5. Let a ray from S is falling on the film and after reflections produce interference pattern.

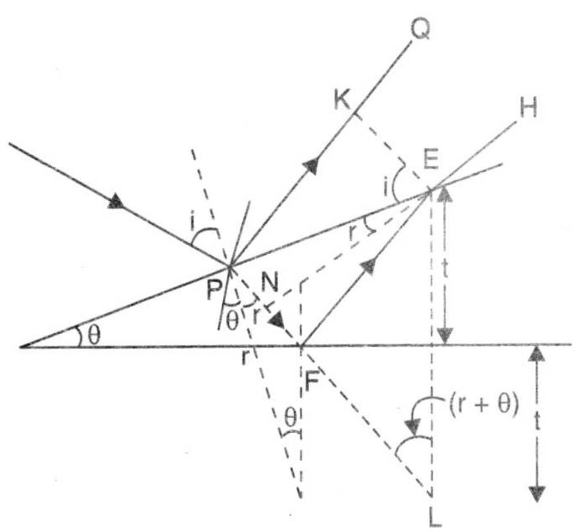

FIGURE 4.5

$$\begin{aligned}\Delta &= [(PF + FE)_{\text{film}} + (PK)_{\text{air}}] \\ &= \mu\,(PF + FE) - (PK) \\ &= \mu\,(PN + NF + FE) - PK \\ &= \mu\,(PN + NF + FE - PN) && [\because\ PK = \mu(PN)] \\ &= \mu\,(NF + FE) && [\because\ FE = FL] \\ &= \mu\,(NF + FL) \\ &= \mu(NL) = 2\,\mu t \cos(r + \theta) && \ldots(13)\end{aligned}$$

Then total path difference considering refraction from denser medium is taking place

$$\boxed{\Delta_T = [2\,\mu t \cos(r+\theta) - \lambda/2]} \qquad ...(14)$$

Condition for maxima

$$2\,\mu t \cos(r+\theta) = (2n+1)\,\lambda/2 \qquad n = 0, 1, 2, ... \qquad ...(15)$$

Condition for minima

$$2\,\mu t \cos(r+\theta) = n\lambda \qquad n = 0, 1, 2, 3, ... \qquad ...(16)$$

Thus the film will appear bright if the thickness t satisfies the condition of maxima

$$t = \frac{(2n+1)\,\lambda}{4\,\mu \cos(r+\theta)} \qquad ...(17)$$

and it will appear dark when

$$t = \frac{n\lambda}{2\,\mu \cos(r+\theta)} \qquad ...(18)$$

Hence, we move along the direction of increasing thickness we observe dark, bright, dark λ ...... fringes. For $t = 0$ *i.e.*, at the edge of film $\Delta = \lambda/2$ so the film will appear dark. Then width of the fringes so observed can be found

$$\beta = \frac{\lambda}{2\,\mu \tan\theta \cos(r+\theta)} \qquad ...(19)$$

In case of normal incidence $r = 0$

$$2\,\mu t = (2n+1)\,\lambda/2 \text{ (maxima)} \qquad ...(20)$$

$$2\,\mu t = n\lambda \text{ (minima)} \qquad ...(21)$$

$$\boxed{\beta = \frac{\lambda}{2\,\mu\theta}} = \text{fringe width} \qquad ...(22)$$

4.10 NEWTON'S FRINGES AND HAIDINGER FRINGES

We have seen that thin film causes a path difference $2\,\mu t \cos r$ between interfering rays. Now, when area of the film is small, the rays from various portions of the film reaching the eye have almost the same inclination. So it is the variation in thickness that gives fringes. Each fringe is the locus of all such points where thickness is same. Such fringes are known as *Newton's fringes or fringes of equal thickness*. On the other hand when film has uniform thickness the path difference changes with r only. Each fringe in this case represent the locus of all points on the film, from which rays are

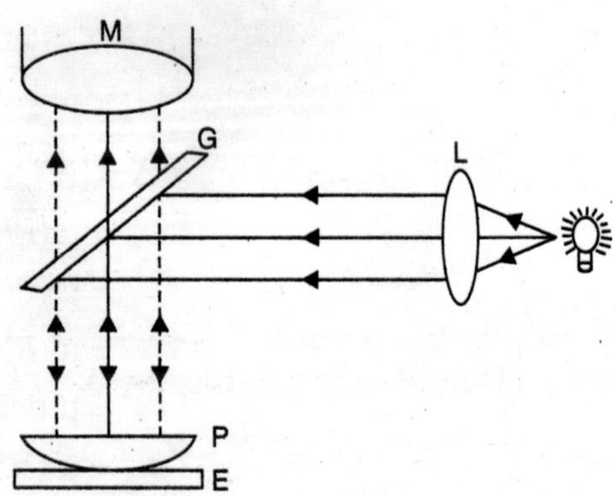

FIGURE 4.6 **Newton's ring apparatus**

equally inclined to the normal. Such fringes are called *Haidinger's fringes* or *fringes of equal inclination. To get such fringes the source must be an extended one.*

Newton's rings: It is a special case of interference in a film of variable thickness such as that formed between a plane glass plate and a convex lens in contact with it. When monochromatic light falls over it normally we get a central dark spot surrounded by alternatively bright and dark circular rings. When white light is used the rings would be coloured.

Experimental Arrangement: Let S be the extended source of light, rays from which after passing through a lens L falls upon a glass plate G at 45°. After partial reflection these rays fall on a plano convex lens P placed on the glass plate E. The interference occurs between the rays reflected from the two surfaces of the air film and viewed through microscope M as shown Figure 4.6.

Theory: The air film formed is of wedge shape so the path difference produced will be

$$\Delta = 2\,\mu t \cos(\theta + r) - \lambda/2 \qquad ...(1)$$

For normal inciden $r = 0$ so

$$\Delta = 2\,\mu t - \lambda/2 \qquad \text{(as } \theta \text{ is very small)}$$

At the point of contact $t = 0$

So $$\Delta = \lambda/2 \qquad ...(2)$$

The central fringe will be dark.

Condition for maxima and minima

$$2\,\mu t = (2n + 1)\,\lambda/2 \qquad ...(3)$$

$$2\,\mu t = n\lambda \qquad n = 0, 1, 2 \qquad ...(4)$$

we get alternatively bright and dark rings.

Diameter of the rings: From the Figure 4.7.

$$DB \times DC = AD \times DG$$

and $DB = DC = r$, radius of ring under consideration.

$CE = DA = BF = t$, thickness of the film at B.

$OG = OA = R$, radius of curved surface of lens.

Then, $$r^2 = t\,(2R - t) = 2Rt - t^2 \simeq 2Rt$$

$$\boxed{t = \frac{r^2}{2R}} \qquad ...(5)$$

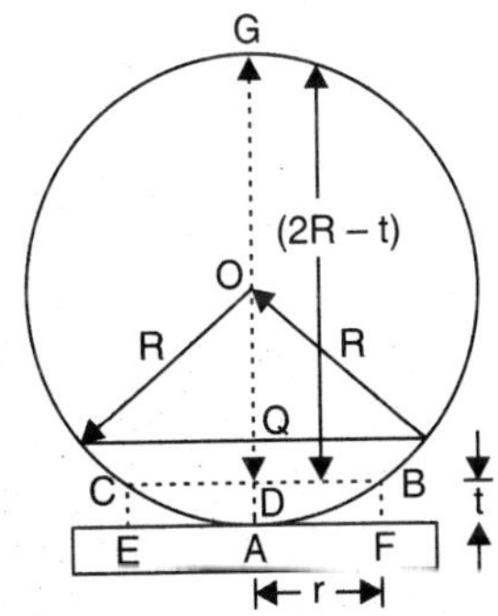

FIGURE 4.7

Diameter of bright rings D_n : From the condition of maxima

$$2\,\mu\,\frac{r_n^2}{2R} = (2n+1)\frac{\lambda}{2},$$

or $$r_n^{\,2} = \frac{(2n+1)\,\lambda R}{2\mu} = \left(\frac{D_n}{2}\right)^2,$$

$$D_n^{\,2} = \frac{2\,(2n+1)\,\lambda R}{\mu},$$

$$= 2\,(2n + 1)\,\lambda R \qquad \text{(For air, } \mu = 1) \qquad \text{...(6)}$$

or

$$\boxed{D_n = \sqrt{(2\lambda R)}\,\sqrt{(2n+1)} = K\,\sqrt{(2n+1)}} \qquad \text{...(7)}$$

$$D_n \propto \sqrt{\text{odd number}}\,.$$

Thus, the diameter of bright rings are proportional to the square root of the odd natural numbers.

Diameter of dark ring D_n : Using the condition for minimum

$$2\,\mu\,\frac{r_n^2}{2R} = n\lambda \quad \text{or} \quad r_n^{\,2} = \frac{n\lambda R}{\mu} \qquad \text{...(8)}$$

which gives

$$(D_n)^2 = 4n\lambda R \qquad \text{(For air } \mu = 1)$$

or

$$\boxed{D_n = \sqrt{(4\,\lambda R)}\,\sqrt{n}} \qquad \text{...(9)}$$

and

$$D_n \propto \sqrt{n}$$

Thus, the diameter of dark rings are proportional to the square root of natural numbers.

4.10.1 Measurement of Wavelength of Light by Newton's Rings

For dark rings we know

$$\frac{(D_n)^2}{n} = 4\lambda R,$$

or

$$\lambda = \frac{(D_n)^2}{4nR} \qquad \text{...(10)}$$

But as the fringes around the centre are not very clear so n cannot be measured correctly. To avoid any mistake one can consider two clear fringes n^{th} and $(n + p)^{th}$

So, $D_n^{\,2} = 4n\lambda R$ and $D_{(n+p)}^2 = 4(n + p)\,\lambda R,$

Then $\left(D_{n+p}^2 - D_n^2\right) = 4p\lambda R,$

or

$$\lambda = \frac{D_{n+p}^2 - D_n^2}{4pR} \qquad \text{...(11)}$$

Thus, measuring the diameters and knowing p and R, λ can be measured.

4.10.2 Measurement of Refractive Index of Liquid by Newton's Rings

For this purpose liquid film is formed between the lens and glass plate.

We have, as above,

$$\left[D_{n+p}^2 - D_n^2\right]_{\text{liquid}} = \frac{4p\lambda R}{\mu} \qquad ...(12)$$

$$\left[D_{n+p}^2 + D_n^2\right]_{\text{air}} = 4p\lambda R \qquad ...(13)$$

which give

$$\boxed{\mu = \frac{\left[D_{n+p}^2 - D_n^2\right]_{\text{air}}}{\left[D_{n+p}^2 - D_n^2\right]_{\text{liquid}}}}$$

One can see that rings contract **with the introduction of liquid.**

4.11 NUMERICALS BASED ON INTERFERENCE OF LIGHT

Problem 1. *A film of thickness 0.6 μm of soap solution having refractive index 1.4 is formed. It is observed at an angle of incidence 4°. Determine that wave lengths of visible light, absent in the reflected light.*

Hint: The condition for a particular wavelength, to be absent from the reflected light is $2\,\mu t \cos r = \lambda, 2\lambda, 3\lambda, ...$

Let these wavelengths be $\lambda_1, \lambda_2, \lambda_3, ...$

$$\mu = \frac{\sin i}{\sin r} \quad \Rightarrow \quad r = \sin^{-1}(\sin 4°/1.4) = 27°20'$$

$$\lambda_1 = 2 \times 1.4 \times 0.6 \times 10^{-6} \times \cos 27°20' = 14923 \text{ Å}$$

$$2\lambda_2 = 2\mu t \cos r \quad \Rightarrow \quad \lambda_2 = 7461 \text{ Å}$$

Similarly $\quad 3\lambda_3 = 2\mu t \cos r \quad \Rightarrow \quad \lambda_3 = 4974 \text{ Å}$

The visible region is 4000 to 7000 Å

So the absent wavelength will be 4974 Å.

Problem 2. *A biprism is placed at a distance of 5 cm in front of a narrow slit, Illuminated by sodium light (λ = 5890 Å) and the distance between the virtual sources is found to be 0.05 cm. Find the width of the fringes observed in eyepiece placed at a distance of 75 cm from the biprism.*

Hint: $\beta = \dfrac{\lambda D}{d} = \dfrac{5890 \times 10^{-10} \times (75+5) \times 10^{-2}}{0.05 \times 10^{-2}}$

$= 9.424$ Å.

Problem 3. *When a thin piece of glass 3.4 × 10^{-4} cm thick is placed in the path of one of the interfering beams in a biprism arrangements, it is found that the central bright fringe shifts through a distance equal to the width of four fringes. Find the refractive index of the piece of glass. Wavelength of light used is 5.46 × 10^{-5} cm.*

Hint: $(\mu - 1)\, t = n\lambda$

$$\boxed{\mu = 1.64}$$

Problem 4. *A surface of refractive index 1.5 is to be coated with a film of magnesium fluoride (μ = 1.38) to minimise the reflection. Calculate the minimum thickness of film for normal incidence of light of wavelength 5000 Å.*

Hint: $$t = \frac{\lambda}{4\mu} = \frac{5000}{4 \times 1.38} = \boxed{906\ \text{Å}}$$

Problem 5. *Light of wavelength 6000 Å falls normally on a thin wedge shaped film of refractive index 1.4, forming fringes that are 2 mm apart. Find the angle of the wedge.*

Hint: $$\beta = \frac{\lambda}{2\theta\mu} \Rightarrow \theta = \frac{\lambda}{2\mu\beta}$$

$$\boxed{\theta = 1.07 \times 10^{-4}\ \text{radian}}$$

Problem 6. *Interference finges are produced by monochromatic light falling normally on a wedge shaped film of cellophane of refractive index 1.4. If angle of wedge is 20 seconds of an arc and the distance between successive fringes is 0.25 cm, calculate the wavelength of light.*

Hint: $$\lambda = 2\theta\mu\beta$$

$$= \frac{2 \times 20 \times \pi}{60 \times 60 \times 180} \times 1.4 \times 0.25$$

$$\lambda = 6790\ \text{Å}$$

Problem 7. *A planoconvex lens of radius 300 cm is placed on an optically flat glass plate and is illuminated by monochromatic light. The diameter of 8th dark ring in the transmitted system is 0.72 cm. Calculate the wavelength of light used.*

Hint: $$r^2 = \frac{(2n-1)\lambda R}{2}$$

$$\boxed{\lambda = 5760\ \text{Å}}$$

Problem 8. *In the Newton's rings arrangement, the radius of curvature of the curved surface is 50 cm. The radii of the 9th and 16th rings are 0.18 cm and 0.2235 cm respectively. Calculate the wavelength.*

Hint: $$D_9 = 2 \times 0.18 = 0.36\ \text{cm}$$

$$D_{16} = 2 \times 0.2235 = 0.447\ \text{cm}$$

$$R = 50\ \text{cm},\ p = 7$$

Then $$\lambda = \frac{D_{n+p}^2 - D_n^2}{4pR} = \boxed{5015 \text{ Å}}$$

Problem 9. *In a Newton's rings experiment the diameter of the 12th ring changes from 1.50 cm to 1.35 when a liquid is introduced between the lens and the plate. Calculate the refractive index of the liquid.*

Hint: For liquid medium $D_1^2 = \dfrac{4n\lambda R}{\mu}$ ...(*i*)

and for air medium $D_2^2 = 4n\lambda R$ ...(*ii*)

Then $$\mu = \left(\frac{D_2}{D_1}\right)^2 = 1.235.$$

Problem 10. *In Newton's rings pattern the diameter of 6th bright ring is 0.5 cm. The wavelength of the light used is 4590 Å. Light enters the film with angle of incidence of 39°. Calculate the radius of curvature of lens.*

Hint: $$2\mu t \cos r = (2n + 1)\frac{\lambda}{2} \text{ and } t = \frac{r^2}{2R}$$

Then $$r_n^2 = \frac{(2n+1)\lambda R}{2\mu \cos r}$$

and $$d_n^2 = \frac{2(2n+1)\lambda R}{\mu \cos r}$$

$$R = 193.66 \text{ cm}.$$

EXERCISES

1. State the basic conditions for observing interference of light. What do you understand by Fresnel's biprism and explain the formation of interference fringes by it? How do you determine the wavelength of monochromatic light by using Fresnel's biprism experiment?
   [*Amity University, Noida, Jan. 2007*]
2. Derive an expression for the diameter of nth dark ring in Newton's ring formation by monochromatic light. [*Amity University, Noida, Dec. 2005*]
3. Describe theory of colours of a thin film, under what conditions the film may appear dark? Using theory of thin films, how would you determine wavelength of given light?
4. Explain formation of Newton's rings in reflected light.
5. Explain effect of introducing a thin plate of mica in the path of one of the interfering beams of Fresnel's biprism. Deduce expression for displacement.

6. Discuss formation of Newton's rings by (*i*) reflected light (*ii*) transmitted light. Derive an expression for radius of nth dark ring in reflected light.
7. Explain formation of Newton's rings. Show how would you use them to determine the wavelength of sodium light. Derive formula used.
8. What are coherent sources? Why two independent sources of light of same wavelength cannot show interference?
9. What is interference of light? Discuss the necessary conditions for interference of light.
10. Why is it necessary to have coherent sources for observing interference of light? Explain why fringes formed by a biprism disappear when the slit is made wide?

SUGGESTION FOR FURTHER READING

1. F.A. Jenkins and H.E. White, Fundamentals of Optics, McGraw Hill Book Company, Singapore, 1981.
2. R.S. Longhurst, Geometrical and Physical Optics, Orient Longman Ltd, Hyderabad, 1999.
3. F.L. Pedrotti, L.S. Pedrotti and L.M. Pedrotti, Introduction to optics, Pearson Education inc., New Delhi, 2007.
4. Aroy Ghatak, Optics, Tata Mc Graw-Hill Publishing Co. Ltd., New Delhi, 2009.
5. E. Hecht and A.R. Ganeshan, Optics, Pearson Education, Singapore, 2009.
6. K.Ghosh and A.Manna, Textbook of Physical Optics, Macmillan India Ltd., Delhi, 2007.
7. R.S. Sirohi, Wave Optics and Its Applications, Orient Longman Ltd., Hyderabad, 2001.
8. Devraj Singh, Engineering Physics, Vol. 1, Dhanpat Rai and Co, (P) Ltd., New Delhi, 2008.
9. Devraj Singh, Fundamentals of Engineering Physics, Vol. 1, Dhanpat Rai and Co. (P) Ltd, New Delhi, 2009.

Chapter 5

DIFFRACTION OF LIGHT

5.1 INTRODUCTION

If an aperture is placed between a source of light and a screen, a sharp illuminated region is obtained on the screen. This indicates that light travels in straight line. Newton's corpuscular theory was able to explain this behaviour. Later on it was proposed that light behaves as a wave. The waves have the property of bending round any corner, as everyone is aware of this property possessed by sound waves. Surely, if light is a wave it must also show this behaviour. The first indication of bending of light round on edge was observed by Grimaldi. He observed that with a small source of light the shadow of light obstacle was larger than given by geometrical construction it was also observed that shadow was not well defined but consisting of some coloured fringes near the outer side of the shadow.

Thus bending of light round the corner or the departure of light path from true rectilinear path was established and phenomenon was called as diffraction. The bright-dark or coloured fringes so obtained are called diffraction pattern.

The diffraction phenomenon could not be explained by Newton's corpuscular theory and Huygen's wave theory. Next attempt was made by Thomas Young. He considered the diffraction as interference between the direct light and light reflected from the edges of obstacle. The idea was not accepted because Fresnel showed that the details of the diffraction pattern do not depend upon the material of the obstacle. The correct explanation was provided by Fresnel who considered that diffraction phenomenon is caused by the interference of the innumerable secondary wavelets produced by the unobstructed position of the same wavefront. He was also able to explain the approximate rectilinear propagation of light.

Diffraction phenomenon can be classified into following two classes only on the basis of positions of source and screen:

(*i*) **Fresnel's diffraction:** In this class either the source or screen or both are at finite distance from the obstacle and thus distances are important. Here the incident wavefronts are either spherical or cylinderical.

(*ii*) **Frounhofer's diffraction:** In this class both the source and the screen are at infinite distance from the obstacle and thus inclination are important not the distances. Here the wavefront is plane one.

5.2 FRESNEL'S THEORY OF DIFFRACTION

On the basis of the following assumptions, Fresnel developed the *theory of diffraction* which explained the various *diffraction effects*.

Assumptions: He assumed that:

(*i*) A wavefront can be divided into a large number of elements called Fresnel's strips or zones whose are is very small.

(*ii*) Each element of wavefront sends secondary waves continuously.

(*iii*) The resultant effect at any point will be combination of the effect of all secondary waves reaching at that point.

(*iv*) The effect at any point due to a particular zone depends on

(*a*) the distance of point from the zone.

(*b*) the inclination of the point with reference to zone under consideration.

(*c*) area of the zone.

5.3 HALF PERIOD ZONES OF PLANE WAVE

Let a source S emits a plane wave front $ABCD$ travelling from left to right and has wavelength λ. Now, we wish to see the effect of wavefront at point P at a distance p from the wavefront. Let us now divide the wavefront into Fresnel's zones with P as a centre and radii equal to $p + n\lambda/2$ (n = 1, 2, 3 ...).

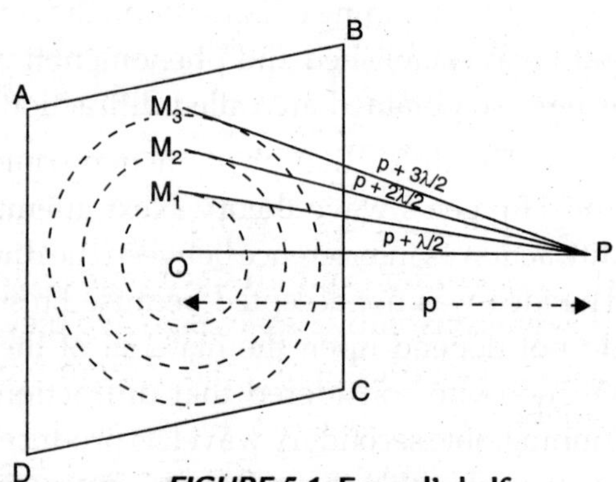

FIGURE 5.1 **Fresnel's half period zone**

Draw concentric spheres on the wavefront as shown in Figure 5.1. The area between two spheres is called zone. The secondary waves from any two consecutive zones reach the point P with a path difference of $\lambda/2$ ($\cong T/2$) that is why the name half period zones. Here T stands for period. The point O is called the pole of the wavefront with respect to point P.

Radii of Half Period Zone: It will be

$$OM_1 = \sqrt{\{(p+\lambda/2)^2 - p^2\}} \simeq \sqrt{(p\lambda)} \quad ...(1)$$

$$OM_2 = \sqrt{\{(p+2\lambda/2)^2 - p^2\}} \simeq \sqrt{(2p\lambda)} \quad ...(2)$$

Similarly,

$$OM_n = \sqrt{\{(p+n\lambda/2)^2 - p^2\}} \simeq \sqrt{(np\lambda)} \quad ...(3)$$

Thus radii are proportional to the square roots of natural numbers.

Area of Half Period Zone: The area of nth zone will be

$$= \pi\left\{(p+n\lambda/2)^2 - p^2\right\} - \pi\left\{((p+(n-1)\lambda/2)^2 - p^2\right\}$$

$$= \pi\left\{p\lambda + \frac{\lambda^2}{4}(2n-1)\right\} \simeq \pi p\lambda$$

which says that area of each half period zone is nearly the same.

The distance of point P from half period zone: It is

$$= \frac{(p+n\lambda/2)+(p+(n-1)\lambda/2)}{2} = p + (2n-1)\lambda/4$$

Amplitude at point P due to one zone: It is given as

$$R_n \propto \frac{\text{area of the zone}}{\text{distance of point } P \text{ from zone}} \times \text{obliquity factor}$$

$$\propto \frac{\pi\left\{p\lambda + \frac{\lambda^2}{4}(2n-1)\right\}}{p+(2n-1)\lambda/4}(1+\cos\theta_n)$$

$$\propto \pi\lambda(1+\cos\theta_n)$$

If n increases, $\cos\theta_n$ decreases and hence R_n also decreases.

Resultant amplitude of point P due to whole wavefront: As the path difference between the two consecutive zones is $\lambda/2$ so they are reaching in opposite phase. If R_1, R_2, R_3 and so on are the amplitudes at point P from various zones then the resultant amplitude at point P will be

$$R = R_1 - R_2 + R_3 - R_4 \ldots (-1)^{n-1} R_n$$

We can have

$$R_2 = \frac{R_1 + R_3}{2}, R_4 = \frac{R_3 + R_5}{2} \text{ and so on}$$

then

$$R = \frac{R_1}{2} + \frac{R_n}{2}, \quad \text{for} \quad n \text{ to be odd}$$

$$= \frac{R_1}{2} + \frac{R_{n-1}}{2} - R_n, \quad \text{for} \quad n \text{ to be even}$$

taking $R_{n-1} \approx R_n$ as n is very large, then

$$R = \frac{R_1}{2} \pm \frac{R_n}{2} \cong \frac{R_1}{2}$$

Thus the amplitude due to a large wavefront at a point is just half that due to first half period Fresnel zone. The intensity will be

$$I \propto R^2,$$

$$\propto \frac{R_1^2}{4}$$

i.e. one fourth that due to the first half period zone.

5.4 ZONE PLATE

It is an interesting optical device, constructed on the basis of Fresnel half period hypothesis, which proves rectilinear propagation of light. To construct a zone plate number of concentric circles with the radii proportional to square root of natural numbers are drawn on a paper. The alternate zones are pointed black and a reduced photograph is taken on glass plate which is known as zone plate. If the odd zones are transparent then it is called positive zone plate and when even zones are transparent it is called negative zone plate as shown in Figure 5.2.

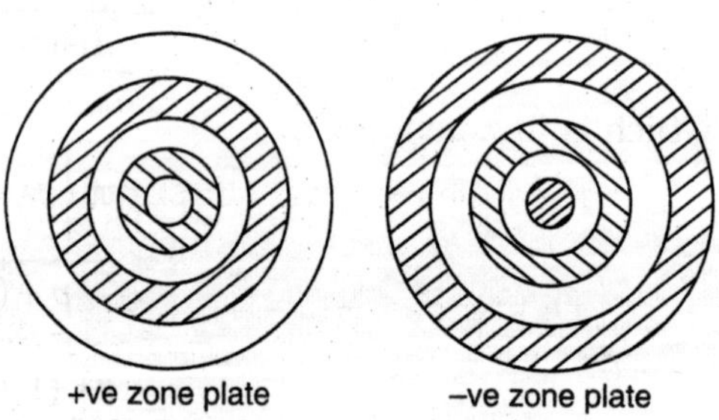

***FIGURE 5.2* Zone plate**

Theory: Let S is the source, P is the point of observation and XY is the edge wise section zone plate as shown in Figure 5.3. Then

$$SM_n = (a^2 + r_n^2)^{1/2}$$

$$= a + \frac{r_n^2}{2a} \qquad \text{...(1)}$$

Similarly $$PM_n = b + \frac{r_n^2}{2b} \qquad ...(2)$$

Then path difference between two rays reaching at P

$$\Delta = (SM_nP - SOP) = [(SM_n + PM_n) - (SO + OP)],$$

$$= \left[a + b + \frac{r_n^2}{2}(1/a + 1/b) - (a + b)\right]$$

$$= \frac{r_n^2}{2}(1/a + 1/b) \qquad ...(3)$$

Here r_n is the radii of nth zone.

Since the difference between two nearby zones is $\lambda/2$ therefore

$$\Delta = \frac{r_n^2}{2}(1/a + 1/b) = n\lambda/2$$

or $$r_n = \sqrt{\left(\frac{ab\lambda}{(a+b)}\right)}\sqrt{n},$$

$$\propto \sqrt{n} \qquad ...(4)$$

The area of nth zone will be

$$= \pi r_n^2 - \pi r_{n-1}^2 = \frac{\pi ab\lambda}{(a+b)} \qquad ...(5)$$

which is independent of n.

Then the resultant amplitude at P, will be

$$R = R_1 - R_2 + R_3 - R_4 + ... = R_1/2 \qquad ...(6)$$

$$R_+ = R_1 + R_3 + R_5 ... \text{ positive plate} \qquad ...(7)$$

$$R_- = -(R_2 + R_4 + R_6 + ...) \text{ negative plate} \qquad ...(8)$$

We find R_+ or R_- are very large in comparison with $1/2\ R_1$ when all zones are exposed.

Further one can write

$$\frac{r_n^2}{2}(1/a + 1/b) = n\lambda/2$$

or $$1/a + 1/b + \frac{n\lambda}{r_n^2} = 0$$

Which can be compared with the lens formula

$$1/u + 1/v = 1/f$$

and hence one can have

$$\frac{1}{f} = \frac{n\lambda}{r_n^2} \quad \text{or} \quad f = \frac{r_n^2}{n\lambda}$$

where f is known as principal focal length of zone plate and thus it acts as a convergent lens with multiple foci.

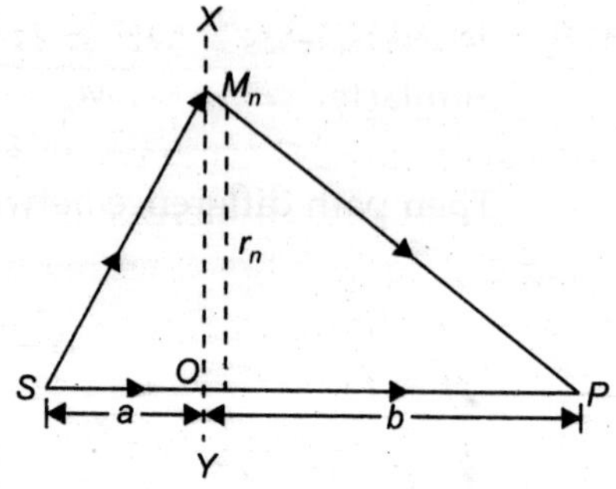

FIGURE 5.3

5.5 EXPLANATION OF RECTILINEAR PROPAGATION OF LIGHT

Consider a circular or aperture on which a plane wavefront is falling normally as shown in the Figure 5.4. Let the screen is placed at same distance and parallel to the aperture. Let there are four points on the screen such as P_1 inside, P_2 outside, P_3 at boundary from inside and P_4 at the boundary of circle from outside. These points have corresponding the poles as O_1, O_2, O_3 and O_4 on the aperture. In case of O_1 sufficient number of zones are exposed so we get resultant amplitude $1/2\, R_1$ at P_1 and so uniform illumination. For O_2 almost all the zones are blocked so we get resultant amplitude zero at P_2 and so complete darkness. For O_3 and O_4 few zones are exposed partially so resultant intensities at P_3 and P_4 is not uniform but decreases as we go from inside to outside.

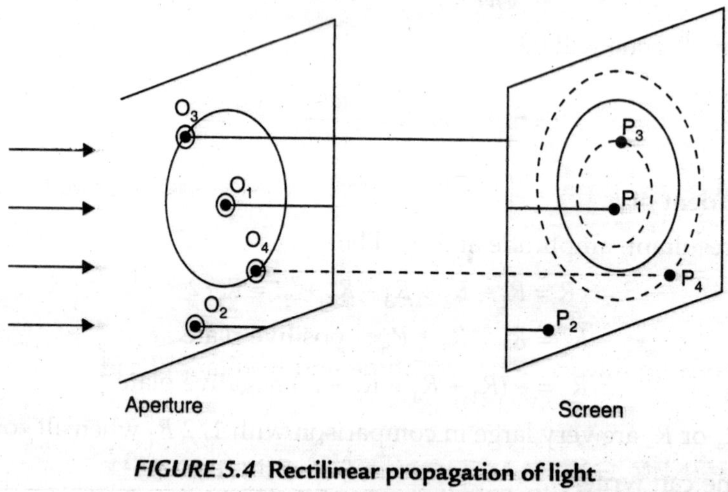

FIGURE 5.4 **Rectilinear propagation of light**

Thus there is uniform illumination inside the circle, complete darkness outside but in between two dotted circle the intensity is not uniform. Hence one can conclude that rectilinear propagation of light is explained on the basis of wave theory but only approximately.

5.6 RESULTANT OF *n* HARMONIC WAVES

Consider n simple harmonic vibrations having equal amplitudes, periods and phases increasing in arithmetic progression by an amount δ. To find the resultant amplitude of such vibrations, a polygon as shown in Figure 5.5 is constructed. The closing side OP will give the resultant amplitude R and a resultant phase ϕ. Resolving the amplitudes parallel and perpendicular to first vibration. We get

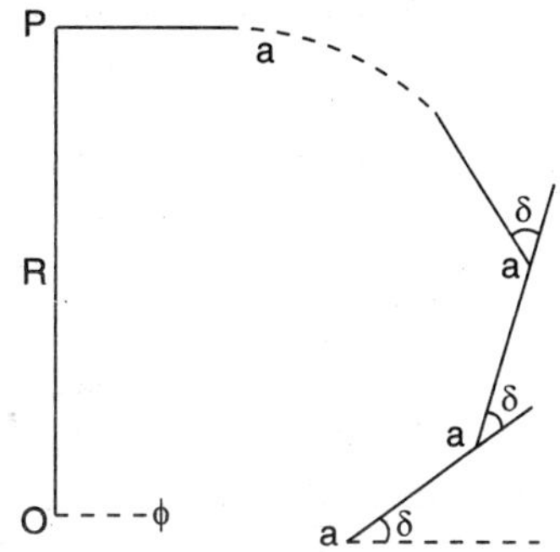

FIGURE 5.5

$$R \cos \phi = a + a \cos \delta + \dots + a \cos (n-1)\,\delta \qquad \dots(1)$$

$$R \sin \phi = 0 + a \sin \delta + \dots + a \sin (n-1)\,\delta \qquad \dots(2)$$

Multiplying equation (1) by (2) sin $\delta/2$ and equation (2) by 2 cos $\delta/2$ one can have respectively:

$$\text{R} \cos \phi = \frac{a \sin n\,\delta/2}{\sin \delta/2} \cos \frac{(n-1)\,\delta}{2} \qquad \dots(3)$$

and

$$R \sin \phi = \frac{a \sin n\,\delta/2}{\sin \delta/2} \sin \frac{(n-1)\,\delta}{2} \qquad \dots(4)$$

Squaring and adding, we get

$$R = \frac{a \sin n\,\delta/2}{\sin \delta/2} \qquad \dots(5)$$

and on dividing

$$\tan \phi = \tan \frac{(n-1)\,\delta}{2} \quad \text{or} \quad \phi = \frac{(n-1)\,\delta}{2} \qquad \dots(6)$$

When a and δ are very small and n is infinitely large such that na remains finite then taking $n\delta = 2a$ and $na = t$, we get

$$R = t\,\frac{\sin \alpha}{\alpha} \quad \text{and} \quad \phi = \frac{n\delta}{2} = \alpha \qquad \dots(7)$$

which are the expression for resultant amplitude and resultant phase.

5.7 FRAUNHOFER DIFFRACTION AT A SINGLE SLIT

Let S is a source of monochromatic light of wavelength λ, L is collimating lens AB is a slit of width a, L' is another conversing lens and XY is the screen light coming out from source and passing through slit is focussed at the screen. A diffraction pattern is obtained on the screen which consists of central bright band having alternate dark and bright bands of decreasing intensity on both the sides. The complete arrangement is shown in Figure 5.6.

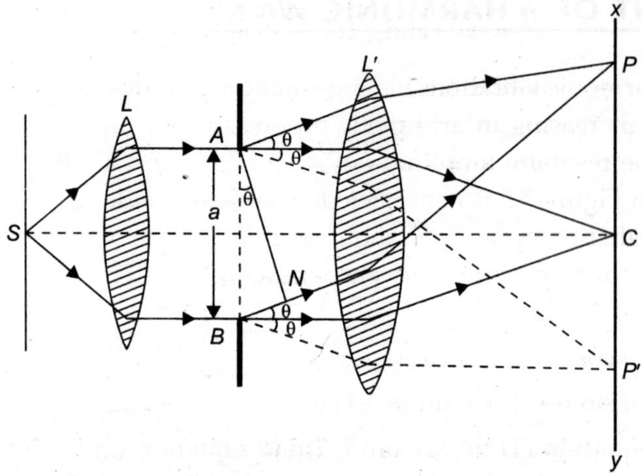

FIGURE 5.6 **Fraunhofer's diffraction at single slit**

Analysis and explanation: According to Huygen's theory a point in *AB* send out secondary waves in all directions. The diffracted ray along the direction of incident ray are focussed at *C* and those at an angle θ and focussed at *P* and *P*′. Being at equidistant from all slits points, secondary wave will reach in same phase at *C* and so the intensity well be maximum.

For the intensity at *P*, let *AN* is normal to *BN*, then path difference between the extreme rays is

$$\Delta = BN = AN \sin\theta = a \sin\theta \qquad ...(1)$$

or phase difference

$$= \frac{2\pi}{\lambda} a \sin\theta \qquad ...(2)$$

which is zero for the ray from *A* and maximum for the ray from *B*. Let *AB* consists of *n* secondary sources then the phase difference between any two consecutive source will be

$$\frac{1}{n}\frac{2\pi}{\lambda} a \sin\theta = \delta \text{ (say)} \qquad ...(3)$$

The resultant amplitude and phase at *P* will be

$$R = \frac{a \sin n\,\delta/2}{\sin \delta/2} = \frac{a \sin \dfrac{\pi a \sin\theta}{\lambda}}{\sin \dfrac{\pi a \sin\theta}{n\lambda}} = \frac{a \sin\alpha}{\sin \alpha/n}, \quad \left[\begin{array}{l} n \text{ is very large} \\ \text{then } \sin\dfrac{\alpha}{n} = \dfrac{\alpha}{n} \end{array}\right]$$

$$= na\frac{\sin\alpha}{\alpha} = A\frac{\sin\alpha}{\alpha} \qquad ...(4)$$

where $$A = na \quad \text{and} \quad \alpha = \frac{\pi a \sin\theta}{\lambda}$$

Corresponding the intensity is

$$I = R^2 = \frac{A^2 \sin^2 \alpha}{\alpha^2} \qquad \text{...(5)}$$

Condition of maxima and minima

$$\frac{dI}{d\alpha} = 0 \;\Rightarrow\; \frac{d}{d\alpha}\left[A^2 \frac{\sin^2 \alpha}{\alpha^2}\right] = 0$$

where

$$(i)\ \frac{\sin\alpha}{\alpha} = 0 \qquad \text{or} \qquad (ii)\ \alpha = \tan\alpha \qquad \text{...(6)}$$

Condition of minimum intensity

Intensity will be 0 when $\frac{\sin\alpha}{\alpha} = 0$ or $\sin\alpha = 0$

$$\alpha = \pm m\pi$$

$$\frac{\pi a \sin\theta}{\lambda} = \pm m\pi$$

$$\boxed{a \sin\theta = \pm m\lambda} \qquad m = 1, 2, 3, \ldots\ldots \qquad \text{...(7)}$$

Condition of maximum intensity: Intensity will be maximum

$$\alpha = \tan\alpha \qquad \text{...(8)}$$

The value of α satisfying this equation are obtained graphically by plotting the curve $y = \alpha$ and $y = \tan\alpha$ on the same graph (Figure 5.7). The point of intersection will give

$$\alpha = 0, \pm 3\pi/2, \pm 5\pi/2, \pm 7\pi/2,$$
$$\simeq 0, \pm 1.43\pi, \pm 2.462\pi, \pm 3.471\pi, \qquad \text{...(9)}$$

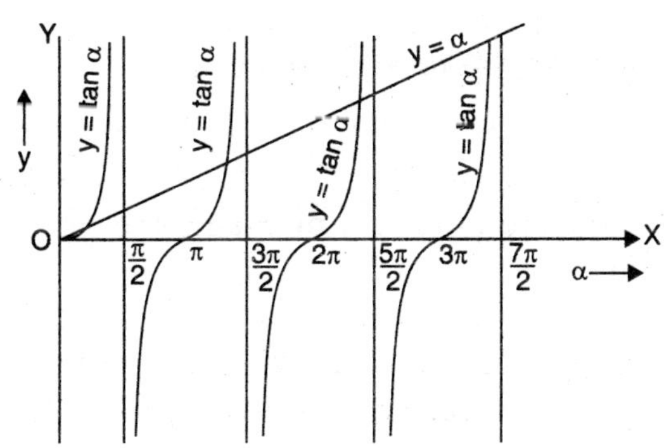

FIGURE 5.7 **Graphical representation of positions of secondary maxima's in the diffraction**

$\alpha = 0$ correspond point to central maximum whose intensity is given as

$$I = \underset{\alpha \to 0}{\text{Lt}} \frac{A^2 \sin^2 \alpha}{\alpha^2} = A^2 = I_0 \qquad \left[\text{as } \underset{\alpha \to 0}{\text{Lt}} \frac{\sin \alpha}{\alpha} = 1\right] \qquad ...(10)$$

The other maxima are given by

$$a \sin \theta = \pm \frac{(2m+1)\lambda}{2} \qquad ...(11)$$

and their intensities as

$$m = 1: \quad I_1 = \frac{A^2 (\sin 3\pi/2)^2}{(3\pi/2)^2} = \frac{4 I_0}{9\pi^2} = \frac{I_0}{22}$$

$$m = 2: \quad I_2 = 4I_0/25\pi^2 = I_0/61$$

$$m_3 = I_3 = 4I_0/49\pi^2 = I_0/121 \text{ and so on}$$

The diffraction pattern consists of a bright central maximum surrounded alternatively by minima maximum.

5.8 A PLANE TRANSMISSION DIFFRACTION GRATING (N-SLITS DIFFRACTION)

A plane diffraction grating is an arrangement consisting of a large number of close, parallel, straight, transparent and equidistant slits, each of equal width a, with neighbouring slits being separated by an opaque region of width b. A grating is made by drawing a series of very fine, equidistant and parallel lines on an optically plane glass plate by means of a fine diamond pen. The light cannot passed through the lines drawn by diamond; while the spacing between the lines is transparent to the light. There can be 15,000 lines per inch or more is such a grating to produce a diffraction of visible light. The spacing $(a + b)$ between adjacent slits is called the **diffraction element** or **grating element**. If the lines are drawn on a silvered surface of the mirror (plane or concave) then light is reflected from the positions of mirrors in between any two lines and it forms a plane concave reflection grating.

Since the original gratings are quite expensive for practical purposes their photographic reproductions are generally used.

The commercial gratings are produced by taking the cast of an actual grating on a transparent film such as cellulose acetate. A thin layer of collodin solution (celluloid dissolved in a volatile solvent) is poured on the surface of ruled grating and allowed to dry. Thin collodin film is stripped off from grating surface. This film, which retains the impressions of the original grating is preserved by mounting the film between two glass sheets. Now-a-days holographic gratings are also produced. Holograpic gratings have a much large number of lines per cm than a ruled grating.

Theory of Grating: Suppose a plane diffraction grating, consisting of large number of N parallel slits each of width a and separation b, is illuminated normally by a plane wave front of monochromatic light of wavelength λ as shown in Figure 5.8. The light diffracted through N slits is focussed by a convex lens on screen XY placed in the focal plane of the lens L. The diffraction pattern obtained on the screen with very large number of slit consists of extremely sharp principle interference maximum; while the intensity of secondary maxima becomes negligibly small so that these are not visible in the diffraction pattern.

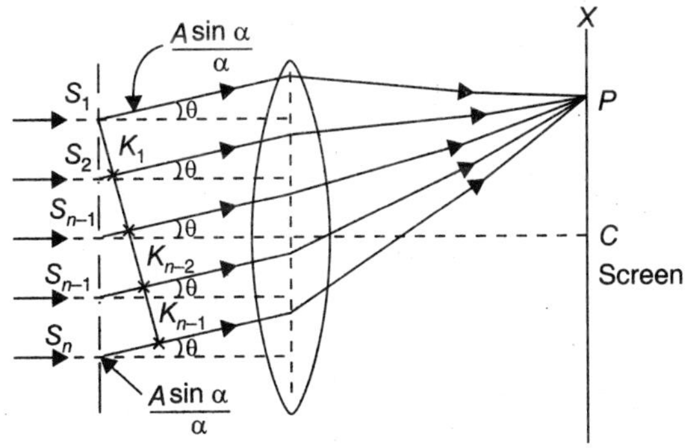

FIGURE 5.8 **Diffraction grating**

According to Huygen-Fresnel principle, all points within each slit becomes the sources of secondary wavelets, which spread out in all directions. From the theory of Fraunhofer diffraction at a single slit, the resultant amplitude R due to all waves diffracted form each slit in the direction θ is given by $R = \dfrac{A \sin \alpha}{\alpha}$, where A is the constant and $\alpha = \dfrac{\pi a \sin \theta}{\lambda}$. Thus the resultant of all the secondary waves may be treated as a single wave of amplitude R and phase α starting from the middle point of each slit and travelling at an angle θ with the normal. Thus the waves diffracted from all the N slits in the direction θ are equivalent to N parallel waves, each starting from the middle points $S_1, S_2, S_3, \ldots S_{n-1}, S_n$ of slits. Draw perpendicular $S_1 K_{n-1}$ from S_1 on $S_n K_{n-1}$, then path difference between the waves originating from the slits S_1 and S_2 is

$$S_2K_1 = S_1S_2 \sin \theta = (a + b) \sin \theta$$

Similarly between rays from S_2 and S_3 will be

$$S_3K_2 = S_2S_3 \sin \theta = (a + b) \sin \theta$$

Thus the path difference between the waves from any two consecutive slits is $(a + b) \sin \theta$.

Therefore the corresponding phase difference is $\dfrac{2\pi}{\lambda} (a + b) \sin \theta = 2\beta$. As we pass from one vibration to another, the path goes on increasing by an amount $\dfrac{2\pi}{\lambda} (a + b) \sin \theta$. Thus the phase increases in arithmetic progression. Therefore, the problem reduces to find the resultant amplitude of N waves of equal amplitude $\left(\frac{A \sin \alpha}{\alpha}\right)$ and period but the phase increasing in the arithmatic progression. In this case the resultant amplitude in the direction θ is given by

$$R' = R\,\frac{\sin N\beta}{\sin \beta} = \frac{A \sin \alpha}{\alpha} \cdot \frac{\sin N\beta}{\sin \beta}$$

The corresponding intensity at P is given by

$$I = R^2 = \frac{A^2 \sin^2 \alpha}{\alpha^2} \cdot \frac{\sin^2 N\beta}{\sin^2 \beta} \qquad ...(1)$$

The first factor of equation (1), that is, $\frac{A^2 \sin^2 \alpha}{\alpha^2}$ represents intensity distribution due to diffraction at a single slit; whereas the second factor, *i.e.* $\frac{\sin^2 N\beta}{\sin^2 \beta}$ gives the distribution of intensity in the diffraction pattern due to interference in the waves from all slits.

Principal Maxima

The condition for principal maxima is

$$\sin \beta = 0 \quad i.e. \quad \beta = \pm n\pi, \quad \text{where} \quad n = 0, 1, 2, 3, 4, ...$$

we have, $\sin N\beta = 0$, so $\frac{\sin N\beta}{\sin \beta} = \frac{0}{0}$ becomes indeterminate. But it may be evaluated by applying the usual method of differentiating the numerator and denominator and applying it limit as β tends to $\pm n\pi$

Thus
$$\lim_{\beta \to \pm n\pi} \frac{\sin N\beta}{\sin \beta} = \lim_{\beta \to \pm n\pi} \frac{\frac{d}{d\beta}(\sin N\beta)}{\frac{d}{d\beta}(\sin \beta)}$$

$$= \underset{\beta \to \pm n\pi}{\text{Lt}} \frac{N \cos N\beta}{\cos \beta} = N$$

Therefore, the intensity of principal maxima is proportional to N^2 and increases as the number of slits increase.

Substituting the value of $\frac{\sin N\beta}{\sin \beta} = N$ in equation (1), we get

$$I_P = \frac{A^2 \sin^2 \alpha}{\alpha^2} N^2 \qquad ...(2)$$

Thus, if we increase the number of slits (N), the intensity of principal maxima increases.

The direction of principal maxima are given by

$$\sin \beta = 0, \quad i.e., \quad \beta = \pm n\pi, \quad \text{where} \quad n = 0, 1, 2, 3, ...$$

or $$\frac{\pi}{\lambda}(a+b)\sin\theta = \pm n\pi \Rightarrow (a+b)\sin\theta = \pm n\lambda \qquad ...(3)$$

If we put $n = 0$ in equation (3), we get $\theta = 0$ and equation (3) gives the direction of zero order principal maximum. The first, second, third, ... order principal maxima may be obtained by putting $n = 1, 2, 3, ...$ in equation (3).

Minima: The intensity is minimum, when

$$\sin N\beta = 0; \quad \text{but} \quad \sin\beta \neq 0$$

Therefore $$N\beta = \pm m\pi$$

or $$N\frac{\pi}{\lambda}(a+b)\sin\theta = \pm m\pi$$

or $$N(a+b)\sin\theta = \pm m\lambda \qquad ...(4)$$

Here m can have all integral values except $0, N, 2N, 3N, ...$ because for these values of m, $\sin\beta = 0$ which gives the positions of principal maxima. Positive and negative signs shows that the minima lie symmetrically on both sides of the central principal maximum.

It is clear from equation (4) that for $m = 0$, we get zero order principal maximum, $m = 1, 2, 3, 4, = (N - 1)$ gives minima governed by equation (4) and then at $m = N$, we get principal maxima of first order. This indicates that, there are $(N - 1)$ equispaced minima between zero and first orders maxima. Thus, there are $(N - 1)$ minimum between two successive principal maxima.

Secondary Maxima: The above study reveals that there are $(N - 1)$ minima between two successive principal maxima. Hence there are $(N - 2)$ other maxima coming alternatively with the minima between two successive principal maxima. These maxima are called **secondary maxima**. To find the positions of the secondary maxima, we first differentiate equation (1) with respect to β and equating to zero

$$\frac{dI}{d\beta} = \frac{A^2\sin^2\alpha}{\alpha^2}.2\left[\frac{\sin N\beta}{\sin\beta}\right]\frac{N\cos N\beta\sin\beta - \sin N\beta\cos\beta}{\sin^2\beta} = 0$$

or $$N\cos N\beta\sin\beta = \sin N\beta\cos\beta = 0$$

or $$\tan N\beta = N\tan\beta \qquad ...(5)$$

To find the intensity of secondary maximum, we make these of the triangle shown in Figure 5.9.

We have $$\sin N\beta = \frac{N\tan\beta}{\sqrt{(1+N^2\tan^2\beta)}}$$

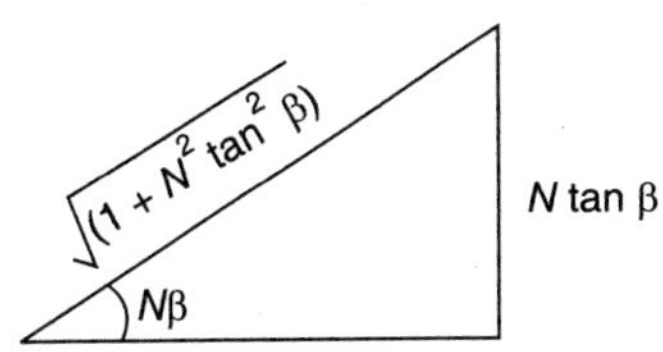

FIGURE 5.9

Therefore $$\frac{\sin^2 N\beta}{\sin^2 \beta} = \frac{(N^2 \tan^2 \beta)}{\sin^2 \beta (1 + N^2 \tan^2 \beta)}$$

or $$\frac{\sin^2 N\beta}{\sin^2 \beta} = \frac{N^2 \tan^2 \beta}{(1 + N^2 \tan^2 \beta) \sin^2 \beta} = \frac{N^2}{1 + (N^2 - 1) \sin^2 \beta}$$

Putting this value of $\frac{\sin^2 N\beta}{\sin^2 \beta}$ in equation (1), we get the intensity of secondary maxima as

$$I_s = \frac{A^2 \sin^2 \alpha}{\alpha^2} \frac{N^2}{\left[1 + (N^2 - 1) \sin^2 \beta\right]} \quad ...(6)$$

This indicates the intensity of secondary maxima is proportional to $\frac{N^2}{\left[1 + (N^2 - 1) \sin^2 \beta\right]}$, whereas the intensity of principal maxima is proportional to N^2.

5.8.1 Absent Spectra with a Diffraction Grating

It may be possible that while the first order spectra is clearly visible, second order may be not be visible at all and the third order may again be visible. It happen when for again angle of diffraction θ, the path difference between the diffracted ray from the two extreme ends of one slit is equal to an integral multiple of λ if the path difference between the secondary waves from the corresponding point in the two halves will be λ/2 and they will can all one another effect resulting is zero intensity. Thus the mining of single slit pattern are obtained in the direction given by.

$$a \sin \theta = m\lambda \quad ...(1)$$

where m = 1, 2, 3, ... excluding zero but the condition for n^{th} order principles maximum in the grating spectrum is

$$(a + b) \sin \theta = n\lambda \quad ...(2)$$

If the two conditions given by equation (2) are simultaneously satisfied then the direction in which the grating spectrum should give us a maximum every slit by itself will produce darkness in that direction and hence the most favourable phase for reinforcement will not be able to produce an illumination *i.e.*, the resultant intensity will be zero and hence the absent spectrum. Therefore dividing equation (2) by equation (1)

$$\frac{(a + b) \sin \theta}{a \sin \theta} = \frac{n}{m}$$

$$\frac{(a + b)}{a} = \frac{n}{m} \quad ...(3)$$

This is the condition for the absent spectra in the diffraction pattern

If $a = b$ *i.e.*, the width of transparent portion is equal to the width of opaque portion then from equation (3) $n = 2m$

i.e., 2nd, 4th, 6th etc., orders of the spectra will be absent corresponds to the minima due to single slit given by $m = 1, 2, 3$ etc.

$$b = 2a$$

$$n = 3m$$

i.e., 3rd, 6th, 9th etc., order of the spectra will be absent corresponding to a minima due to a single slit given by $m = 1, 2, 3$ etc.

5.8.2 Number of Orders of Spectra with a Grating

The number of spectra that are visible in a given grating can be easily calculated with the help of the equation.

$$(a + b) \sin \theta = n\lambda$$

$$n = \frac{(a + b) \sin \theta}{\lambda}$$

Here $(a + b)$ is the grating element and is equal to $1/N = 2.54\ N$ cm, N being number of lines per inch in the grating. Maximum possible value of the angle of diffraction θ is 90°, Therefore $\sin \theta = 1$ and the maximum possible order of spectra.

$$\eta_{max} = \frac{(a + b)}{\lambda}$$

If $(a + b)$ is between λ and 2λ *i.e.*, grating element $(a + b) < 2\lambda$ then,

$$\eta_{max} < \frac{2\lambda}{\lambda} < 2$$

and hence only the first order of spectrum is seen if $(a + b)$ is between 2λ and 3λ first two order will obtained and so on.

5.9 RESOLVING POWER OF OPTICAL INSTRUMENTS

When two objects are very class to each other, it may not be possible for our eye to see them separately. If we wish to see them separately, then we will have to make use of some optical instruments like microscope, telescope, grating, prism etc. The ability of an optical instrument to form distinctly separate images of two objects, very close to each other is called the resolving power of instrument. A lens system like microscope and telescope gives us a geometrical resolution while a grating or a prism gives a spectral resolution. In fact the image of a point object or line is not simply a point or line but what we get is a diffraction pattern of decreasing intensity. For a two point system two diffraction patterns are obtained which may and may not overlap depending upon their separation. The minimum separation between two objects that can be resolved by an optical instrument is called resolving limit of that instrument. The resolving power is inversely proportional to the resolving limit.

5.9.1 Rayleigh Criterion of Resolution

According to Lord Rayleigh's arbitrary criterion two nearby images are said to be resolved if the position of central maximum of one coincides with the first minima of the other or vice versa.

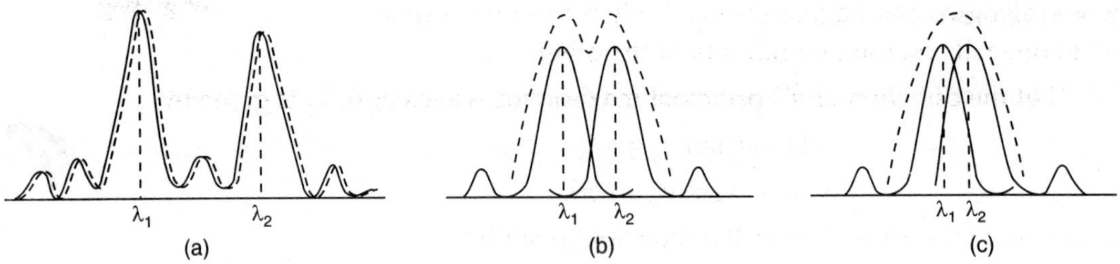

FIGURE 5.10 **Rayleigh criterian**

To illustrate this let us consider the diffraction patterns due to two wavelengths λ_1 and λ_2. There may be three possibilities. First let the difference $(\lambda_1 \sim \lambda_2)$ is sufficiently large so that central maximum are quite separate, this situation is called *well resolved*.

Secondly consider that $(\lambda_1 \sim \lambda_2)$ is such that central maximum due to one falls on the first minima of the other. The resultant intensity curve shows a distinct dip in the middle of two central maxima. This situation is called *just resolved* as the intensity of the dip can be resolved by our eyes.

$$I_{dip} = 0.81\, I_{max} \qquad \text{...(1)}$$

Thirdly let the $(\lambda_1 \sim \lambda_2)$ is very small such that they come still closes as shown in Figure 5.10. The intensity curves have sufficient overlapping and two images cannot be distinguished separately. The resultant curve almost appears as one maxima. This case is known as *unsolved*. Thus the minimum limit of resolution is that when two patterns are *just resolved*.

5.10 RESOLVING POWER OF PLANE DIFFRACTION GRATING

We know that the diffraction grating has ability to produce spectrum *i.e.*, to separate the lines of nearly equal wavelengths and therefore it has resolving capability. The resolving power of a grating may be defined as its ability to form separate diffraction maxima of two wavelengths which are very close to each other. If λ is the mean value of the two wavelengths and $d\lambda$ is the difference between two then resolving power may be defined as

$$\text{resolving power} = \lambda / d\lambda$$

Expression for resolving power: Let a beam of light having two wavelengths λ_1 and λ_2 is falling normally on a grating *AB* which has $(a + b)$ grating element and *N* number of slits as shown in Figure 5.11. After passing through grating rays forms the diffraction patterns which can be seen through telescope.

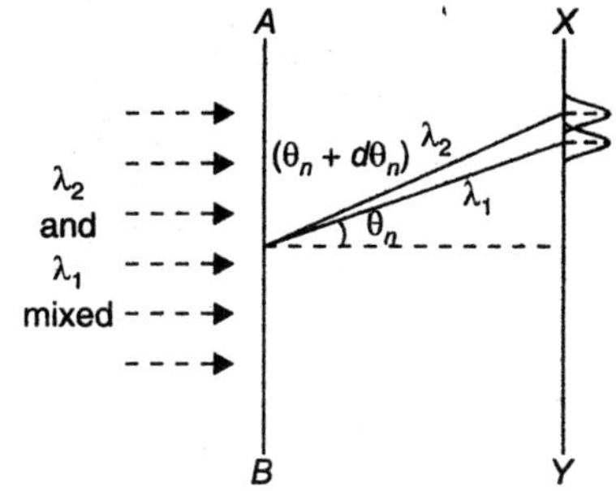

FIGURE 5.11 **Resolving power of grating**

Now, if these patterns are very close together they overlap and cannot be seen separately.

However, if they satisfy the Rayleigh criterion, that is the wavelengths can be just resolved when central maxima due to one falls on the first minima of the other.

Let the direction of nth principal maxima for wavelength λ_1 is given by

$$(a + b) \sin \theta_n = n\lambda_1$$

or

$$N (a + b) \sin \theta_n = Nn\lambda_1$$

and the first minima will be in the direction given by

$$N (a + b) \sin (\theta_n + d\theta_n) = m\lambda_1$$

where m is an integer except 0, N, $2N$ ..., because at these values condition of maxima will be satisfied. The first minima adjacent to the nth maxima will be in the direction $(\theta_n + d\theta_n)$ only when $m = (nN + 1)$. Thus

$$N (a + b) \sin (\theta_n + d\theta_n) = (nN + 1)\, \lambda_1$$

For just resolution, the principal maxima for the wavelength λ_2 must be formed in the direction $(\theta_n + d\theta_n)$.

Therefore $\quad (a + b) \sin (\theta_n + d\theta_n) = n\lambda_2$

or

$$N (a + b) \sin (\theta_n + d\theta_n) = Nn\lambda_2$$

Now equating the two equations

$$(nN + 1)\, \lambda_1 = Nn\lambda_2$$

$$(nN + 1)\, \lambda = Nn\, (\lambda + d\lambda)$$

$$\lambda = Nn\, d\lambda$$

where

$$\lambda_1 = \lambda, \quad \lambda_2 - \lambda_1 = d\lambda, \quad \lambda_2 = \lambda + d\lambda$$

Thus resolving power of grating is found as

$$\boxed{\text{R.P.} = \frac{\lambda}{d\lambda} = nN}$$

Resolving power = order of spectrum × total number of lines on grating

which can also be written as

$$= \frac{N (a + b) \sin \theta_n}{\lambda} = \frac{\omega \sin \theta_n}{\lambda}$$

where, $\omega = N(a + b)$ is the total width of lined space in grating.

$$\text{R.P.}_{max} = \frac{N(a+b)}{\lambda} = \frac{\omega}{\lambda}$$

when $\theta_n = 90°$.

5.11 RELATION BETWEEN RESOLVING POWER AND DISPERSIVE POWER OF A GRATING

We know that resolving power

$$\text{R.P.} = \frac{\lambda}{d\lambda} = nN$$

and dispersive power $\quad \text{D.P.} = \frac{d\theta}{d\lambda} = \frac{n}{(a+b)\cos\theta_n}$

Therefore, $\quad \frac{\lambda}{d\lambda} = nN = N(a+b)\cos\theta_n \frac{n}{(a+b)\cos\theta_n}$

$$\boxed{\frac{\lambda}{d\lambda} = A \times \frac{d\theta}{d\lambda}}$$

Resolving power = total aperture of telescope objective × dispersive power.

The resolving power of a grating can be increased by

(*i*) Increasing the number of lines on the grating N.

(*ii*) Increasing the sides of spectrum n.

(*iii*) Increasing the total width of grating 'ω', for which one has to make use of whole aperture of telescopes objective.

5.12 NUMERICAL PROBLEMS BASED ON DIFFRACTION OF LIGHT

Problem 1. *Light of 5000 Å is incident on a circular hole of radius (i) 1 cm and (ii) 1 mm. How many half period zones are contained in the circle if the screen is placed at a distance of 1 m to observe the diffraction?*

Hint: (*i*) Area of the circular hole = $\pi r^2 = \pi$ cm^2

Area of each half period zone = $\pi b\lambda = 5 \times 10^{-5} \times \pi$ cm^2

No. of half period zones = $\frac{\pi}{5 \times 10^{-5}\pi} = 20{,}000$

(*ii*) Area of circular hole = $\pi \times (0.1)^2$ cm^2

$$\text{No. of half period zones} = \frac{\pi \times (0.1)^2}{5 \times 10^{-5} \times \pi} = 0.2 \times 10^3 = 200$$

Problem 2. *Find the radius of first zone is a zone plate of focal length 25 cm for light of wavelength 5000 Å.*

Hint:
$$f_n = \frac{r_n^2}{n\lambda} \Rightarrow f_1 = \frac{r_1^2}{\lambda} \quad \text{(for first zone } n = 1)$$

or
$$r_1 = \sqrt{f_1 \lambda} = 0.0354 \text{ cm.}$$

Problem 3. *The spectral lines of sodium D_1 and D_2 have wavelengths of approximately 5890 Å and 5896 Å. A sodium lamp sends incident plane wave onto a slit of width 2 μm. A screen is located 2 m from the slit. Find the spacing between the first maxima of the two sodium lines as measured on the screen.*

Hint: Distance of the secondary maximum from the centre of the screen is given by

$$\sin\theta = \frac{x}{D} = \frac{3}{2}\frac{\lambda}{a} \Rightarrow x = \frac{3}{2}.\frac{D\lambda}{a}$$

For the two wavelength $x_1 = \frac{3}{2}\frac{D\lambda_1}{a}$ and $x_2 = \frac{3}{2}\frac{D\lambda_2}{a}$

$$\text{Spacing } (x_2 - x_1) = \frac{3D}{2a}(\lambda_2 - \lambda_1) = 0.9 \text{ mm.}$$

Problem 4. *In a double slit Fraunhofer diffraction pattern, the screen is 160 cm away from the slits. The slit width are 0.08 mm and they are 0.4 mm apart. Calculate the wavelength of light if the fringe spacing is 0.25 cm. Also find missing orders.*

Hint: The fringe spacing $\beta = \frac{\lambda D}{2d} \Rightarrow \lambda = \frac{\beta(2d)}{D} = 6250 \times 10^{-8}$ cm

For missing order $\frac{a+b}{a} = \frac{n}{m}$

$$b = 0.04 \text{ cm and } a = 0.008 \text{ cm}$$

$$n = 6m, \qquad \text{where } m = 1, 2, 3, 4, \ldots\ldots$$

$$n = 6, 12, 18, \ldots\ldots$$

Problem 5. *A plane transmission grating has 6000 lines/cm. Calculate the highest order spectrum which can be seen with a light of wavelengths 4000 Å.*

Hint: $(a + b)\sin\theta_{max} = n_{max}\lambda$ $\quad [\because \sin\theta_{max} = 1]$

$$\Rightarrow \quad n_{max} = \frac{(a+b)}{\lambda} = 4.$$

Problem 6. *The number of lines in a grating X is N_1 and the number of lines in a grating Y is N_2, N_1 is less than N_2. The total ruled width of the two gratings X and Y is same. Compare their resolving powers.*

Solution: Resolving power of a grating is given by R.P. $= \frac{\lambda}{d\lambda} = nN$ ...(*i*)

According to grating equation $(a + b) \sin \theta = n\lambda$ or $n = \frac{(a+b)\sin\theta}{\lambda}$ ...(*ii*)

From equations (*i*) and (*ii*)

$$\text{R.P.} = \frac{N(a+b)\sin\theta}{\lambda}$$

But $W = N(a + b)$, where W = width of the ruled surface

$$\Rightarrow \qquad \text{R.P.} = \frac{W\sin\theta}{\lambda} \qquad \text{...(iii)}$$

From equation (*iii*), it is clear that the resolving power of both the gratings is same and it is independent of the number of lines for a given width of ruled surface.

Problem 7. *What other spectral lines in the range 4000 Å will coincide with the fifth order line of 6000 Å in a grating spectrum.*

Hint: $(a + b) \sin \theta = n\lambda = 5 \times 6000 \text{ Å} \Rightarrow \frac{5\times 6000}{n} \text{ Å}$

For $n = 6$, $\lambda = 5000$ Å, and for $n = 7$, $\lambda = 4285.7$ Å.

Problem 8. *A diffraction grating which has 4000 lines per cm is used at normal incidence. Calculate the dispersive power of the grating in the third spectrum in the wavelength region 5000 Å.*

Hint: $$\frac{d\theta}{d\lambda} = \frac{n}{(a+b)\cos\theta}$$

Also $(a + b) \sin \theta = n\lambda \Rightarrow \sin\theta = \frac{n\lambda}{a+b}$

$$\cos\theta = \sqrt{(1-\sin^2\theta)}, \text{ then D.P. } \frac{d\theta}{d\lambda} = 15{,}000.$$

Problem 9. *What is the minimum number of lines per cm in a 2.5 cm wide grating which will just resolve the two sodium lines (5890 Å and 5896 Å) in the second order spectrum.*

Hint: $$\text{R.P. } \frac{\lambda}{d\lambda} = nN \Rightarrow N = \frac{\lambda}{d\lambda}.\frac{1}{n} \cong 491$$

Number of lines per cm $= \frac{491}{2.5} = 196.$

Problem 10. *How many orders will be visible if the wavelength of the incident radiation is 5000 Å and the number of lines on the grating is 2620 in one inch.*

[Amity University, Noida, Nov. 2009]

Hint: $(a+b)\sin\theta = n\lambda \quad (a+b) = \frac{2.54}{2620}$ cm.

$$n = \frac{(a+b)}{\lambda} \qquad \left[\text{as } \theta = \frac{\pi}{2}\right]$$

$$= \mathbf{19.}$$

EXERCISES

1. Discuss the phenomenon of Fraunhofer diffraction at a single slit and obtain condition for first minimum from central maximum.
2. What is the difference between Fresnel diffractioin and Fraunhofer diffraction? Derive an expression for the angular width of central diffraction maximum in the case of Fraunhofer diffraction at a single slit. [*Amity Univ., Noida, Dec. 2005*]
3. Give the construction of a plane diffraction and for *N* number of slits in the grating, obtain the condition for first secondary minimum after the central maxima. Also write in general number of secondary minimum and number of secondary maxima in between two primary maxima.
4. (*a*) What will be the effect on a diffraction pattern of a grating if the grating element is of the order of λ or less?

   (*b*) How would you increase resolving power of a diffraction grating?

   (*c*) What is Rayleigh criterion of resolution?

   (*d*) Is there any difference in the diffraction pattern due to double slit and interference due to two sources?
5. What is Rayleigh criterion of resolution? Deduce an expression for resolving power of the diffraction grating.
6. Discuss the phenomenon of Fraunhofer's diffraction at a single slit and show that the intensities of the successive maximum are nearly in the ratio of $\frac{4}{9\pi^2} : \frac{4}{25\pi^2} : \frac{4}{49\pi^2} : ...$

   [*Amity Univ., Noida Jan. 2007*]
7. What is meant by diffraction of light? Distinguish between Fresnel and Fraunhofer classes of diffraction.
8. Explain Rayleigh criterion of resolution obtain an expression of resolving power of grating.
9. What do you understand by the resolving power of a grating? Derive necessary expression.
10. Give the construction and theory of plane transmission grating and explain the formation of spectra by it. Explain what are absent spectra.

SUGGESTION FOR FURTHER READING

1. J. Beynon, Introductory University Optics, Prentice-Hall of India Pvt. Ltd., New Delhi, 1998.
2. R.S. Longhurst, Geometrical and Physical Optics, Orient Longman Ltd., Hyderabad, 1999.
3. K. Ghosh and A. Manna, Textbook of Physical Optics, Macmillan India Ltd., New Delhi, 2007.
4. Devraj Singh, Engineering Physics, Dhanpat Rai and Co., New Delhi, 2008.
5. E-Hecht, Optics, Pearson Education, New Delhi, 2009.
6. F.L. Pedrotti, L.S. Pedrotti, Principles of Optics, Prentice Hall International, New York, 1987.
7. Devraj Singh, Fundamentals of Engineering Physics, Dhanpat Rai and Co., New Delhi, 2009.
8. J.W. Goodman, International Trends in Optics, Academic Press, New York, 1991.

Chapter **6**

POLARISATION OF LIGHT

6.1 INTRODUCTION

The phenomenon of interference and diffraction reveals that light has wave nature like sound. But wave may be classified into two classes : (*i*) Longitudinal wave (*ii*) Transverse wave. Sound waves are longitudinal in nature. Now we shall investigate the nature of light waves.

In order to investigate the nature of light we first demonstrate which is explained below. Take a stretched string *AB* one end (*A*) of which is free while other end (*B*) is fixed. Let it passes through two slits S_1 and S_2.

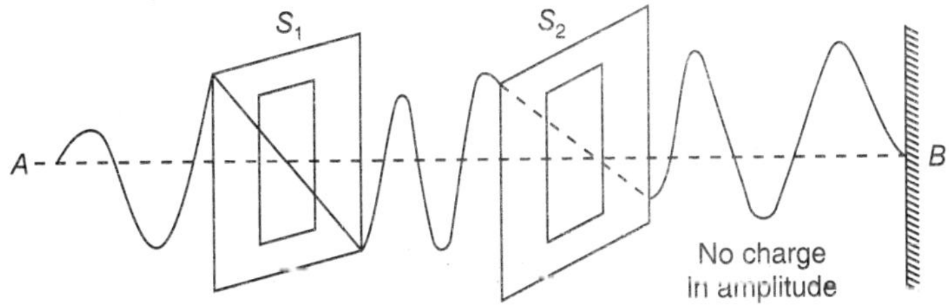

FIGURE 6.1

Now set up a longitudinal wave by moving end *A* forward and backward along the string. If slit S_2 is rotated in any direction then there is no charge in the amplitude of the wave. Now, set up transverse wave by moving the end *A* perpendicular to the length of the string.

When S_1 and S_2 are parallel to each other, then there is no charge in the amplitude of wave coming out of slit. If slit S_2 is rotated slowly either in clockwise or in anti-clockwise, then amplitude of the wave coming out of slit S_2 diminishes. If S_2 is perpendicular to S_1 then there is no wave coming out of S_2. Thus the transverse wave can pass through the slits when these slits are parallel to each other as shown in Figure 6.1.

Same type of behaviour is shown by the light when it passes through *tourmaline crystal* cut parallel to optic axis.

If we take two tourmaline crystals *P* and *Q* which are parallel to each other then we get light coming out of crystal *Q* as shown in Figure 6.2.

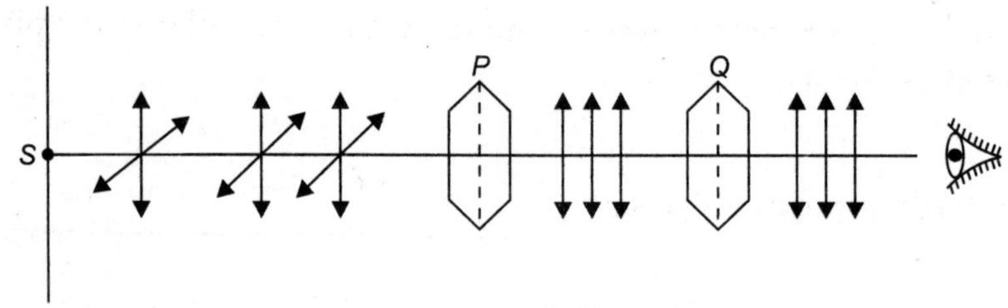

FIGURE 6.2

If we rotate the crystal *Q* either in clockwise or anti-clockwise, then intensity of light emerging out of tourmaline crystal *Q* decreases. When the axis of two crystals become perpendicular to each other then intensity of light emerging out of *Q* is zero. It is also noted that only vibrations parallel to crystallographic axis are allowed to pass through the crystal. Thus the vibrations are confined only in a particular direction.

Thus on the basis of above experiment one can say that light waves are transverse in nature *i.e.*, the displacement associated with light waves is at right angle to the direction of propagation of the wave. The tourmaline crystal *P* is known as polariser while *Q* is known as analyser.

6.2 ORDINARY OR UNPOLARISED LIGHT

Lord Maxwell explained the nature of displacement associated with light wave. In his well known electromagnetic theory he points out that light waves are electromagnetic which consists of a vibrating electric and magnetic field. Both the fields are mutually perpendicular to each other and both are perpendicular to the direction of propagation as shown in Figure 6.3.

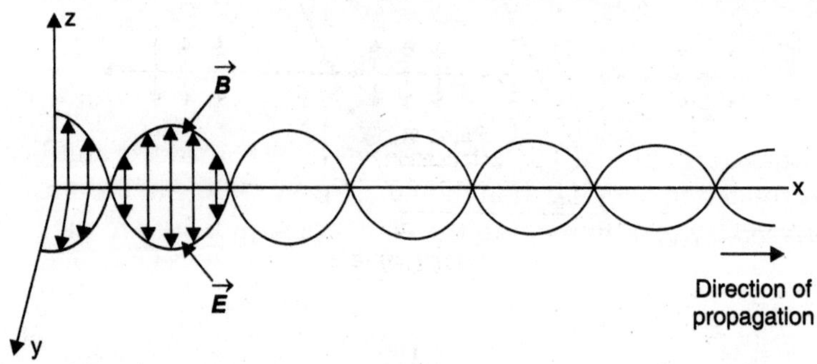

FIGURE 6.3

In the figure **E** represents the electric field vector and **B** represents the magnetic field vector. It is the electric field vector which produces all the observed effect of light. So, vibration mean vibrations of electric field vector.

In any light source, light is emitted by atoms or molecules. As source of light contains million atoms and million excited atoms oriented at random which emit light for a very short duration. Thus ordinary light will have vibrations in all directions perpendicular to the direction of propagation. The unpolarised light is generally represented by a star ✳.

It is to be noted that a vibration which is in plane of the paper, is represented by a double head arrow. In contrary to it, if the vibration is perpendicular to the plane of paper, it is then represented by a dot.

6.3 PLANE POLARISED LIGHT

As it has been already explained the light coming out of tourmaline crystal *P* has vibrations only in one direction parallel to the crystallographic axis. This type of light which attains one sidedness is called linearly polarised or plane polarised light. If the vibration is parallel to the plane of paper, it is represented by double head arrow ↕ while the vibrations are perpendicular to the plane of paper, it is represented by a dot. When an ordinary light is passed through a tourmaline crystal, then light is polarised and vibrations are confined to only one direction perpendicular to the direction of propagation of light. This type of light is called plane polarised light.

6.4 PLANE OF POLARISATION AND PLANE OF VIBRATION

The plane of polarisations is that plane in which there is no vibration. In Figure 6.4 *PQRS*, is the plane of polarisation. The plane containing the vibration and direction of propagation is called plane of vibration. In Figure 6.4 *ABCD* is the plane of vibration.

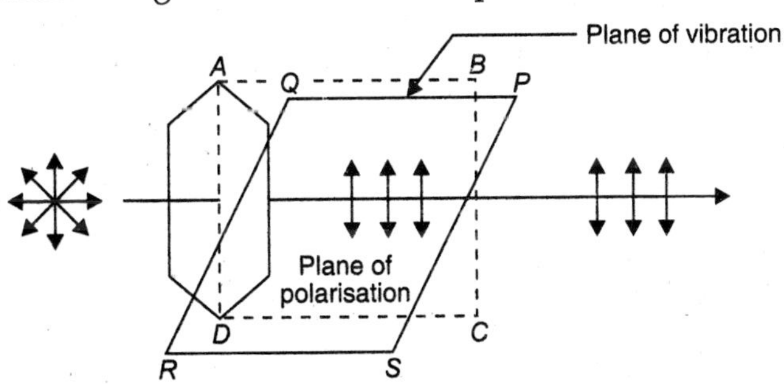

FIGURE 6.4

6.5 DIFFERENT METHODS FOR PRODUCTION OF POLARISED LIGHT

One can produce polarised light by following methods:

(*i*) Polarisation by reflection
(*ii*) Polarisation by refraction
(*iii*) Polarisation by double refraction.

6.5.1 Polarisation by Reflection (Brewster's Law)

Louis Malus in 1808 discovered that when unpolarised light is incident on the surface of any transparent medium then the reflected and refracted beams are partially polarised. Degree of polarisation depends upon the angle of incidence. At a certain angle of incidence the reflected light is completely polarised, called the angle of polarisation. It is to be noted that refracted ray is partially polarised.

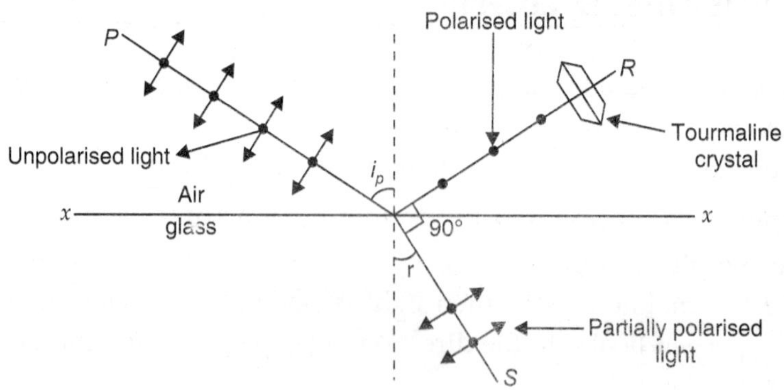

FIGURE 6.5

Note. Unpolarised light is either represented by ↕ or ✱.

In 1811, Sir Brewster and his co-workers showed that the tangent of the angle of polarisation is numerically equal to the refractive index of the medium.

i.e.,
$$\boxed{\mu = \tan i_p}$$

where μ is the refractive index and i_p is the polarising angle.

Above relation is called *Brewster's law.* As the refractive index of the medium depends upon the wavelength of light so polarising angle varies with the wavelength. Thus complete polarisation is possible only with monochromatic light.

When light is incident at the polarising angle, the reflected and refracted rays are mutually perpendicular to each other. This can be easily shown with the help of Brewster's law.

$$\mu = \tan i_p = \frac{\sin i_p}{\cos i_p} \quad ...(1)$$

Also, from Snell's Law,

$$\mu = \frac{\sin i_p}{\sin r} \quad ...(2)$$

From equations (1) and (2),

$$\Rightarrow \quad \frac{\sin i_p}{\cos i_p} = \frac{\sin i_p}{\sin r}$$

$$\Rightarrow \quad \cos i_p = \sin r = \cos (90° - r)$$

$$\Rightarrow \quad i_p = 90° - r$$

$$\Rightarrow \quad \boxed{i_p + r = 90°.}$$

Polarisation of light by reflection can be verified by observing reflected light through a tourmaline crystal. It has been found that light is obtained only for a particular position of crystal while it diminishes when the crystal is rotated slowly and at a particular position the crystal refuses to pass the light through itself. These observations enable us to conclude that light after reflection is completely polarised.

Note. For glass angle of polarisation is 57.5°.

6.5.2 Polarisation by Refraction

According to Brewster's law, when the beam of light is incident on the surface of a transparent material at polarising angle then the reflected light is completely plane polarized having vibrations perpendicular to the plane of incidence. However, the transmitted light is not completely plane polarised. The intensity of the reflected beam is very weak as only 8% of the incident light is reflected at each reflection. Thus to increase the intensity of plane polarised reflected beam one should increase the number of reflections.

The light incident at the polarising angle on a pile of thin parallel placed plates placed one on top of the other separated by thin sheet of papers with a central aperture otherwise plates will make good contact with each other and would act practically as a single slab. Then at each reflection more of the component vibrations perpendicular to the plane of incidence are filtered out with the result that the transmitted light becomes richer and richer in plane polarised light having vibrations in the plane of incidence. The apparatus has generally ten thin plates at an angle of 32.5° with the axis of the wooden tube as shown in Figure 6.6.

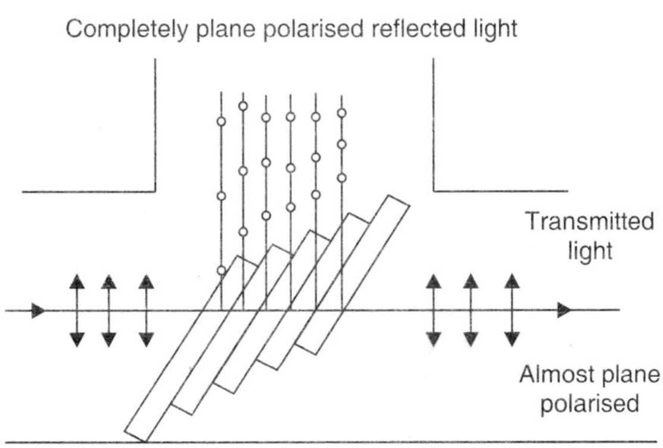

FIGURE 6.6

Law of Malus: According to law of Malus, when a beam of completely plane polarised light is incident on an analyser, the intensity (I) of light transmitted through analyser varies directly as square of cosine of angle (θ) between the plane of transmission of the analyser and polariser.

Mathematically, $I \propto \cos^2 \theta$

Let a be the amplitude of vibrations transmitted by the polariser, along OP. The plane of polariser OA makes an angle θ with OP. Resolving amplitude of vibrations into rectangular components.

(*i*) $a \cos \theta$ along OA parallel to plane of transmission of analyser

(*ii*) $a \sin \theta$ along OB, perpendicular to plane of transmission of analyser.

As component transmitted through analyser is $a \cos \theta$ only, so intensity of transmitted light is

$$I \propto (a \cos \theta)^2$$

$$I = K a^2 \cos^2 \theta$$

$$I = I_0 \cos^2 \theta \qquad (K^2 = I_0)$$

where I_0 is intensity of light transmitted by polariser which is constant.

Note. When an unpolarised light is passed through a polariser, it get polarised. It can be proved that intensity of polarised light is half of the intensity of unpolarised light.

$$\boxed{I_{pol} = \frac{1}{2} I_{unpol}.}$$

In unpolarized light, the vibrations are in all probable directions in a plane perpendicular to the direction of propagation. So,

$$[\cos^2 \theta]_{av} = \frac{1}{2\pi} \int_0^{2\pi} \cos^2 \theta \, d\theta$$

$$[\cos^2 \theta]_{av} = \frac{1}{2\pi} \int_0^{2\pi} \left[\frac{1 + \cos 2\theta}{2}\right] d\theta$$

$$[\cos^2 \theta]_{av} = \frac{1}{4\pi} \left[\theta + \frac{\sin 2\theta}{2}\right]_0^{2\pi}$$

$$[\cos^2 \theta]_{av} = \frac{1}{2}$$

Using law of Malus

$$I = I_o \cos^2 \theta$$

$$\boxed{I_{pol} = I_{unpol} \times \frac{1}{2}.}$$

6.5.3 Polarisation by Double Refraction (Birefringence)

When a ray of light enters from one medium to another medium, it deviates from its path either toward or away from the normal obeying Snell's law. If medium is isotropic, refraction takes place only in one direction, whereas in case of anisotropic medium, Erasmus Bartholinus in 1669 discovered that the refraction is not confined to single direction. He found that when a ray of ordinary light is incident on calcite or quartz crystals, it splits up into two polarised refracted rays. This phenomenon is called **double refraction** and crystal is known as **doubly refracting crystals**. The phenomenon of double refraction can be explained by a simple experiment. Mark an ink dot on a piece of paper and place a calcite crystal over this dot. Two images of ink mark will be observed. On rotating the calcite crystal either clockwise or anti-clockwise, it is found that one image rotates with the rotation of crystal and other image remains stationary as shown in Figure 6.7. The stationary image is known as **ordinary image** and rotating image is known as **extraordinary image**. The refracted ray which produces ordinary image is known as **O-ray *i.e.*, ordinary rays** and obey ordinary laws of refraction, and the refracted ray which produces extraordinary image is known as **extraordinary ray (E-ray)** and does not obeys ordinary laws of refraction.

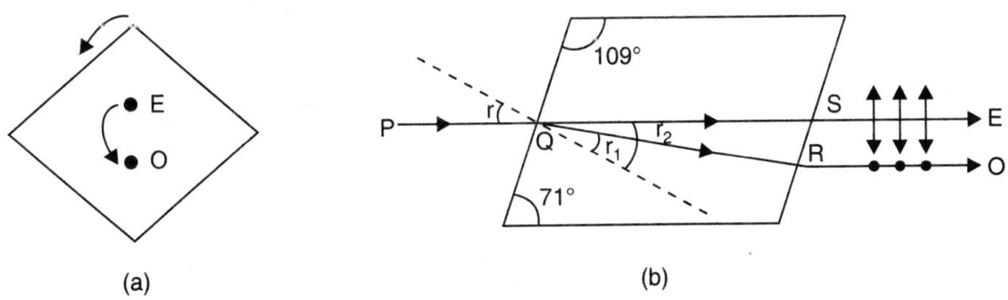

FIGURE 6.7

Consider a narrow beam of light *PQ* incident on a calcite crystal making an angle *i* with the normal, it is refracted along two directions *QR* and *QS* at angles r_1 and r_2 respectively, as shown in Figure 6.7(*b*). These two rays finally emerges as *O*-ray and *E*-ray which are parallel to each other as well as parallel to direction of incident beam.

For *O*-ray, $$\mu_O = \frac{\sin i}{\sin r_1}$$

For *E*-ray, $$\mu_E = \frac{\sin i}{\sin r_2}$$

As $r_1 < r_2$, therefore $\mu_O > \mu_E$ or $v_O < v_E$. This is because

$$\mu = \frac{\sin i}{\sin r} = \frac{\text{Velocity of light in air}}{\text{Velocity of light in medium}} = \frac{v_{air}}{v_{med.}}$$

Birefringence: The difference between refractive index for *O*-Ray and *E*-Ray is called birefringence *i.e.*,,

$$\boxed{\mu_o \approx \mu_E = \text{birefringence}}$$

Thus in case of calcite crystal the velocity *O*-ray is less than that of *E*-Ray *i.e.,* in crystal of calcite extraordinary ray travels faster than that of ordinary ray. Further it has been found that, for *O*-ray, the refractive index is same for all angle of incidence whereas for *E*-ray, the refractive index is different for different angle of incidence. Thus *O*-ray travels with same speed in all direction whereas *E*-ray travels with different speed in different directions.

It has been observed that both *O*-ray and *E*-ray are plane polarised. The vibrations of ordinary ray are perpendicular to principal section of the crystal and vibrations of extraordinary ray are along the principal section of the crystal.

(*a*) **Geometry of Calcite Crystal**

The calcite crystal, also known as Iceland spar ($CaCO_3$) is a colourless crystal, transparent to visible as well as to ultraviolet light. It is available in different shapes and can be easily reduced to rhombohedron bounded by six parallelograms with angle 102° and 78° as shown in Figure 6.8(*a*). At corners *A* and *H* all the faces make equal obtuse angles, are known as *blunt corners*. A line passing through one of the blunt corners and is equally inclined to all the three edges meeting over there gives the direction of optic axis. Any line parallel to this line is also known as optic axis.

A line joining two opposite blunt corners is not an optic axis, but only in case of cubic crystal where all the three edges are equal, a line joining two opposite blunt corners will be an optic axis.

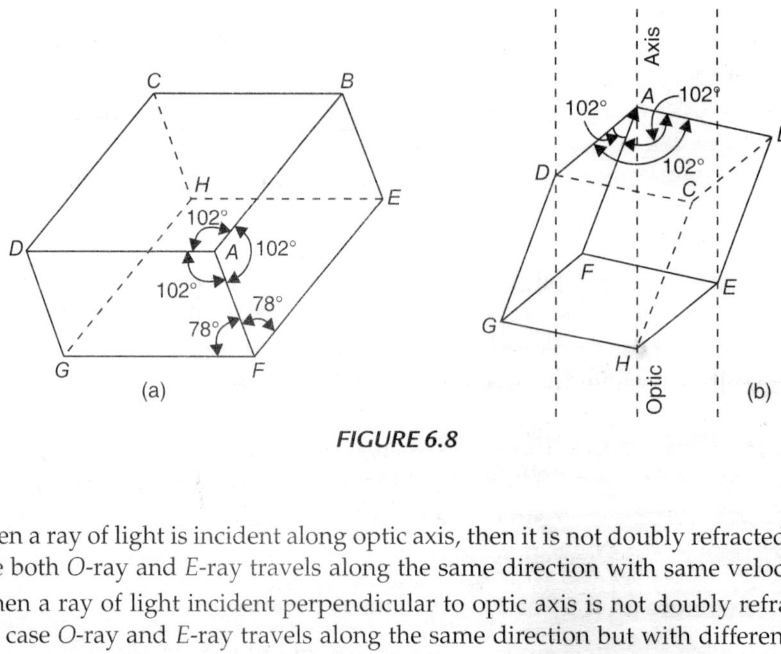

FIGURE 6.8

Note:

(*i*) When a ray of light is incident along optic axis, then it is not doubly refracted, because in this case both *O*-ray and *E*-ray travels along the same direction with same velocities.

(*ii*) When a ray of light incident perpendicular to optic axis is not doubly refracted because in this case *O*-ray and *E*-ray travels along the same direction but with different velocities.

(*b*) **Principal Section of the Crystal**

A plane containing optic axis and perpendicular to the opposite faces of the crystal is called *principal section of the crystal*. As there are six faces in a crystal so for every point, there are three principal sections passing through any point inside the crystal, one corresponding to each pair of opposite faces. A principal section always cuts the surface of calcite crystal in parallelogram with angles 109° and 71°.

(*c*) **Principal Plane of the Crystal**

For ordinary ray, principal plane is the plane drawn through optic axis and ordinary ray and for extraordinary ray, principal plane is the plane drawn through optic axis and extraordinary ray. The ordinary ray always lie in the plane of incidence, whereas for extraordinary ray this is not generally true. The incidence plane of two rays do not coincide but in particular case, when the plane of incidence is a principal section then principal section of crystal and principal planes of ordinary and extraordinary rays coincide.

(*d*) **Types of Crystal**

Uniaxial Crystals: Those crystals in which there is a single direction called optic axis along which *O*-rays and *E*-rays are transmitted with same velocity and along any other direction they have different velocities are called uniaxial crystal *e.g.*, calcite, quartz, tourmaline etc.

Biaxial Crystals: Those crystals in which there are two directions along which *O*-rays and *E*-rays are transmitted with same velocity (*i.e.*, they have more than one optic axis) are called biaxial crystals *e.g.*, borax, mica, topaz etc.

6.6 PRODUCTION OF POLARISED LIGHT OF DOUBLE REFRACTION

In the phenomenon of double refraction using calcite crystal, it has been found that *O*-ray and *E*-ray are plane polarised, because when a tourmaline crystal is placed in the path of *O*-ray and *E*-ray, then on rotating the crystal, intensity of both the images changes. If intensity of one image (extraordinary) decreases then intensity of other image (ordinary) increases. In complete rotation, both the images alternatively can be extinguished at two places. It clearly shows that both *O*-ray and *E*-ray are plane polarised with their planes perpendicular to each other.

Huygen's Experiment:

Huygen's in 1678, demonstrated polarisation of light by double refraction. In his experiment, he passed a beam of light through a pair of calcite crystals and made the following observations:

(*i*) When principal sections of both the crystals are parallel *i.e.*, $\theta = 0$ as shown in Figure 6.9*a*. Two images O_1 and E_1 are formed due to a ray falling normally on the surface *AB* of first crystal. This is because, a ray falling normally on first crystal gets splitted into *O*-ray and *E*-ray. The *O*-ray transverse second crystal without any deviation whereas *E*-ray transverse second crystal along the path parallel to that in the first crystal. Thus two rays represented by O_1 and E_1 emerge exactly parallel with the separation twice as in case of first crystal.

(*ii*) If the second crystal is rotated, each of the two rays *O*-ray and *E*-ray suffers double refraction in the second crystal and four images are observed. The old images O_1 and E_1 become dimmer and in between them two new faint images O_2 and E_2 are formed. On rotation of second crystal, the images O_1 and O_2 remains stationary while E_1 and E_2 rotate in a circular path around O_1 and O_2 respectively. In addition to it, intensity of old image O_1 and E_1 decreases whereas that of new images O_2 and E_2 increases. When principal section of second crystal makes an angle of 45° with the principal section of first crystal, all the images are of equal intensity as shown in Figure 6.9(*b*).

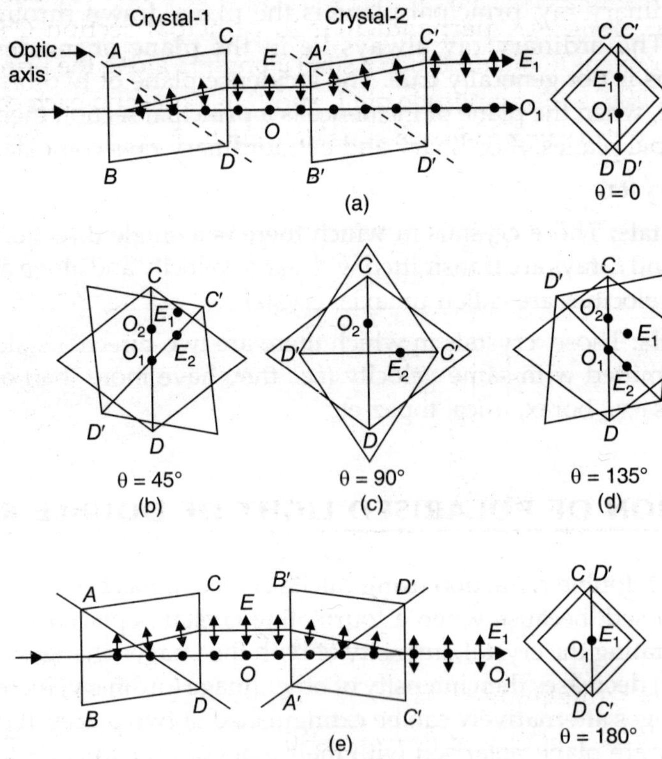

FIGURE 6.9

(*iii*) When $\theta = 90°$, the old images O_1 and E_1 disappear and new images O_2 and E_2 acquires maximum intensity as shown in Figure 6.9(*c*).

(*iv*) On further rotation, when $\theta = 135°$, again four images of equal intensity are observed as shown in Figure 6.9(*d*).

(*v*) At $\theta = 180°$, the principal section of two planes are parallel but their optic axis are oriented in opposite direction as shown in Figure 6.9(*e*), so the images O_2 and E_2 disappear and images O_1 and E_1 will superimpose to form a single.

(*vi*) If rotation continues from 180° to 360°, all the above changes takes place in reverse direction.

Thus, it has been derived experimentally that first crystal provides polarised lights as *O*-ray and *E*-ray.

Explanation: The above observations can be explained physically as:

Let θ be the angle between principal section *CD* and *C'D'* of two crystals at any instant. When an ordinary ray enters first crystals, it splits up into two plane polarised component as *O*-ray and *E*-ray. The vibrations of *O*-ray are perpendicular to principal section *CD*. In Figure 6.10,

O-ray and E-ray are represented by PO and PE respectively and each having same amplitude say a. Both O-ray and E-ray enter second crystal and gets splitted further into two components each due to double refraction. The components of O-ray are: ordinary component O_1 of amplitude $a \cos \theta$ along PO_1 perpendicular to principal section $C'D'$ of crystal-2 and extraordinary component E_2 of amplitude $a \sin \theta$ along PE_1 along the principal section $C'D'$ of crystal-2. From the above the amplitude of O_1 and E_1 is same and each equal to $a \cos \theta$ while those of O_2 and E_2 is also same and each equal to $a \sin \theta$.

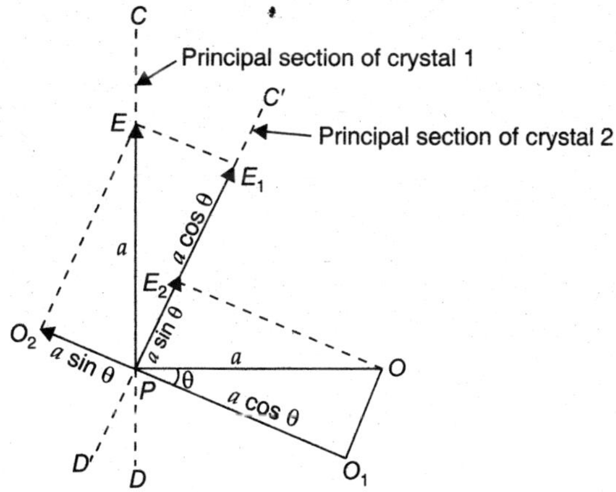

FIGURE 6.10

As Intensity = (Amplitude)^2

So Intensity of each of O_1 and $E_1 = a^2 \cos^2 \theta$

and Intensity of each of O_2 and $E_2 = a^2 \sin^2 \theta$

When $\theta = 0$ or $\theta = 180°$

$a^2 \cos^2 \theta = a^2$ *i.e.*, O_1 and E_1 has maximum intensity.

and $a^2 \sin^2 \theta = 0$ *i.e.*, O_2 and E_2 disappear.

When $\theta = 45°$ or $\theta = 135°$

and $\left.\begin{aligned} a^2 \cos^2 \theta &= \frac{a^2}{2} \\ a^2 \sin^2 0 &= \frac{a^2}{2} \end{aligned}\right\}$ *i.e.*, O_1E_1 and O_2E_2 all have same intensity.

When $\theta = 90°$

$a^2 \cos^2 \theta = 0$ *i.e.*, O_1 and E_1 disappear.

and $a^2 \sin^2 \theta = a^2$ *i.e.*, O_2 and E_2 has maximum intensity.

Thus all the observations made by Huygen can be explained physically and for all positions the sum of intensities of two components is $a^2 \cos^2 \theta + a^2 \sin^2 \theta = a^2$, which is equal to intensity of incident beam.

6.7 NICOL PRISM

William Nicol in 1928 invented an optical device used for producing and analysing plane polarised light. The device was named as Nicol prism after his name.

Principal: It is based on the phenomenon of double refraction. *i.e.*, when a ray of light is passed through calcite crystal, it splits up into two rays (*i*) the ordinary rays which is plane polarised with its vibrations perpendicular to direction of the principal axis of the crystal (*ii*) *E*-ray which is also plane polarised with its vibrations parallel to direction of principal axis of the crystal. In order to have a plane polarised beam one of the two rays has to be eliminated. In Nicol prism, ordinary ray has been eliminated by total internal reflection and ray emerging through the crystal is only extraordinary ray which is plane polarised.

Construction: Nicol prism is prepared by taking a crystal *ADEGFHBC* whose length is three times its breadth and having *ABCD* as principal section with angle $\angle ABC = 71°$ as shown in Figure 6.11. The end faces *AB* and *CD* are cut in such a way that they make angles 68° and 112° instead of 71° and 109° as shown in Figure 6.11.

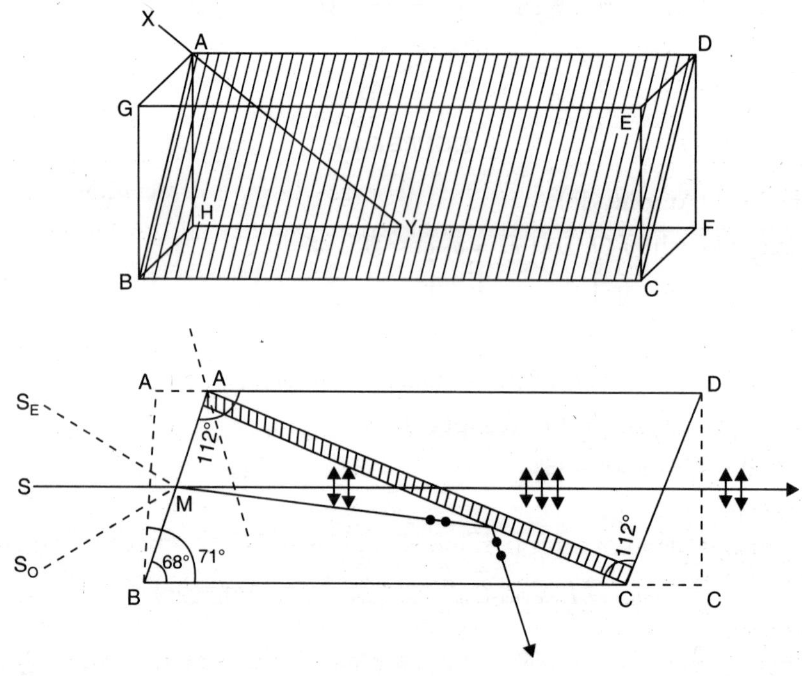

FIGURE 6.11

The resulting crystal is then cut into two pieces along the plane *A′C′* passing through opposite blunt corners and perpendicular to principal section. The cut surfaces are ground and polished optically flat and then cemented together with the layer of Canada Balsam. The Canada Balsam is a transparent liquid having refractive index midway between the refractive indices of crystal of *O*-ray and *E*-ray. For sodium light the refractive indices are

$$\mu_O = 1.65837 \simeq 1.66$$
$$\mu_B = 1.55$$
$$\mu_E = 1.48641 \simeq 1.48$$

Thus Canada Balsam is optically denser than calcite of *E*-ray and rarer from *O*-ray. The sides of crystal are coated lamp Black and enclosed in a brass tube. The ends are kept open for incidence and emergence of light.

Working: When ray *SM* of unpolarised light is incident on face *A'B*, it splits into two refracted rays viz *O*-ray and *E*-ray. The ordinary ray goes from calcite to Canada Balsam is travelling from optically denser medium to rarer medium can be totally internally refracted and extraordinary ray is travelling from optically rarer to denser medium is transmitted. The critical angle for ordinary ray will be

$$\sin C = \frac{\mu_O}{\mu_B} = \frac{1.55}{1.66} = 0.933$$
$$C = 69°.$$

The angle of incidence on Canada Balsam depends upon the angle which *A'B* makes with blunt edge *BC'* and also on breadth of length ratio of the crystal. This was the only reason that length of crystal is chosen thrice of breadth and natural angle 71° is reduced to 68°. Because by doing so, *O*-ray falls on Canada Balsam layer at an angle greater than critical angle *C*, so it is totally internally reflected and absorbed, whereas *E*-ray is transmitted. The transmitted extraordinary ray is plane polarised having vibrations parallel to principal section of the Nicol prism. Thus, Nicol prism act as a polariser.

Drawbacks of Nicol Prism

1. Nicol prism can act as polariser effectively only if incident beam is slightly convergent or slightly divergent and fails, if incident beam is highly convergent or divergent.
2. If angle of incidence of incident ray S_0M at the crystal surface is increased, the angle of incidence at the Canada Balsam surface decreases. If angle SMS_0 is greater than 14°, the angle of incidence at the Canada Balsam surface is less than 69° and ordinary ray is also transmitted through the Nicol prism. Hence emergent ray from the Nicol will be mixture of *O*-ray and *E*-ray *i.e.*, will not be plane polarised.
3. The refractive index of calcite crystal is different for different direction of *E*-ray, being minimum when it is travelling at right angle to optic and maximum when it travelling along optic axis.

Because along optic axis *E*-ray and *O*-ray travels with same speed for intermediate angles it is between 1.486 and 1.658. For a particular value of angle of incidence of ray S_EM, μ_E may be more than μ_B and *E*-ray will also be totally internally reflected and no light emerges from the Nicol.

Thus a nicol can polarize light if it is confined within an angle of 14° on either side of *SM*.

6.7.1 Nicol Prism as an analyser (or Parallel and Crossed Nicols)

A Nicol prism can be used as a polariser and an analyser. When two Nicol prisms are placed co-axially as shown in Figure 6.12, then Nicol prism *P* acts as polariser and Nicol prism *A* acts as analyser. Such an arrangement is known as polariscope.

When principal section of both the Nicols are parallel, then emitted *E*-ray from polariser *P* has vibrations parallel to principal section of analyser *A*, so get freely transmitted through it. In this setting of Nicols the intensity of emitted light is maximum. This position and the position when the angle between the principle sections of two prisms is 180° is known as "Parallel Nicols".

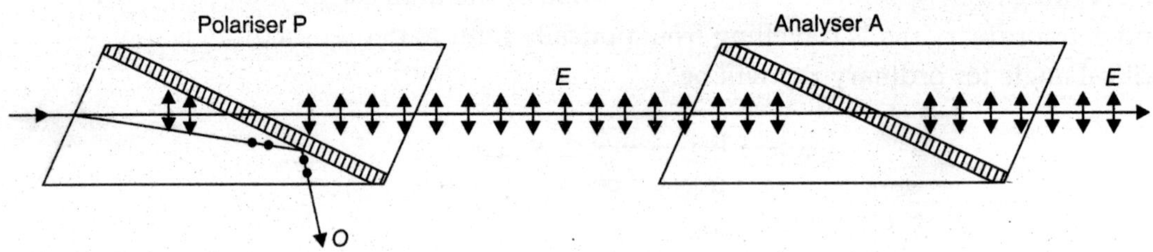

(a)

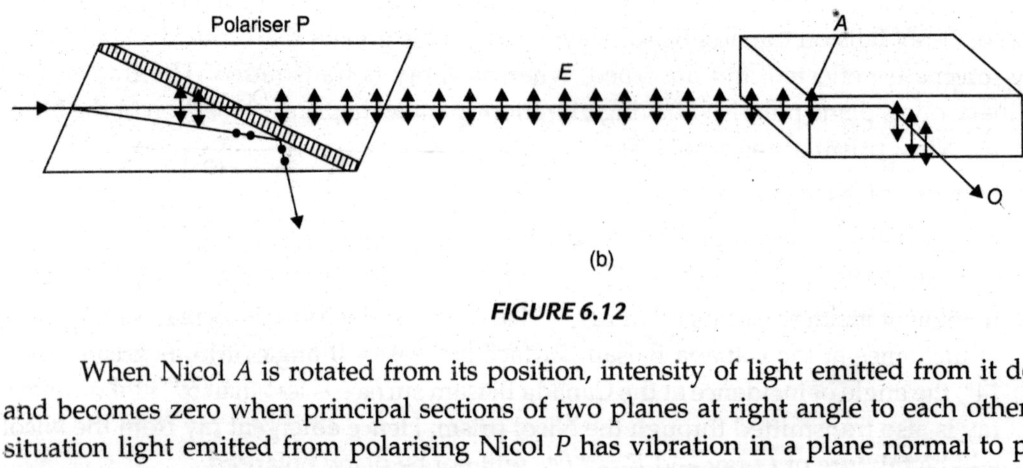

(b)

FIGURE 6.12

When Nicol *A* is rotated from its position, intensity of light emitted from it decreases and becomes zero when principal sections of two planes at right angle to each other. In this situation light emitted from polarising Nicol *P* has vibration in a plane normal to principal section of analysing Nicol *A* and is totally internally reflected back from Canada Balsam layer and no light is emitted. In this setting, two Nicols are said to be "*Crossed Nicols*".

For all other intermediate positions between parallel and crossed, the *E*-ray emitted by polariser falls on analyser and get splitted into two components: one having vibrations in the principal section of analyser and other having vibrations perpendicular to principal section of analyser. The latter is totally internally reflected by Canada Balsam layer while the former is freely emitted. The intensity of emitted light is given by Malus law *i.e.*, $I \propto \cos^2 \theta$, where θ is angle between principal sections of analyser or polariser.

6.8 THEORY FOR PRODUCTION OF CIRCULARLY AND ELLIPTICALLY POLARISED LIGHT

Consider a plane polarised light of wavelength λ and amplitude a from Nicol N is incident normally on a doubly refracting calcite crystal cut with its faces parallel to optic axis and is oriented in such a way that vibration in plane polarised light makes an angle θ with optic axis of the crystal as shown in Figure 6.13 (b).

On entering the crystal the light gets splitted into two components called ordinary and extraordinary and represented as O-ray and E-ray respectively. The O-ray has vibrations perpendicular to optic axis with amplitude $a \sin \theta$ and E-ray has vibrations along optic axis with amplitude $a \cos \theta$ as shown in Figure 6.13 (c). Both the components travels in same direction but with different velocities. In case of calcite E-ray travels faster than O-ray ($\mu_O > \mu_E$) whereas in case of quartz the situation is reversed.

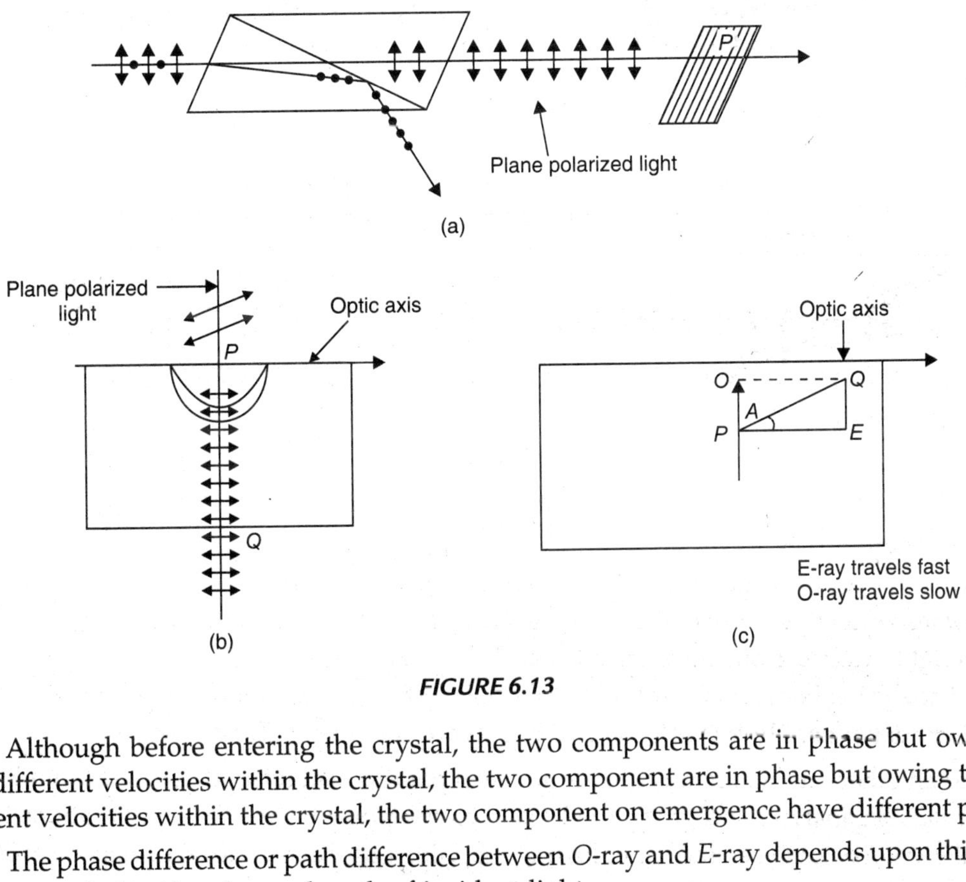

FIGURE 6.13

Although before entering the crystal, the two components are in phase but owing to their different velocities within the crystal, the two component are in phase but owing to their different velocities within the crystal, the two component on emergence have different phases.

The phase difference or path difference between O-ray and E-ray depends upon thickness and nature of crystal and wavelength of incident light.

Thus incident plane polarised light after passing through the crystal consist of two simple harmonic vibration, in two mutually perpendicular planes having same period but different amplitude and phases. They superimpose and result into a single motion which can be linear, circular or elliptical polarised depending upon value of θ and phase difference.

To calculate phase difference between *O*-ray and *E*-ray

By definition, $$\mu = \frac{\lambda_{air}}{\lambda_{med}}$$

so $$\lambda_E = \frac{\lambda}{\mu_E} \text{ and } \lambda_O = \frac{\lambda}{\mu_O}$$

Number of waves within the crystal = $\dfrac{\text{thickness}}{\text{wavelength}}$.

For *E*-ray, $$n_E = \frac{t}{\lambda_E} = \frac{t\mu_E}{\lambda}$$

For *E*-ray, $$n_O = \frac{t}{\lambda_O} = \frac{t\mu_O}{\lambda}$$

As we have considered calcite crystal *i.e.*, negative crystal, so $\mu_O > \mu_E$. Therefore the number of waves by which *E*-ray leads *O*-ray can be given as

$$\frac{1}{\lambda}(\mu_O - \mu_E)$$

As before entering both the rays were in same phase and each wave corresponds to a phase angle of 2π radians, therefore phase difference between *E*-ray and *O*-ray after emergences will be

$$\delta = 2\pi \frac{t}{\lambda}(\mu_O - \mu_E)$$

Let displacements of *E*-ray and *O*-ray along and perpendicular to optic axis may be represented as

$$x = a\cos\theta \sin\omega t \qquad \text{For } E\text{-ray} \qquad ...(1)$$

$$y = a\sin\theta \sin(\omega t + \delta) \qquad \text{For O-ray} \qquad ...(2)$$

Equations (1) and (2) may be written as

$$x = A\sin\omega t \qquad ...(3)$$

$$y = B\sin(\omega t + \delta) \qquad ...(4)$$

where $A = a\cos\theta$ and $B = a\sin\theta$

From equation (4), we get

$$\frac{y}{B} = \sin(\omega t + \delta)$$

or $$\frac{y}{B} = \sin\omega t\cos\delta + \cos\omega t\sin\delta \qquad ...(5)$$

From equation (3), we get

$$\sin\omega t = \frac{x}{A} \text{ therefore, } \cos\omega t = \sqrt{1 - \frac{x^2}{A^2}}$$

Put sin ω*t* and cos ω*t* in equation (5), we get

$$\frac{y}{B} = \frac{x}{A}\cos\delta + \sqrt{1-\frac{x^2}{A^2}}\sin\delta$$

or
$$\frac{y}{B} - \frac{x}{A}\cos\delta = \sqrt{1-\frac{x^2}{A^2}}\sin\delta$$

Squaring both sides, we have

$$\frac{y^2}{B^2} + \frac{x^2}{A^2}\cos^2\delta - \frac{2xy}{AB}\cos\delta = \left(1-\frac{x^2}{A^2}\right)\sin^2\delta$$

or
$$\frac{x^2}{A^2} + \frac{y^2}{B^2} - \frac{2xy}{AB}\cos\delta = \sin^2\delta \qquad ...(6)$$

This is general equation of ellipse.

Special case:

1. When δ = 0, sin δ = 0 and cos δ = 1

From equation (6),

$$\frac{x^2}{A^2} + \frac{y^2}{B^2} - \frac{2xy}{AB} = 0$$

or
$$\frac{x}{A} - \frac{y}{B} = 0$$

$$y = \frac{B}{A}x$$

This is equation of a straight line. Thus emergent light will be plane polarised.

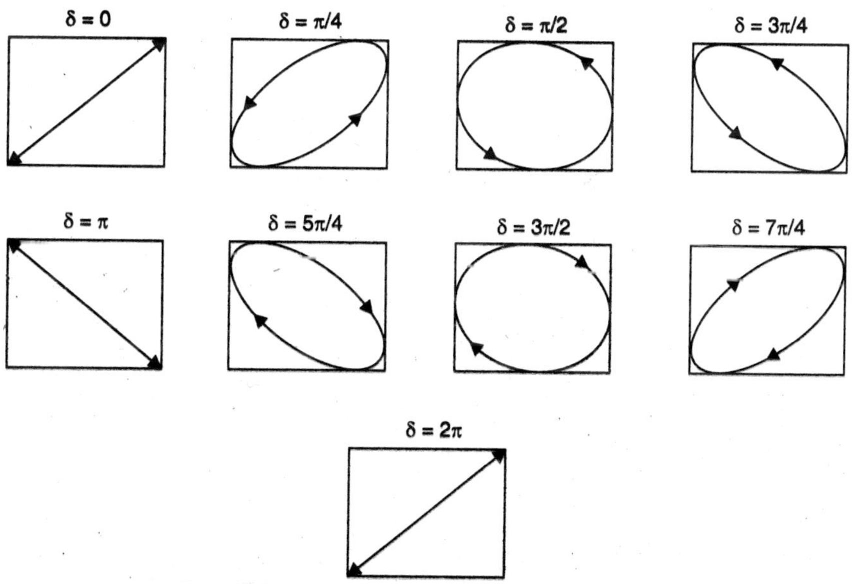

FIGURE 6.14

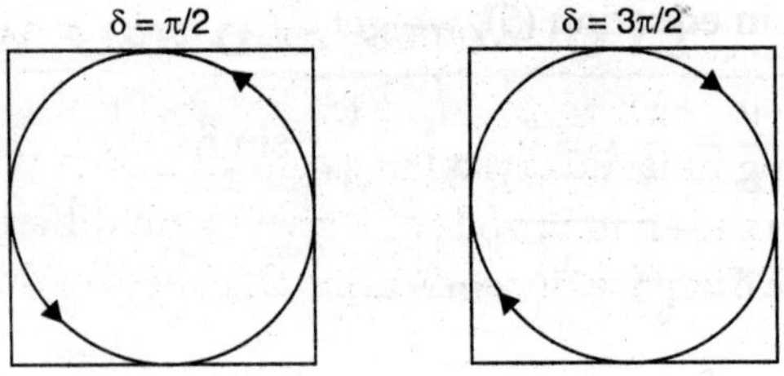

(a) (b)

FIGURE 6.15

2. When $\delta = \frac{\pi}{2}, \frac{3\pi}{2} \; .... \; \frac{(2n+1)\pi}{2}$

From equation (6),

$$\frac{x^2}{A^2} + \frac{y^2}{B^2} = 1$$

This is the equation of symmetrical ellipse. Hence emergent light will be *elliptically polarised.*

3. When $\delta = \frac{\pi}{2}, \frac{3\pi}{2} \; .... \; \frac{(2n+1)\pi}{2}$ and $a = b$

From equation (6),

$$x^2 + y^2 = A^2$$

This is the equation of circle of radius a. Hence emergent ray will be *circularly polarised.*

Note. For $A = B$, the vibration of the incident light makes an angle 45° with the optic axis.

4. When $\delta = \pi, 3\pi$ .... $\cos\delta = -1$ and $\sin\delta = 0$

From equation (6),

$$\frac{x^2}{A^2} + \frac{y^2}{B^2} + \frac{2xy}{AB} = 0 \text{ or } \frac{x}{A} + \frac{y}{B} = 0$$

$$y = -\frac{B}{A}x$$

which is again equation of a straight line but with negative slope.

5. When $\delta = \frac{\pi}{4}, \frac{3\pi}{4}$ .... $\sin\delta = \cos\delta = \frac{1}{\sqrt{2}}$

From equation (6),

$$\frac{x^2}{A^2} + \frac{y^2}{B^2} \mp \sqrt{2}\,\frac{xy}{AB} = \frac{1}{2} \qquad \text{–ve sign for } \delta = \frac{\pi}{4}$$

$$\text{+ve sign for } \delta = \frac{3\pi}{4}.$$

This equation of ellipse in which major axis is inclined to positive and negative x-axis respectively for $\delta = \frac{\pi}{4}$ and $\delta = \frac{3\pi}{4}$ respectively.

6.9 RETARDATION PLATE : QUARTER AND HALF WAVE PLATE

The plates of a doubly refracting uniaxial crystal of calcite or quartz whose refracting faces are cut parallel to optic-axis and are used to introduce a given path difference between *O*-ray and *E*-ray. The magnitude of path difference depends upon thickness of doubly refracting uniaxial plate.

6.9.1 Quarter Wave Plate (QWP)

A plate of doubly refracting uniaxial crystal of calcite or quartz whose refracting faces are cut parallel to optic axis and can introduce a path difference of $\lambda/4$ or phase difference $\pi/2$ between *O*-ray and *E*-ray is known *as quarter wave plate.*

Let t be the thickness of a doubly refracting uniaxial quarter wave plate of calcite or quartz. Then path travelled within the plate by ordinary and extraordinary rays will be equivalent to path $\mu_O t$ and $\mu_E t$ in air, so path difference introduced between the two rays is given as :

For calcite (negative crystal), path difference = $(\mu_O - \mu_E)t$

For quartz (positive crystal), path difference = $(\mu_E - \mu_O)t$

For quarter wave plate this path difference is $\lambda/4$, therefore for calcite or negative crystal

$$(\mu_O - \mu_E)t = \frac{\lambda}{4}$$

or

$$t = \frac{\lambda}{4(\mu_O - \mu_E)}$$

For quartz or positive crystal,

$$t = \frac{\lambda}{4(\mu_E - \mu_O)}.$$

Figure 6.16 shows action of quarter wave plate. A plane polarised light with its plane of vibration inclined at an angle θ with the optic axis of plates fall on it.

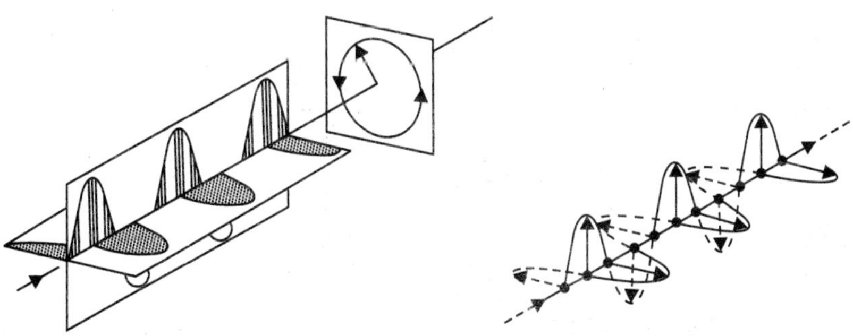

(a) (b)

FIGURE 6.16

If θ has arbitrary value except (0°, 45°, 90°), the emergent light will be elliptically polarised, for θ = 0° or 90°, the emergent light will be plane polarised, and for θ = 45° the emergent ray will be circularly polarised.

6.9.2 Half Wave Plate (HWP)

A plate of doubly refracting uniaxial crystal of calcite or quartz whose refracting faces are cut parallel to optic axis and can introduce a path difference of $\lambda/2$ or phase different of π between *O*-ray and *E*-ray is known as *half wave plate.*

Let *t* be the thickness of a doubly refracting uniaxial half wave plate of calcite or quartz. Then path travelled within the plate by ordinary and extraordinary rays will be equivalent to path $\mu_O t$ and $\mu_E t$ in air, so path difference introduced between the two rays is given as :

For calcite (negative crystal), path difference = $(\mu_O - \mu_E)t$

For quartz (positive crystal), path difference = $(\mu_E - \mu_O)t$

For half wave plate this path difference is $\lambda/2$,

therefore for calcite or negative crystal

$$(\mu_O - \mu_E)t = \frac{\lambda}{2}$$

or
$$t = \frac{\lambda}{2(\mu_O - \mu_E)}$$

For quartz or positive crystal

$$t = \frac{\lambda}{2(\mu_E - \mu_O)}$$

AB = incident polarised light

$A'B'$ = emergent polarised light

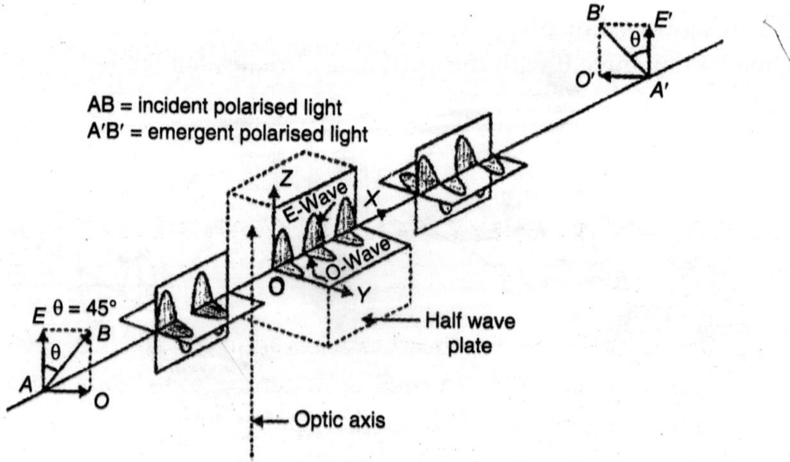

FIGURE 6.17

Figure 6.17 shows action of half wave plate. A plane polarised light with its plane of vibration inclined at an angle $\theta = 45°$ with the optic axis of plate fall on it. The plate introduces a path difference of $\lambda/2$ between ordinary and extraordinary waves. The emergent light is plane polarised and direction of polarisation of linear incident light get rotated through $2\theta = 90°$.

6.10 PRODUCTION OF PLANE, CIRCULAR AND ELLIPTICAL POLARISED LIGHT

(*i*) *Plane polarised light*. For producing plane polarised light, a beam of unpolarised light is passed through doubly refracting Nicol prism. Within the prism the beam gets split into ordinary and extraordinary rays. The ordinary ray is totally internally reflected back by Canada Balsam and only extraordinary ray passes through the Nicol prism. The emergent extraordinary ray is plane polarised having vibrations along the optic axis of Nicol prism.

(*ii*) *Circularly polarised light*. When two plane polarised light beams having equal amplitudes and plane of vibrations in two mutually perpendicular directions with path difference $\lambda/4$ or phase difference of $\pi/2$ superimpose on each other, form a circularly polarised light. For this, an ordinary monochromatic light is passed through Nicol prism, it gives polarised light. The plane polarised light is then allowed to fall normally on a quarter wave plate in such a manner that vibration of polarised light makes an angle of 45° with the optic axis of quarter wave plate. In this situation, the light emerging out from quarter wave plate is circularly polarised.

(*iii*) *Elliptically polarised light*. When two plane polarised light beams of unequal amplitudes and having plane of vibrations with path difference of $\lambda/4$ or phase difference of $\pi/2$ superimpose on each other, form an *elliptically polarised light*. For this, an ordinary monochromatic light is passed through Nicol prism, it gives *plane polarised light*. The plane polarised light is then allowed to fall on a quarter wave plate in such a manner that vibration of plane polarised light makes angle other than 45° (because when $\theta = 45°$, the amplitudes of *O*-rays and *E*-rays will become equal). In this situation, the light emerging out from quarter wave plate is elliptically polarised.

6.11 DETECTION OF PLANE, CIRCULARLY AND ELLIPTICALLY POLARISED LIGHT

(*i*) *Plane Polarised Light:* The light beam is allowed to fall on Nicol prism. If on rotation of Nicol prism, intensity of emitted light can be completely extinguished at two places in each rotation, then light is plane polarised.

(*ii*) *Circularly Polarised Light:* The light beam is allowed to fall on a Nicol prism. If on rotation of Nicol prism the intensity of emitted light remains same, then light is either circularly polarised or unpolarised. To differentiate between unpolarised and circularly polarised light, the light is first passed through quarter wave plate and then through Nicol prism. Because if beam is circularly polarised then after passing through quarter wave-plate an extra difference of $\lambda/4$ is introduced between ordinary and extraordinary component and gets converted into plane polarised. Thus on rotating the Nicol, the light can be extinguished at two plates. If, on the other hand, the beam is unpolarised, it remains unpolarised after passing through quarter wave plate and on rotating the Nicol, there is no change in intensity of emitted light (Figure 6.18).

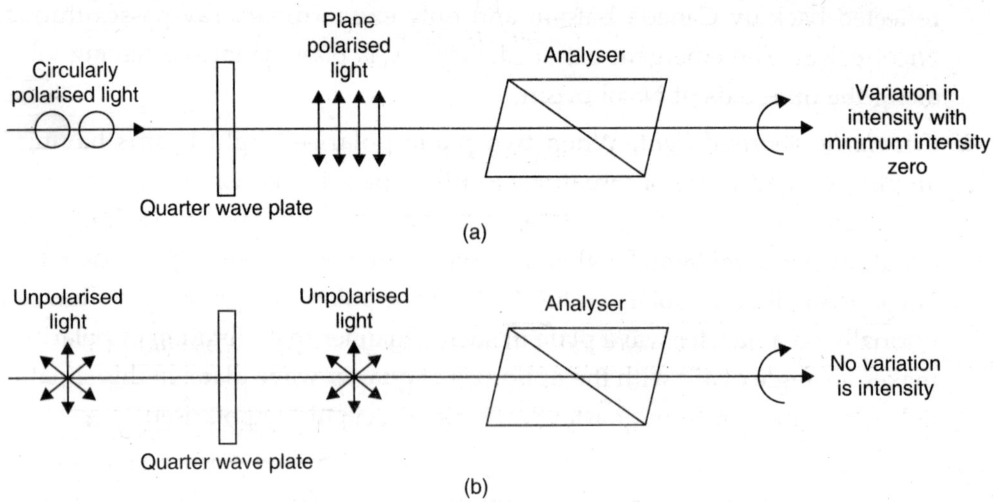

FIGURE 6.18

(*iii*) *Elliptically Polarised Light.* The light beam is allowed to fall on Nicol prism. If on rotation of Nicol prism, the intensity of emitted light varies from maximum to minimum, then light is either elliptically polarised or a mixture of plane polarised and unpolarised. To differentiate between the two, the light is first passed through quarter wave plate and then through Nicol prism. Because, if beam is elliptically polarised, then after passing through quarter wave plate, an extra path difference of $\lambda/4$ is introduced between *O*-ray and *E*-ray and get converted into plane polarised. Thus, on rotating the Nicol, the light can be extinguished at two places. If, on the other hand, beam is mixture of polarised and unpolarised it remains mixture after passing through quarter wave plate and on rotating the Nicol intensity of emitted light varies from maximum to minimum (Figure 6.19).

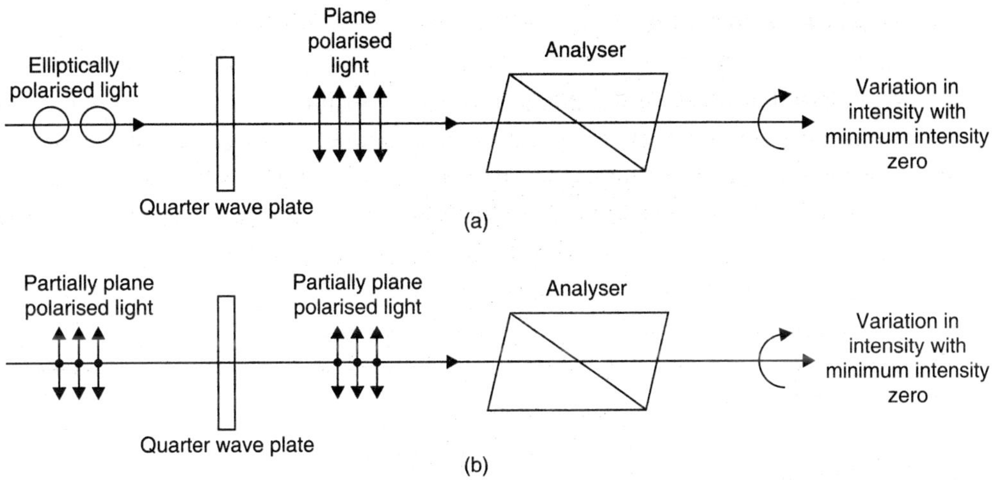

FIGURE 6.19

Summary of detection of polarisation of light:

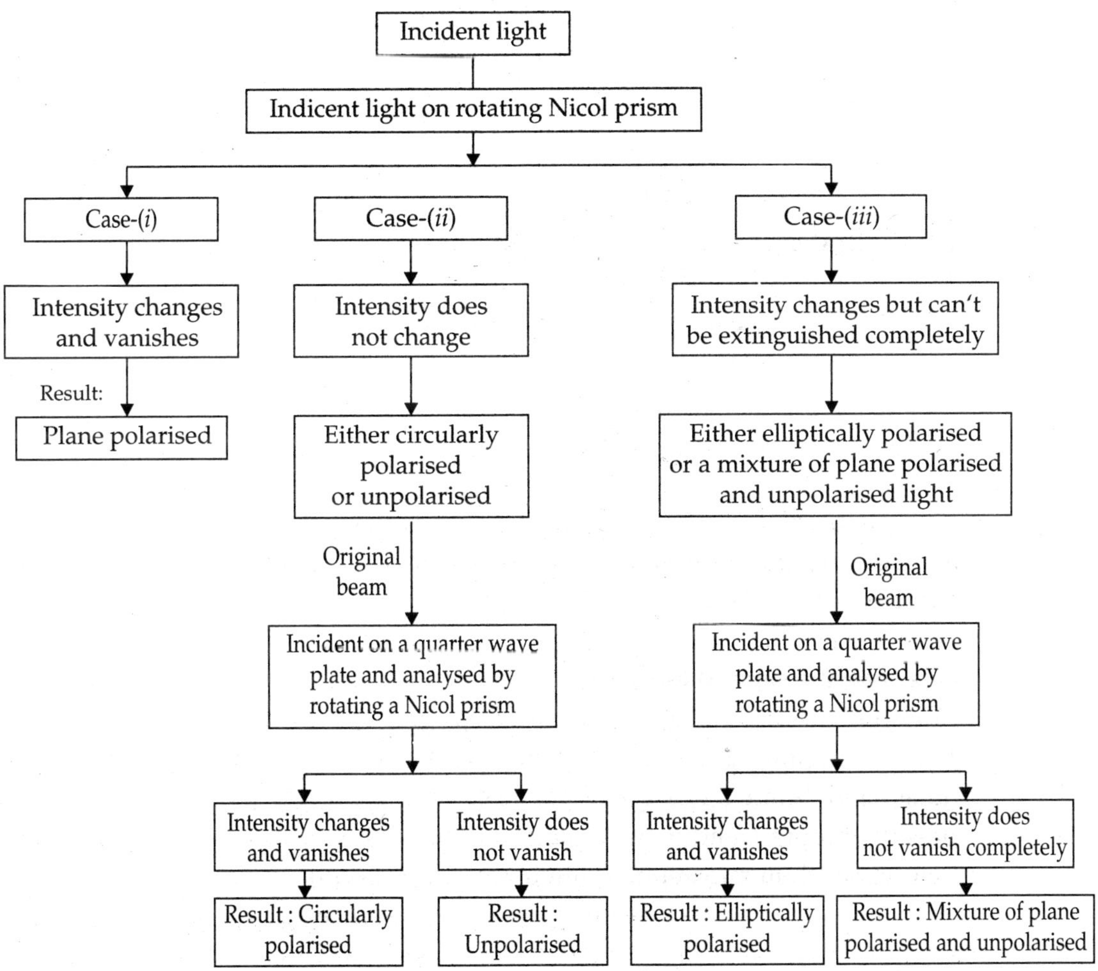

6.12 OPTICAL ROTATION

When a plane polarised light is passed through certain substance, the plane of polarisation of the emergent light get rotated through certain angle about the direction of propagation of light. This phenomenon is called *'Optical rotation'* or *'Optical activity'* and the substance which rotates plane of polarised light are called *optically active medium e.g.*, Sugar solution, turpentine, sodium chlorate, cinnabar etc.

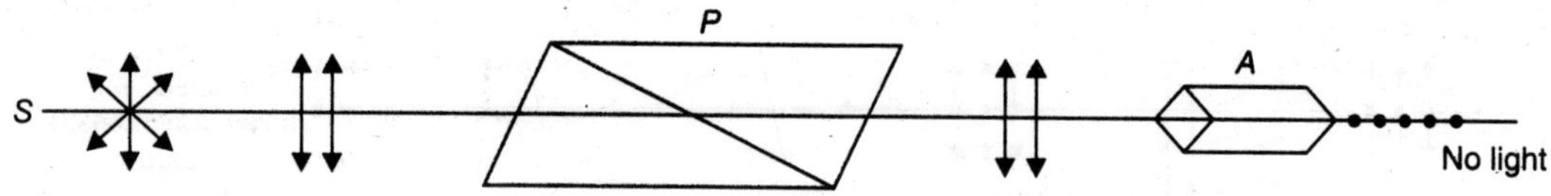

(a)

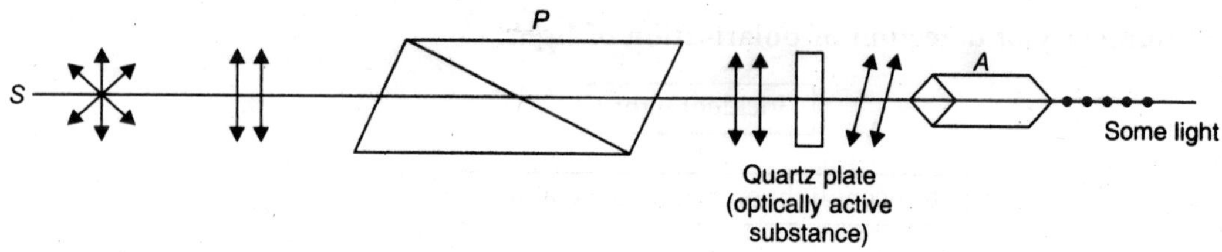

(b)

FIGURE 6.20

The optically active substances are of two types:

(*i*) Dextro rotatory or right handed.

(*ii*) Laevo rotatory or left handed.

(*i*) **Dextro rotatory or right handed.** The substance which rotates plane of polarisation of light towards right or in clockwise direction are called *dextro rotatory* or *right handed.*

(*ii*) **Laevo rotatory or left handed.** The substance which rotates plane of polarisation of light towards left or in anti-clockwise directions are called *laevo rotatory* or *left handed.*

6.12.1 Fresnel's Theory of Optical Rotation

According to Fresnel's a linearly plane polarised light can be supposed to be consists of two circularly polarised vibrations rotating in opposite directions. When a plane polarised light enters a crystal of calcite (which is optically inactive substance) along optical axis, it is resolved into two circularly polarised vibration rotating in opposite direction with same angular frequency or velocity such that resultant of these two vibrations at all point of time is along the *optic axis.* Vibrations rotating in clockwise (in Figure 6.21 (*a*)), direction are represented by *OR* and vibration rotating in anticlockwise direction are represented by *OL*. The resultant *OR* and OL at all point of time will be along AB. In case of Quartz (which is optically active), a linearly polarised light on entering the crystal is resolved into two circularly polarised vibrations rotating in opposite direction with different angular velocity or frequency. In case of left handed optically active quartz crystal, anticlockwise vibrations travel faster, while in case of right handed optically active, quartz crystal, clockwise vibrations travel faster.

Consider a right handed quartz crystal in which clockwise component travels faster than left handed component. Suppose at any instant of time, right handed component traverse angle δ greater than left handed component as shown in Figure 6.21 (*b*). The new position of resultant of OL' and OR' will be along CD *i.e.*, plane of vibration of light has been rotated through angle $\delta/2$ towards right after passing through quartz crystal. The angle $\delta/2$ depends upon thickness of crystal.

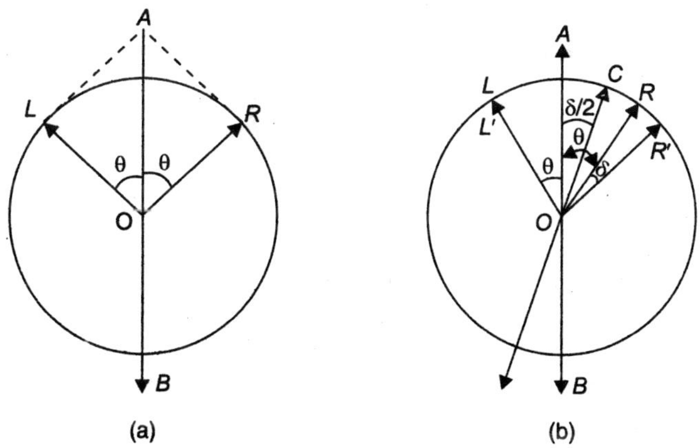

FIGURE 6.21

Mathematically,

(*a*) *For calcite crystal which is optically inactive*: When linearly plane polarised light enters a calcite crystal it get resolved into two circularly polarised vibrations. One is moving anticlockwise with same angular frequency or velocity. As each circularly polarised vibration further consist of two rectangular components having phase differences $\pi/2$, so,

for clockwise circular vibration

$$x_1 = a \sin \theta = a \sin \omega t.$$

$$x_2 = a \cos \theta = a \cos \omega t.$$

for anticlockwise circular vibration

$$x_2 = -a \sin \theta = -a \sin \omega t$$

$$y_2 = a \cos \theta = a \cos \omega t$$

From above, the resultant displacement of vibrations along x-axis and y-axis respectively are given as,

$$x = x_1 + x_2 = a \sin \theta - a \sin \theta = 0$$

$$y = y_1 + y_2 = a \cos \omega t + a \cos \omega t = 2\,a \cos \omega t$$

Hence resultant vibration has amplitude $2a$ and is plane y-axis *i.e.*, along original direction AB. Hence two oppositely circularly polarised vibrations give rise to a plane polarised vibrations.

(*b*) *For quartz crystal which is optically active:* When linearly plane polarised light enters a quartz crystal (right handed), it gets resolved into circularly polarised vibrations moving in opposite direction with different angular frequency or velocity. In case of right handed crystal clockwise vibrations travel faster than anticlockwise vibrations. Let at any instant of time anticlockwise vibrations has traversed angle θ and clockwise vibrations has traversed angle (θ + δ). Therefore, component of two circular vibrations at that instant of time will be,

for clockwise vibration

$$x_1 = a \sin (\omega t + \delta)$$

$$y_1 = a \cos (\omega t + \delta)$$

for anticlockwise circular vibration

$$x_2 = - a \sin \omega t$$

$$y_2 = a \cos \omega t$$

From resultant displacement of vibrations along x-axis and y-axis respectively are given as,

$$x = x_1 + x_2 = a \sin (\omega t + \delta) - a \sin \omega t$$

$$= 2a \sin \delta/2 \cos (\omega t + \delta/2)$$

$$y = y_1 + y_2 = a \cos (\omega t + \delta) + a \cos \omega t$$

$$= 2a \cos \delta/2 \cos (\omega t + \delta/2).$$

The resultant vibration along x-axis and y-axis are in same phase, so resultant of these vibrations is plane polarised and makes an angle of δ/2 with original direction AB. Thus, plane of vibrations get rotated through angle δ/2 towards right after passing through a right handed quartz crystal.

From above, we get

$$\tan \delta/2 = y/x$$

Angle of Rotation

If μ_R, μ_L be the refractive indices of quartz crystal for right handed and left handed vibrations respectively, then optical path difference on passing through a quartz crystal slab of thickness 't' is given as,

Path difference = $(\mu_L - \mu_R)t$

If λ be the wavelength of light used, then phase difference,

$$\delta = 2\pi/\lambda\, (\mu_L - \mu_R)t.$$

Angle of rotation

$$\boxed{\frac{\delta}{2} = \frac{\pi}{\lambda} (\mu_L - \mu_R)t.}$$

6.12.2 Specific Rotation

When a linearly plane polarised light is passed through an optically active medium/substance, the plane of linearly polarised light get rotated through certain angle either towards left or right. The angle through which plane polarised light get rotated depends upon

(*i*) thickness of the medium

(*ii*) density of active substance or concentration of solution

(*iii*) Wavelength of light

(*iv*) Temperature.

The rotation produced by 1 decimeter ($\lambda = 10$ cm) long solution of concentration 1 gm/cc at given temperature for given wavelength is known as *specific rotation* and given as,

$$[S]_{\lambda}^{t} = \frac{10\theta}{LC}$$

where, θ = rotation in degree, L = length of tube in cms

C = concentration of solution in gm/cc

t = temperature and λ = wavelength of linearly polarised light

The rotation produced by optically active/medium/substance can be measured by polarimeter.

6.13 POLARIMETER

It is a device which is used to measure the optical rotation produced by an optically active substance. By measuring the angle θ the specific rotation of an optically active substance can be determined.

Two types of polarimeters are generally used in the laboratory now a days:

(*a*) Laurent's Half Shade Polarimeter

(*b*) Biquartz Polarimeter

6.13.1 Laurent's Half Shade Polarimeter

Construction: It consists of a monochromatic source S which is placed at focal point of a convex lens L. Just after the convex lens there is a Nicol Prism P which acts as a polariser. H is a half shade device which divides the field of polarised light emerging out of the Nicol P into two halves generally of unequal brightness. T is a glass tube in which optically active solution is filled. The light after passing through T is allowed to fall on the analyzing Nicol A which can be rotated about the axis of the tube. The rotation of analyser can be measured with the help of a scale C. Laurent's half shade polarimeter is shown in Figure 6.22.

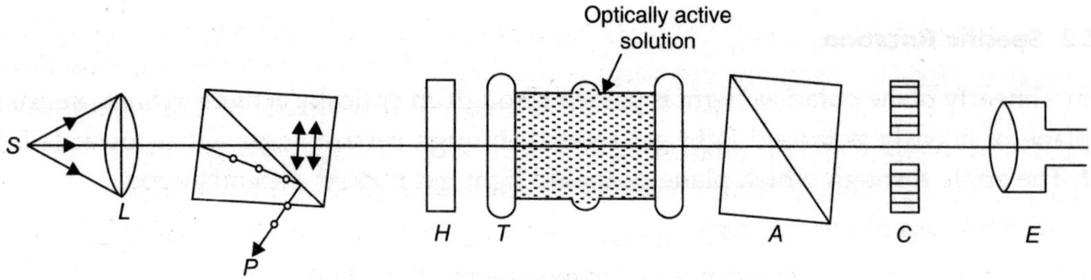

FIGURE 6.22

Working: In order to understand the need of a half shade device, let us suppose that half shade device is not present. The position of the analyzer is so adjusted that the field of view is dark when tube is empty. The position of the analyzer is noted on circular scale. Now the tube is filled with optically active solution and it is set in its proper position. The optically active solution rotates the plane of polarization of the light emerging out of the polariser *P* by some angle. So the light is transmitted by analyzer *A* and the field of view of telescope becomes bright. Now the analyzer is rotated by a finite angle so that the field of view of telescope again become dark. This will happen only when the analyzer is rotated by the same angle by which plane of polarization of light is rotated by optically active solution.

The position of analyzer is again noted. The difference of the two readings will give you angle of rotation of plane of polarization (θ).

A difficulty is faced in the above procedure that when analyzer is rotated for the total darkness, then it is attained gradually and hence it is difficult to find the exact position correctly for which complete darkness is obtained.

To overcome above difficulty half shade device is introduced between polariser *P* and glass tube *T*.

Half Shade Device

It consist of two semicircular plates *ACB* and *ADC*. One half *ACB* is made of glass while other half is made of quartz. Both the halves are cemented together. The quartz is cut parallel to the optic axis. Thickness of the quartz is selected in such a way that it introduces a path difference of $\lambda/2$ between ordinary and extraordinary ray. The thickness of the glass is selected in such a way that it absorbs the same amount of light as is absorbed by quartz half.

Let us consider that the vibration of polarisation is along *OP*. On passing through the glass half the vibrations remain along *OP*. But on passing through quartz half these vibrations will split into *O*- and *E*-components. The *E*-components are parallel to the optic axis while *O*-component is perpendicular to optic axis. The *O*-component travels faster in quartz and hence an emergence *O*-component will be along *OD* instead of along *OC*. Thus components *OA* and *OD* will combine to

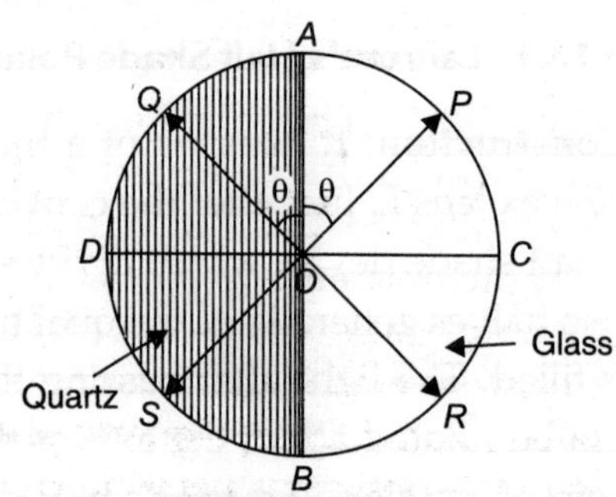

FIGURE 6.23

form a resultant vibration along OQ which makes same angle with optic axis as OP. Now if the Principal plane of the analyzing Nicol is parallel to OP then the light will pass through glass half unobstructed. Hence glass half will be brighter than quartz half or we can say that glass half will be bright and the quartz half will be dark. Similarly if principal plane of analyzing Nicol is parallel to OQ then quartz half will be bright and glass half will be dark.

When the principal plane of analyzer is along AOB then both halves will be equally bright. On the other hand if the principal plane of analyzer is along DOC then both the halves will be equally dark.

Thus it is clear that if the analyzing Nicol is slightly disturbed from DOC then one half becomes brighter than the other. Hence by using half shade device, one can measure angle of rotation more accurately.

Determination of Specific Rotation

In order to determine specific rotation of an optically active substance (say sugar) the polarimeter tube T is first filled with pure water and analyzer is adjusted for equal darkness (Both the halves should be equally dark) point. The position of the analyzer is noted with the help of scale. Now the polarimeter tube is filled with sugar solution of known concentration and again the analyser is adjusted in such a way that again equally dark point is achieved. The position of the analyzer is again noted. The difference of the two readings will give you angle of rotation θ. Hence specific rotation S is determined by using the relation.

$$[S]_{\lambda}^{t} = \frac{\phi}{LC}$$

The above procedure may be repeated for different concentration.

6.13.2 Biquartz's Polarimeter

It is a polarimeter which can be used to determine specific rotation more accurately. The arrangement in biquartz polarimeter is same as in Laurent's half shade polarimeter except half shade device and source of light *i.e.*, half shade device is replaced by biquartz plate and monochromatic source is replaced by white light.

Biquartz plate consists of two semicircular halves AYB and $AY'B$. One half AYB is dextrorotatory while other half is made of laevorotatory quartz as shown in Figure 6.24. Both the halves are cut with their refracting faces perpendicular to the optic axis. The thickness of each half is approximately 3.75 mm so that each half rotates the plane of polarization of yellow light by 90°. One rotates in clockwise while other in anticlockwise. Since white light is falling normally on biquartz plate and quartz crystal is cut with the optic axis perpendicular to the refracting face so

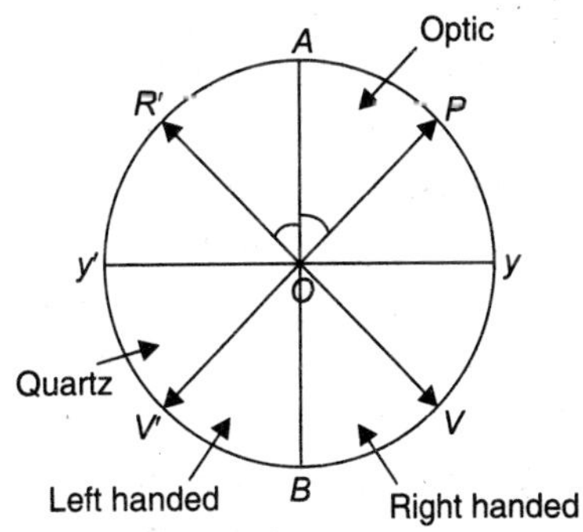

FIGURE 6.24

light does not split into two components and only rotatory dispersion occurs. Red colour is rotated by minimum angle while violet colour by maximum angle. *OR*, *OY*, *OV*, and *OR′*, *OY′*, *OV′* are the vibrations of red, yellow and violet colours in two halves of the Biquartz plate as shown in Figure 6.24.

If the principal plane of analyzing nicol is along *AOB* then colour of both the halves will be greyish violet colled *tint of passage*. If analyzing Nicol is rotated slightly on either side of this position then one half becomes red and other bluish. Thus we can measure the angle of rotation and hence specific rotation more accurately.

Determination of Specific Rotation

In order to determine specific rotation of an optically active solution (say sugar solution) polarimeter tube is first filled with pure water and analyzer is adjusted in such a way position that we get tint of passage. The position of the analyser is noted. Now the polarimeter is filled with sugar solution of known concentration and again the analyzer is adjusted in a position that *tint of passage* is achieved. The position of analyzer is again noted. The difference of the two readings will give you angle of rotation ϕ. Putting this value in equation

$$S = \frac{\phi}{L \times C}, \text{ one get value of specific rotation}$$

6.14 RELATIVE MERITS OF HALF SHADE AND BIQUARTZ POLARIMETER

Half shade polarimeter is very sensitive for those persons who are suffering from colour blindness. It is only suitable for a particular wavelength since the quartz half will work as a half wave plate for a particular wavelength.

In contrary to it biquartz polarimeter is more accurate than half shade polarimeter and equally works with white light.

A person suffering from colour blindness can not work with biquartz polarimeter. It is difficult to adjust the analyzer in a position for which the tint in both the halves is same.

6.15 NUMERICAL PROBLEMS BASED ON POLARISATION OF LIGHT

Problem 1. *A glass plate refractive index 1.54 is used as a polariser. Find angle of polarisation and angle of refraction for it.*

Hint:
$$\mu = \frac{\sin i_p}{\sin r}$$

$$\therefore \quad \sin r = \frac{\sin i_p}{\mu}$$

$$r = \sin^{-1}\left[\frac{\sin 57^\circ}{1.54}\right] = 33^\circ.$$

Problem 2. *The angle of polarisation for flint glass is found to 62°24′. Calculate its refractive index.*

Hint:

$$\mu = \tan i_p$$

$$\mu = \tan 62° \, 24' = 1.912.$$

Problem 3. *A ray of light is incident on a glass surface of refractive index 1.732 at polarizing angle. Calculate the angle of refraction of the ray.*

Hint:

$$\tan i_p = 1.732, \; i_p + r = 90°$$

$$i_p = \tan^{-1}(1.732) = 60°$$

$\therefore$ Angle of refraction $r = 30°$.

Problem 4. *A plane polarized light is incident on a piece of a cut parallel to the axis. Find the least thickness for which the O-Ray and E-Ray combine to form plane polarized light. Given that* $\mu_O = 1.5442$; $\mu_E = 1.5533$ *and* $\lambda = 5 \times 10^{-5}$ *cm.*

Hint:

$$t = \frac{\lambda}{2(\mu_E - \mu_O)} \Rightarrow t = \frac{5 \times 10^{-5}}{2(1.5533 - 1.5442)} = 2.75 \times 10^{-3} \text{ cm.}$$

Problem 5. *If the plane of vibration of incident beam makes an angle of 30° with the optic axis, compare the intensities of ordinary and extraordinary rays.*

Hint:

$$I = I_o \cos^2 \theta$$

For ordinary light	For extra-ordinary light
$I_O = a^2 \sin^2 \theta = \frac{a^2}{4}$	$I_E = a^2 \cos^2 \theta = \frac{3a^2}{4}$

Hence,

$$\frac{I_O}{I_E} = \frac{a^2/4}{3a^2/4} = \frac{1}{3}$$

$$I_E = 3 I_O.$$

Problem 6. *Determine the specific rotation of a given sample of sugar solution if the plane of polarisation is turned through 13.2°. The length of the tube containing 10% sugar solution is 20 cm.*

Hint:

$$[S]_\lambda^t = \frac{10\theta}{l.C} = \frac{10 \times 13.2}{20 \times 0.1} = 66°.$$

Problem 7. *Calculate the thickness of mica sheet required for making a quarter wave plate for* $\lambda = 5460$ *A°. The indices of refraction for the ordinary and extraordinary rays in mica are 1.586 and 1.592.*

Hint:

$$t = \frac{\lambda}{4(\mu_E - \mu_O)}$$

$$= \frac{5460 \times 10^{-8}}{4(1.592 - 1.586)} = 2.275 \times 10^{-2} \text{ cm.}$$

Problem 8. *Calculate the thickness of a double refracting plate capable of producing a path difference of λ/4 between ordinary and extraordinary waves. (λ = 5890 Å, μ_O = 1.53 and μ_E = 1.54)*

Hint:
$$t = \frac{\lambda}{4(\mu_E - \mu_O)} = \frac{5.890 \times 10^{-5}}{4(1.54 - 1.53)} = 1.47 \times 10^{-3} \text{ cm.}$$

Problem 9. *A tube of 20 cm long with solution of 15 gm of cane sugar in 100 cc of water is placed in the path of polarization if the specific rotation of cane sugar is 66°.*

Hint:
$$[S]_\lambda^T = \frac{10\,\theta}{L.C}$$

$$\theta = \frac{SLC}{10} = \frac{66 \times 20 \times 15}{100 \times 10} = 19.8°.$$

Problem 10. *On introducing a polarimeter tube 25 cm long and containing sugar solution of known strength, it is found that the plane of polarization is rotated through 10°. Find the strength of sugar solution in g/cm³. Given that specific rotation of sugar solution is 60° per decimeter per unit concentration.*

Hint:
$$[S]_\lambda^T = \frac{10\,\theta}{L.C} = \frac{10 \times 10}{25 \times C} = 60°$$

$$C = \frac{10\,\theta}{L.C} = \frac{10 \times 10}{25 \times 60} = \frac{1}{15} = 0.0679 \text{ g/m}^3.$$

Problem 11. *Calculate thickness of quarter wave plate for light of wavelength 5000 λ. Given μ_O = 1.54 and ratio of velocity extraordinary to ordinary wave is 1.006.*

Hint:
$$t = \frac{\lambda}{4(\mu_E - \mu_O)} = \frac{5000 \times 10^{-10}}{4(1.54 - 1.53)}$$

$$t = 1.25 \times 10^{-5} \text{ cm.}$$

EXERCISES

1. Show that light waves are transverse in nature.
2. Define plane of vibration and plane of polarisation.
3. Explain, light can be polarised by reflection.
4. Write note on Malus Law.
5. Explain Double reflection.
6. Give Huygen's explanation to show that light can be polarised by double refraction.
7. Write note on Nicol's Prism.
8. Explain, how a Nicol prism can be used as polariser and analyser.

9. Give theory of production of circular and elliptical polarised light.
10. Explain quarter and half wave-plate.
11. Explain optical rotation and give Fresnel's theory of optical rotation.
12. What do you understand by a half and quarter wave-plate? Give the theory and construction of Laurent's half shade polarimeter.
13. What is double refraction? How Nicol prism polarises and analyses the light?
14. What do you understand by optical rotation? Explain Fresnel's theory of the rotation of the plane of polarization. How would you increase the sensitivity of a pair of crossed Nicols?
15. Define specific rotation. Describe the construction and working of a Laurent's half shade polarimeter. Explain fully the action of the half shade device. How would you use it to determine the specific rotation of sugar solution?
16. Describe a Nicol prism and show how it can be used for the study of polarization of light. Explain how you can distinguish between different kinds of polarized light and also a mixture of natural and plane polarized light. Describe the instrument you would use for the purpose.
17. Describe the construction and working of a Nicol Prism to generate plane polarised light. What is the difference between ordinary ray and extraordinary ray? *(Amity University, Dec. 2005)*
18. What do you understand by

    (*i*) plane polarised light (*ii*) circularly polarised light

    (*iii*) elliptically polarised light ? How would you experimentally differentiate between them ?

    (Amity University, Dec. 2005)

SUGGESTION FOR FURTHER READING

1. Ajoy Ghatak, Optics, Tata Mc Graw Hill Publishing Co. Ltd., New Delhi, 2009.
2. F.L. Pedrotti et al., Introduction to Optics, Pearson Education, Delhi, 2007.
3. E. Hecht and A.R. Ganeshan, Optics, Pearson Education, Delhi, 2009.
4. F.A. Jenkins and H.E. White, Fundamentals of Optics, Mc Graw Hill Co. Singapore, 1981.
5. G.B. Deodhar, Introduction to Optics, The Indian Press Ltd., Allahabad, 1949.
6. H.B. Lal, S.R. Tripathi and Kanchan Gaur, Essence of Optics, United Book Depot, Allahabad, 1998.
7. D.P. Khandelwal, Textbook of Optics and Atomic Physics, Shiva Lal Agrawala & Company, Agra, 1979.
8. Devraj Singh, Engineering Physics, Vol. 1, Dhanpat Rai & Co. (P) Ltd., Delhi, 2008.
9. Devraj Singh, Fundamentals of Engineering Physics, Dhanpat Rai & Co. (P) Ltd., Delhi, 2009.
10. N.K. Sethi and S.B. Raizada, Textbook of Optics, Premier Publishing Co., Delhi, 1953.

Chapter 7

ELECTROMAGNETISM

7.1 INTRODUCTION

In 1820, Oersted showed that a conductor carrying current produces a magnetic field around it. If a magnetic needle is placed below a conductor carrying current, it gets deflected. If we make reverse direction of current, deflection and the magnetic field produced are reversed. This is called the *magnetic effect of current*. Electric current flowing in a conductor is equivalent to charges (electrons) in motion. Charges at rest produce electrostatic interaction, whereas charges in motion produce magnetic interaction. The magnetic field produced by a current carrying conductor interacts with a *magnetic interaction.* The magnetic field produced by a current carrying conductor interacts with a magnetic needle placed closed to it which gets deflected. If such a conductor carrying current is placed in place of the magnetic needle, then the magnetic field produced by the two interact with each other. If in place of the magnetic needle, we consider an electron beam it is also deflected. We have learnt about the steady state problems in electrostatics and magnetostatics considering electric and magnetic phenomena independent of each other. A stationary charge produce an electrostatic field whereas a moving charge is associated with an electric and a magnetic fields, *such fields have space and time variation*. This phenomenon is said to be **electromagnetism**.

7.2 SCALAR AND VECTOR FIELDS

A point function is a physical quantity, which may be expressed as a continuous function of the position of a point in a region of space.

The region in which the physical quantity is specified is known as *field.*

Scalar field: When a scalar physical quantity is expressed by a continuous scalar point function $\phi\ (x, y\ , z)$, then the region (x, y, z) specified by ϕ is the *scalar field* and the function ϕ is a *scalar point function* or *scalar field function.* Now if the temperature of a room varies from one position to another inside the room, then it may be expressed as a function of space coordinates (x, y, z). This temperature is the scalar field and may be expressed as

$$T = T\ (x, y, z) = T\ (\mathbf{r})$$

where **r** is the position vector of a point (x, y, z).

Vector field. When a vector physical quantity is expressed by a continuous vector point function $\mathbf{A}\ (x, y\ , z)$ then the region specified by this function is a *vector field.* For example, the velocity **v** of the wind if varying from place to place, then

$$\mathbf{v} = \mathbf{v}\ (x, y\ , z) = \mathbf{v}\ (\mathbf{r})$$

is known as a *vector field.*

7.3 GRADIENT OF A SCALAR FIELD

As we known that the value of a scalar function is constant at a fixed point in space, so the gradient of a scalar function ϕ is determined as

$$\begin{aligned} \text{grad}\ \phi &= i\,\frac{\partial \phi}{\partial x} + j\,\frac{\partial \phi}{\partial y} + k\,\frac{\partial \phi}{\partial z} \\ &= \left(i\frac{\partial}{\partial x} + j\frac{\partial}{\partial y} + k\frac{\partial}{\partial z}\right)\phi \\ &= \nabla\ \phi \qquad \text{...(1)} \end{aligned}$$

where ∇ is an operator 'del'.

Physical Significance of Gradient

A scalar field may be represented by a series of level surfaces each having a constant value of scalar point function ϕ. Examples of these surfaces is isothermal, equidensity and equipotential surfaces. The ϕ changes by a constant value as we move from one surface to another. These surfaces are known as Gaussian surfaces. Since ϕ is a single valued function, the two surfaces cannot intersect each other. Now let the two such surfaces are very close together, be represented by two scalar point functions ϕ and $(\phi + d\phi)$. Let **r** and $(\mathbf{r} + d\mathbf{r})$ be the position vectors of points A and B on the surfaces ϕ and $(\phi + d\phi)$ respectively with respect to an origin O as shown in Figure 7.1. Clearly the vector AB will be $d\mathbf{r}$. Let the minimum distance between the two surfaces dn be in the direction of unit normal vector $\hat{n}$ at A.

From Figure 7.1 it is clear that

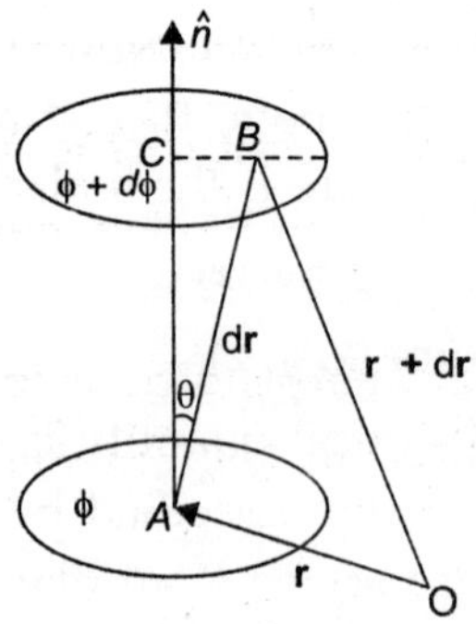

FIGURE 7.1

$$dn = dr \cos \theta$$

$$= |\hat{n}|\,|dr| \cos \theta = \hat{n}.d\mathbf{r}$$

$$\therefore \qquad d\phi = \frac{\partial \phi}{\partial n} dn = \frac{\partial \phi}{\partial n} \hat{n}.d\mathbf{r} \qquad \text{...(1)}$$

Since the continuous scalar function defining the level surfaces (Gaussian surfaces) has a value ϕ at point A (x, y, z) and $(\phi + d\phi)$ at point $(x + dx, y + dy, z + dz)$, we have

$$d\phi = \frac{\partial \phi}{\partial x} dx + \frac{\partial \phi}{\partial y} dy + \frac{\partial \phi}{\partial z} dz$$

$$= \left(i\frac{\partial \phi}{\partial x} + j\frac{\partial \phi}{\partial y} + k\frac{\partial \phi}{\partial z} \right) . (idx + jdy + kdz)$$

$$= \nabla\phi . \, d\mathbf{r} \qquad \text{...(2)}$$

From equations (1) and (2), equating the values of $d\phi$,

we obtain $\nabla\phi . \, d\mathbf{r} = \dfrac{\partial \phi}{\partial n} \hat{n}.d\mathbf{r}$

As $d\mathbf{r}$ is any arbitrary vector, we have

$$\nabla\phi = \frac{\partial \phi}{\partial n} \hat{n}$$

or

$$\boxed{\text{grad } \phi = \frac{\partial \phi}{\partial n} \hat{n}}$$

But $\frac{\partial \phi}{\partial n}\hat{n}$ is the maximum rate of increase for a scalar function ϕ. Therefore, the gradient of a scalar field at any point is a vector field, the magnitude of which is equal to the maximum rate of increase of ϕ at that point and the direction of it is same as that of normal to the level surface at that point.

7.4 EQUIPOTENTIAL SURFACES

All points on an equipotential surface have the same electric potential (i.e., the same charge).

- The electric force neither helps nor hinders motion of an electric charge along an equipotential surface.
- Electric field lines are always perpendicular to an equipotential surface.
- Electric potential is analogous to altitude; one can make maps of each in every similar ways.

An **equipotential surface** is just what its name declares: a surface-real or imaginary-on which all points have the same electric potential.

(*i*) All points on a sphere around a point charge have the same distance r, so all have same potential.

$$V = \frac{K e q}{r}$$

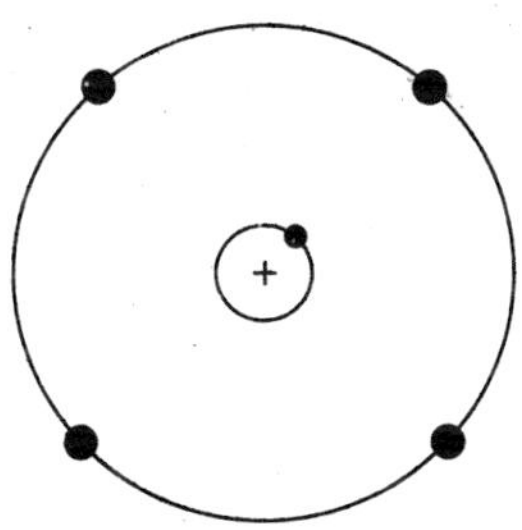

FIGURE 7.2 **Sphere around a point charge**

(*ii*) All points on a plane parallel to the plates a conductor have the same potential (Figure 7.3).

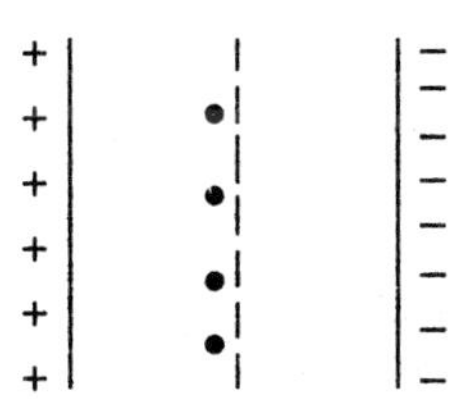

FIGURE 7.3 **Plates of conductor**

What's so great about equipotential surfaces?

— the electric field is always perpendicular to an equipotential surface.

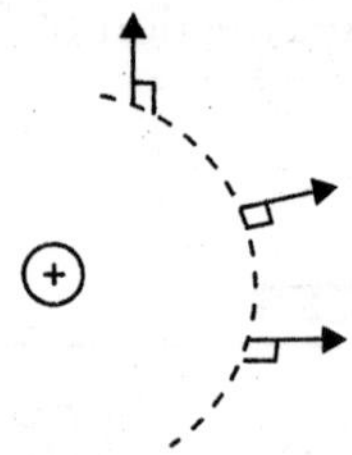

FIGURE 7.4 **Electric field on equipotential surface**

Which implies that a charged particle moving along an equipotential surface always moves perpendicular to the electric force so

— the electric force does no work on particle moving along an equipotential surface. and conversely, one need not do any work to move the particle *i.e.,* particles may move freely.

The Gravity Analogy (again)

In the earth's gravitational field, equipotential surfaces are simply flat, level areas: area of constant **altitude**.

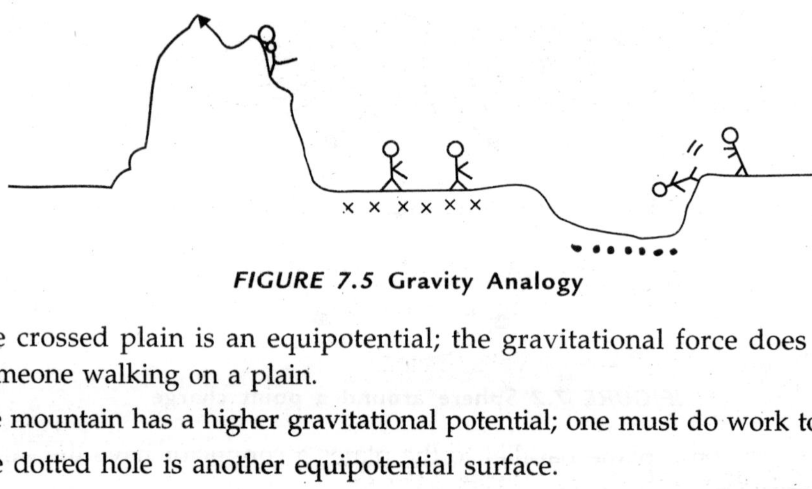

FIGURE 7.5 **Gravity Analogy**

— the crossed plain is an equipotential; the gravitational force does no work on someone walking on a plain.

— the mountain has a higher gravitational potential; one must do work to climb up it.

— the dotted hole is another equipotential surface.

Topographic maps use lines of constant altitudes (gravitational equipotentials) to show the "shape" of the landscape. (Figure 7.6)

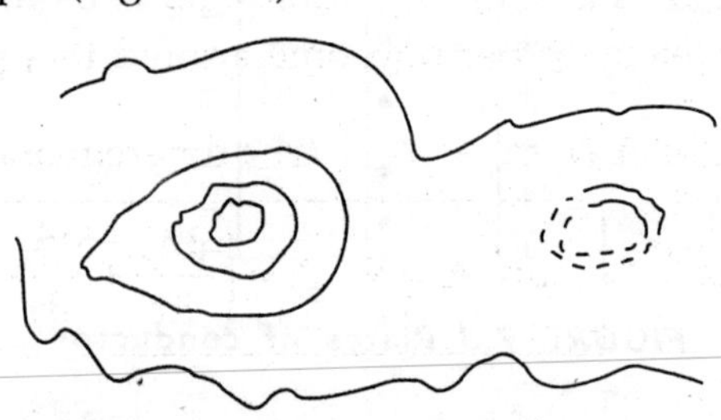

FIGURE 7.6 **Topographic maps**

One can also make maps the electric potential, using lines of equipotential. (Figure 7.7)

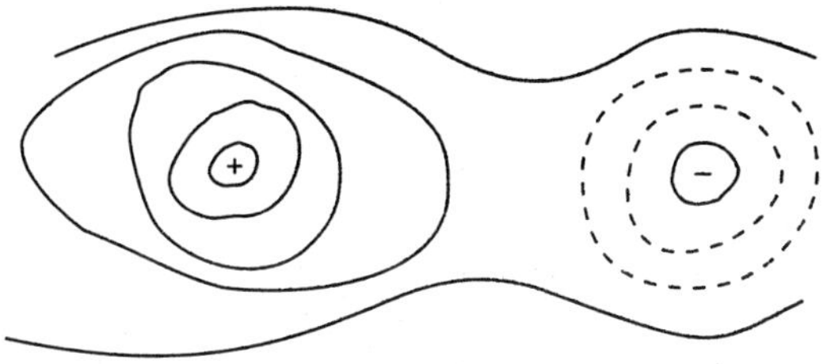

FIGURE 7.7 **Maps of Electric Potential**

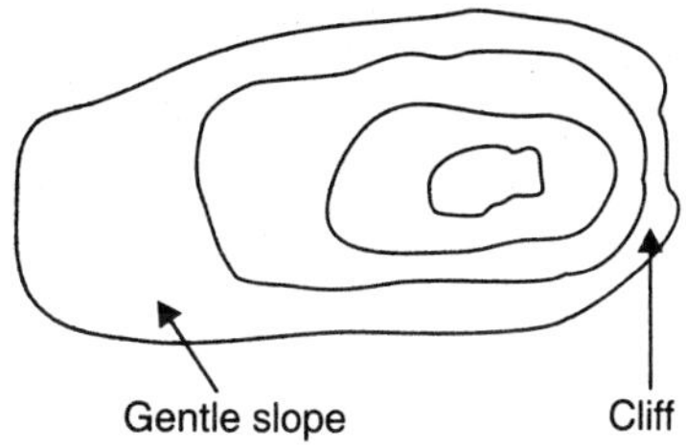

FIGURE 7.8 **(*a*) Topo map**

For small values of radius the equipotential cylindry about each line charge are nearly concentric, with line charges as centre.

On a topo map, lines close together indicate a sharp change in altitude. One can measure the steepness of ground by

$$\text{gradient} = \frac{\text{Change in altitude}}{\text{distance travelled}}$$

On a map of electric field, lines close together indicate a strong electric field. (Figure 7.8 *b*)

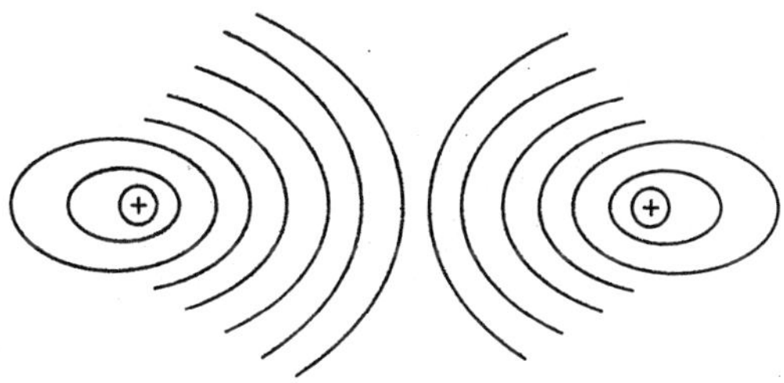

FIGURE 7.8 **(*b*) Closer lines indicates strong electric field**

$$\text{electric field} = \frac{\text{Change in electric potential}}{\text{distance travelled}}$$

$$= \frac{\Delta V}{\Delta x} \text{ Volts/meter.}$$

7.5 DIVERGENCE OF A VECTOR FIELD

The divergence of a vector field at a point is a scalar and is defined as the amount of flux diverging from a unit volume element per unit time around that point.

The divergence of a vector **A** (= $iA_x + jA_y + kA_z$) differentiable at each point (x, y, z) in a region of space is defined as

$$\boxed{\text{div } \mathbf{A} = \nabla \cdot \mathbf{A} = \left(\frac{\partial A_x}{\partial x} + \frac{\partial A_y}{\partial y} + \frac{\partial A_z}{\partial_z}\right).}$$

Physical Significance of Divergence or **Divergence in Cartesian Coordinates (or Expression for the Divergence of a Vector Field).**

Let there be a small rectangular parallelopiped *ABCDEFGH* with its centre at *O* (x, y, z) and the sides having lengths dx, dy and dz are parallel to x, y and z-axes as shown in Figure 7.9.

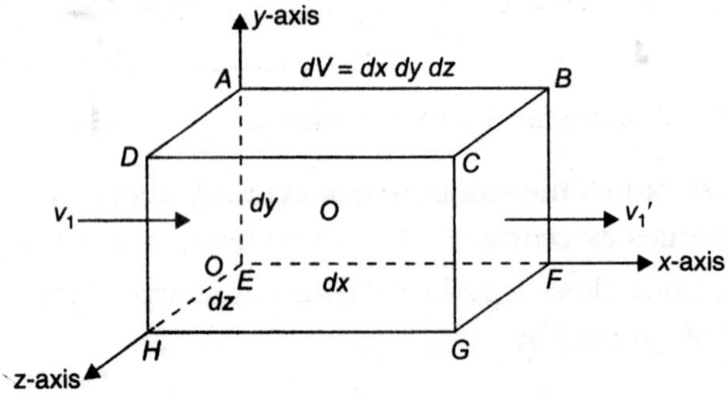

FIGURE 7.9

Let a vector $\mathbf{v} = i\, v_x + j\, v_y + k\, v_z$ represent the velocity of the fluid at the centre *O* of parallelopiped. The rate of change of flow of the liquid in a direction of x-axis may be given as $\dfrac{\partial v_x}{\partial x}$.

Thus the magnitude of component of v along x-axis at the centre of face *BFGC*, *i.e.*, v_1' is given by

$$v_1' = v_x + \frac{1}{2}\frac{\partial v_x}{\partial x}dx \qquad ...(1)$$

Similarly the magnitude of the component of v along x-axis at the centre of the face *AEHD*, *i.e.*, v_1 is given by

$$v_1 = v_x - \frac{1}{2}\frac{\partial v_x}{\partial x}dx \qquad ...(2)$$

Here negative sign is taken because the face *AEHD* is in negative direction of the centre *O* and the magnitude of v along x-axis is v_x.

But the volume of the fluid passing through a face per unit time is defined as,

Flow of volume per second = (Normal component of velocity to the face) × (Area of face)

Therefore, the net fluid entering the face *AEHD* per unit time

$$= \left(v_x - \frac{1}{2}\frac{\partial v_x}{\partial x}dx\right) dy\, dz$$

Similarly the fluid leaving the face *BFGC* per unit time

$$= \left(v_x + \frac{1}{2}\frac{\partial v_x}{\partial x}dx\right) dy\, dz$$

Thus the excess of fluid leaving the parallelopiped along x-axis is

$$= \left(v_x + \frac{1}{2}\frac{\partial v_x}{\partial x}dx\right) dydz - \left(v_x - \frac{1}{2}\frac{\partial v_x}{\partial x}dx\right) dydz$$

$$= \frac{\partial v_x}{\partial x}\, dxdydz$$

In a similar manner, one can find the values of the fluid leaving parallelopiped along y and z-axes as

$$= \frac{\partial v_y}{\partial y}\, dxdydz \text{ and } \frac{\partial v_z}{\partial z}\, dxdydz$$

Therefore, the net volume of fluid leaving or diverging or moving out of parallelopiped per unit time.

$$= \left(\frac{\partial v_x}{\partial x} + \frac{\partial v_y}{\partial y} + \frac{\partial v_z}{\partial z}\right) dxdydz$$

where $dxdydz$ be the volume of parallelopiped. Thus the amount of diverging per unit volume, which is defined as divergence of **v** may be given as

$$\text{div } \mathbf{v} = \frac{\partial v_x}{\partial x} + \frac{\partial v_y}{\partial y} + \frac{\partial v_z}{\partial z}$$

We know the differential operator ∇ is defined as

$$\nabla = i\frac{\partial}{\partial x} + j\frac{\partial}{\partial y} + k\frac{\partial}{\partial z}$$

Therefore, the divergence of a vector field **v** may also be defined as

$$\text{div } \mathbf{v} = \nabla.\mathbf{v} = \left(i\frac{\partial}{\partial x} + j\frac{\partial}{\partial y} + k\frac{\partial}{\partial z}\right).\ (i\, v_x + j\, v_y + k\, v_z)$$

$$= \frac{\partial v_x}{\partial x} + \frac{\partial v_y}{\partial y} + \frac{\partial v_z}{\partial z}$$

The divergence being the scalar product of two vectors is scalar. The following conclusions may be drawn from the above expression:

(*i*) When the div **v** or ∇.**v** is positive then either the fluid is undergoing expansion or point itself is the source of fluid.

(*ii*) When the div **v** is negative then either the fluid is undergoing contraction or the point is the sink of the fluid.

(*iii*) When div **v** = 0, then the volume of the fluid entering and leaving are same and hence we can say that the fluid is incompressible.

7.6 THE CURL OF A VECTOR FIELD

The vector product or cross product of the differential operator ∇ with a vector is known as the curl of that vector,

Thus,

$$\text{curl } \mathbf{A} = \nabla \times \mathbf{A} = \begin{vmatrix} i & j & k \\ \dfrac{\partial}{\partial x} & \dfrac{\partial}{\partial y} & \dfrac{\partial}{\partial z} \\ A_x & A_y & A_z \end{vmatrix} \quad ...(1)$$

where, $\mathbf{A} = i\,A_x + j\,A_y + k\,A_z$.

Hence, curl of a vector point function **A** is the vector product of the del operator with **A**.

The curl of a vector **A** may be expressed as

$$\text{curl } \mathbf{A} = \sum i \times \frac{\partial \mathbf{A}}{\partial x} = i \times \frac{\partial \mathbf{A}}{\partial x} + j \times \frac{\partial \mathbf{A}}{\partial y} + k \times \frac{\partial \mathbf{A}}{\partial z}$$

If curl of a vector field **F** about a point may also be defined as the circulation per unit surface as the surface shrinks to zero. In terms of line integral it is defined as

$$\text{Curl } \mathbf{F}.\hat{n} = \underset{\Delta S \to 0}{\text{Lt}} \frac{\oint \mathbf{F}.d\mathbf{l}}{\Delta S}$$

where $\hat{n}$ is the unit vector normal to area ΔS.

This equation gives the component of the curl which is in the direction of unit normal to ΔS. The net circulation or rotation of the field will be a vector resultant of the circulation in the three orthogonal planes.

Physical Significance of Curl/The Curl in a Cartesian Co-ordinates

Now considering an area of simple shape, *ABCD* in the *x-y* plane. Let the sides

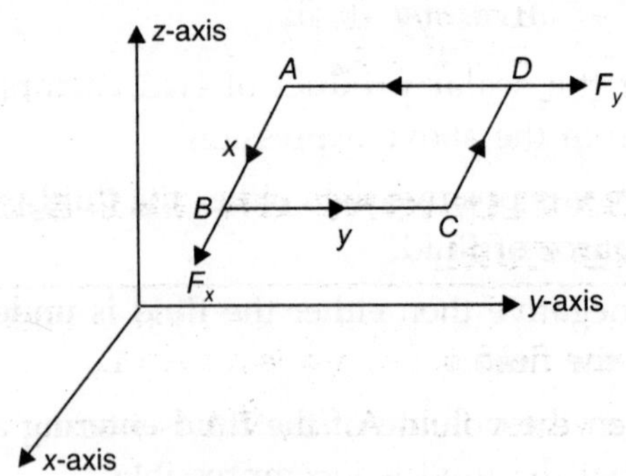

FIGURE 7.10

$$AB = CD = \Delta x \text{ and } BC = AD = \Delta y$$

Let F_x and F_y be the components of F along AB and AD respectively. Then,

$$(F_x)_{DC} = (F_x)_{AB} + \frac{\partial F_x}{\partial y}\Delta y$$

Similarly,
$$(F_y)_{BC} = (F_y)_{AD} + \frac{\partial F_y}{\partial x}\Delta x$$

The line integral over a closed cycle $ABCD$, therefore, is given by

$$\oint_{C(x,y)} \mathbf{F}.d\mathbf{l} = (F_x)_{AB}\,\Delta x + (F_y)_{BC}\,\Delta y + (F_x)_{CD}\,\Delta x + (F_y)_{DA}\,\Delta y$$

or,
$$\oint_{C(x,y)} \mathbf{F}.d\mathbf{l} = F_x\,\Delta x + \left(F_y + \frac{\partial F_y}{\partial x}\Delta x\right)\Delta y - \left(F_x + \frac{\partial F_x}{\partial y}\Delta y\right)\Delta x - F_y\,\Delta y$$

or,
$$\oint_{C(x,y)} \vec{\mathbf{F}}.d\vec{\mathbf{l}} = \left(\frac{\partial F_y}{\partial x} - \frac{\partial F_x}{\partial y}\right)\Delta x\,\Delta y$$

or,
$$(\text{curl } \mathbf{F})_z = \frac{\partial F_y}{\partial x} - \frac{\partial F_x}{\partial y}$$ [From equation (1)]

Similarly,
$$(\text{curl } \mathbf{F})_x = \frac{\partial F_z}{\partial y} - \frac{\partial F_y}{\partial z}$$

and
$$(\text{curl } \mathbf{F})_y = \frac{\partial F_x}{\partial z} - \frac{\partial F_z}{\partial x}$$

Combining the components,

$$(\text{curl } \mathbf{F}) = i\left(\frac{\partial F_z}{\partial y} - \frac{\partial F_y}{\partial z}\right) + j\left(\frac{\partial F_x}{\partial z} - \frac{\partial F_z}{\partial x}\right) + k\left(\frac{\partial F_y}{\partial x} - \frac{\partial F_x}{\partial y}\right)$$

or
$$\nabla \times \mathbf{F} = \begin{vmatrix} i & j & k \\ \dfrac{\partial}{\partial x} & \dfrac{\partial}{\partial y} & \dfrac{\partial}{\partial z} \\ F_x & F_y & F_z \end{vmatrix}.$$

7.7 IMPORTANT CONCLUSIONS ON THE BASIS OF GRADIENT, DIVERGENCE AND CURL

- The gradient of scalar field ϕ, *i.e.*, $\nabla\phi$ is a vector.
- The divergence of a vector field **A**, *i.e.*, $\nabla.\mathbf{A}$ is a scalar.
- The curl of a vector field $\vec{\mathbf{A}}$, *i.e.*, $\nabla \times \mathbf{A}$ is a vector.

- The divergence of the gradient of a scalar field ϕ may be given as

$$\text{div (grad } \phi) = \nabla.(\nabla\, \phi) = \nabla^2\phi = \left(\frac{\partial^2}{\partial x^2} + \frac{\partial^2}{\partial y^2} + \frac{\partial^2}{\partial z^2}\right)\phi$$

where ∇^2 is known as the Laplacian operator.

- The curl of the gradient of a scalar field is zero, *i.e.*,

$$\text{curl (grad } \phi) = \nabla \times (\nabla\, \phi) = 0.$$

It is because the cross-product of two similar vectors (∇) is zero.

- The divergence of curl of a vector field is zero, *i.e.*,

$$\text{div (curl } \mathbf{A}) = \nabla \,.\, (\nabla \times \mathbf{A}) = 0$$

It is because the vector $\nabla \times \mathbf{A}$ is perpendicular to ∇ and therefore the dot product of ∇ with a vector perpendicular to it will be zero.

- The curl of the curl of a vector field is defined as

$$\text{curl (curl } \mathbf{A}) = \nabla \times (\nabla \times \mathbf{A}) = \nabla\,(\nabla.\mathbf{A}) - \nabla^2\mathbf{A}$$
$$= \text{grad (div } \mathbf{A}) - \nabla^2\,\mathbf{A}$$

- **Solenoidal Vector Point Function**

When the divergence of a vector point function **A** is zero, *i.e.*,

$$\text{div } \mathbf{A} = \nabla.\mathbf{A} = 0$$

then it is said to be solenoidal vector point function. Thus the vector point function will be solenoidal over a region, if the flux across any closed surface in that region is zero.

- **Irrotational Vector Point Function**

If the curl of a vector point function is zero it is said to be irrotational *i.e.*, when

$$\text{curl } \mathbf{A} = \nabla \times \mathbf{A} = 0$$

then **A** is irrotational.

7.8 LINE, SURFACE AND VOLUME INTEGRALS

Line Integral: *The integration of a vector along a curve is called its line integral.* As shown in Figure 7.11, let *MN* is a curve drawn between two points *M* and *N* in vector field. Let $d\mathbf{l}$ is an element of length along the curve *MN* at *O*. Suppose **A** is the vector at *O*, making an angle θ with the direction of $d\mathbf{l}$.

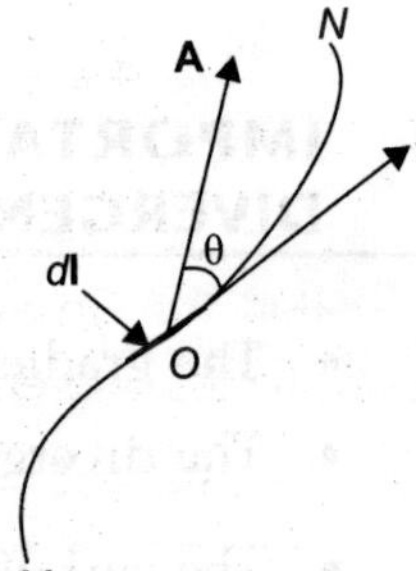

FIGURE 7.11

Then $$\mathbf{A}\,.d\mathbf{l} = A\, dl \cos\theta$$
$$= dl\,(A\cos\theta) \qquad ...(1)$$

Equation (1) shows that the value of **A**.d**l** at any point of the curve is equal to the product of small element dl and the component (A cos θ) of **A** along the direction of d**l**. The value of **A**.d**l** for the complete curve MN can be obtained by integrating Equation (1). Hence

$$\int_M^N \mathbf{A}.d\mathbf{l} = \int_M^N dl\,(A\cos\theta) \qquad ...(2)$$

Integral $\int_M^N \mathbf{A}.d\mathbf{l}$ is defined as the line integral of **A** along the curve MN.

Examples.

(*i*) *If* **A** *represents the force acting on a particle moving along the curve from M to N, then the line integral* $\int_M^N \mathbf{A}.d\mathbf{l}$ *represents the total workdone by the force during the motion of the particle over its entire path from M to N.*

(*ii*) *If* **A** *denotes the electric field intensity at any point, then the line integral represents the potential difference between M and N.*

Surface Integral

Consider a simple surface S in a vector field bounded by a curve as shown in Figure 7.12. Let d**S** be an infinitesimal element of the surface. The surface element of area can be represented by area vector d**S**. If $\hat{n}$ be a unit positive vector (drawn outward the surface) in the direction of dS, then

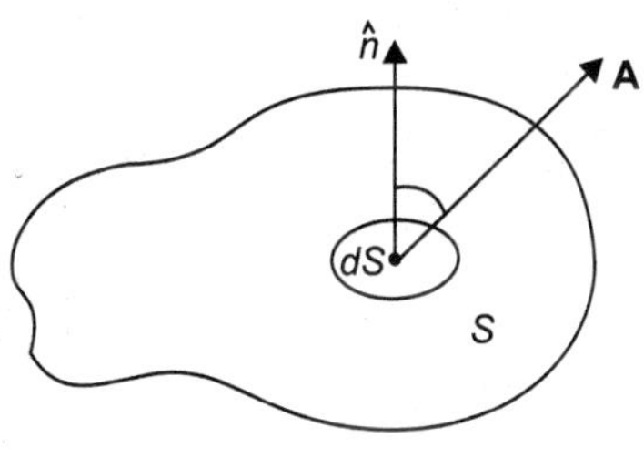

FIGURE 7.12

$$d\mathbf{S} = \hat{n}\,dS$$

Let **A** be a vector at middle of the element dS in the direction making an angle θ with $\hat{n}$.

Now the scalar product

$$\vec{\mathbf{A}}.d\vec{\mathbf{S}} = \vec{\mathbf{A}}.\hat{n}\,dS = AdS\cos\theta$$

is called the flux of vector field $\vec{\mathbf{A}}$ across the area element dS. The total flux of the vector field across the entire surface S is given by

$$\iint_S \mathbf{A}.d\mathbf{S} = \iint_S \mathbf{A}.\hat{n}\,dS = \iint_S A\cos\theta\,dS$$

This is defined as surface integral.

Volume Integral

Consider a closed surface in space enclosing the volume V. If **A** be a vector point function at a point in a small element of volume dV, then the integral

$$\iiint \mathbf{A}\, dV$$

is called the volume integral of vector **A** over the surface.

7.9 GAUSS DIVERGENCE THEOREM

Statement: *"The volume integral of the divergence of a vector field* **A** *taken over any volume V bounded by a closed surface S is equal to the surface integral of* **A** *over the surface S."*

Or

"The volume integral of the divergence of a vector field over the volume enclosed by surface S is equal to the flux of that vector field taken over that surface S."

Hence, this theorem is used to convert volume integral into surface integral.

Mathematically,

$$\iiint_V \text{div}\, \mathbf{A}\, dV = \iiint_V (\nabla.\mathbf{A})\, dV = \iint_S \mathbf{A}.d\mathbf{S}$$

Proof : Let a volume V be enclosed a surface S of any arbitrary shape.

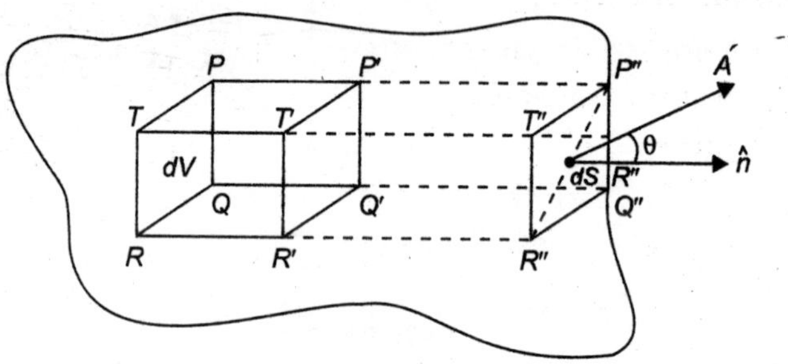

FIGURE 7.13

Let a small volume element $PQRTT'P'Q'R'$ of volume dV lies within surface S as shown in Figure 7.13. As we know that flux diverging per unit volume per second is given by div **A**; therefore, for volume element dV the flux diverging will be div **A**dV. If V be the volume enclosed by the surface S, then the total flux diverging through volume V will be equal to the volume integral.

$$\iiint_V \text{div}\, \mathbf{A}\, dV \qquad ...(1)$$

Now consider a surface element dS and $\hat{n}$ a unit vector normal to dS. Let θ be the angle between **A** and $\hat{n}$ at dS, then **A**.d**S** will give the flux through the surface element dS. Where $d\mathbf{S} = \hat{n}\, dS$ = area vector along $\hat{n}$. Therefore, the total flux passing through the surface S may be obtained by the integral.

$$\iint_S \mathbf{A}.\hat{n}\,dS = \iint_S \mathbf{A}.d\mathbf{S} \quad ...(2)$$

But the total flux through the entire surface S [given by Equation (2)] must be equal to the total flux diverging from the volume V enclosed by surface S [given by Equation (1)] and therefore

$$\boxed{\iiint_V \text{div}\,\mathbf{A}\,dV = \iint_S \mathbf{A}.d\mathbf{S}.}$$

7.10 STOKE'S CURL THEOREM

Statement: *"The surface integral of the curl of a vector field* **A** *taken over any surface S is equal to the line integral of* **A** *around the closed curve forming the periphery of the surface S."*

Or

"The flux of the curl of a vector field **A** *over any closed surface S of any arbitrary shape is equal to the line integral of vector* **A**, *taken over the boundary of that surface."*

Mathematically it may be represented as

$$\iint_S (\nabla \times \mathbf{A})\,.\,d\mathbf{S} = \iint_S \text{curl}\,\mathbf{A}\,.\,d\mathbf{S} = \oint \mathbf{A}\,.\,d\mathbf{l}$$

Hence this theorem is used to convert surface integral into line integral.

Proof: Let us consider a closed curve $C_1C_2C_3C_4C_1$ enclosing a surface area S in a vector field **A** as shown in Figure 7.14.

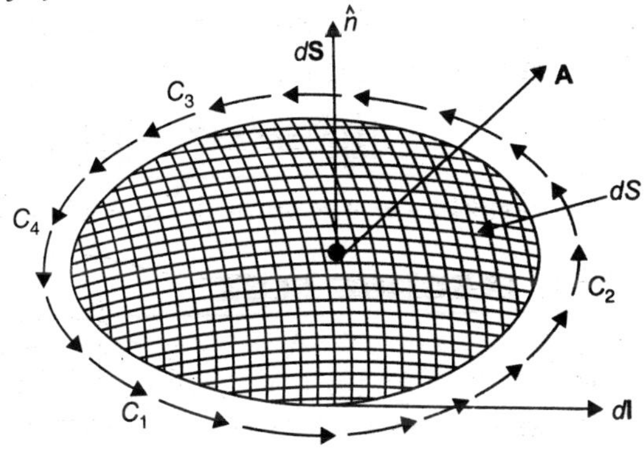

FIGURE 7.14

The line integral of **A** over the boundary of the closed curve $C_1C_2C_3C_4C_1$ may be given as

$$\oint \mathbf{A}\,.\,d\mathbf{l} \quad ...(1)$$

where dl is the small line element of the boundary of curve $C_1C_2C_3C_4C_1$.

Now when the whole surface is divided into large number of infinitesimal rectangular clements each of surface area dS, then the vector area of each element will be $\hat{n}\,dS = d\mathbf{S}$, where $\hat{n}$ is unit vector along the normal on dS.

But the curl of a vector field **A** at any point in the field is equal to the *line integral of that field per unit area taken along the boundary of an infinitesimal area drawn around that point. The direction of the line integral is normal to the surface.*

Therefore, the line integral of **A** along the boundary of each of the rectangular element dS may be written as

$$(\text{curl}\,\mathbf{F})\,.\,d\mathbf{S}$$

But the line integrals taken along the common sides of the continuous element (dS) mutually cancel each other due to being equal in magnitude and opposite in direction as shown in Figure 7.15, therefore, if we want to find out total line integral for all dS elements, then only the sides of the elements lying on the boundary or periphery of the surface S will contribute to the line integral.

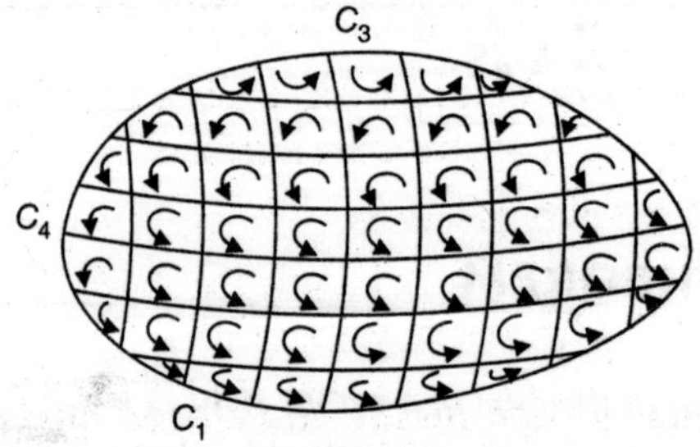

FIGURE 7.15

Thus the total line integral of the vector **A** along the boundary $C_1C_2C_3C_4C_1$ of the surface S

i.e., $\iint_S (\text{curl}\,\mathbf{F}).d\mathbf{S}$ will be equal to the $\oint \mathbf{F}\,.\,d\mathbf{l}$,

i.e.,
$$\iint_S (\text{curl}\,\mathbf{F}).d\mathbf{S} = \iint_S (\nabla \times \mathbf{F}).d\mathbf{S} = \oint \mathbf{F}\,.\,d\mathbf{l}.$$

7.11 ELECTRIC FLUX

The electric flux (ϕ) is defined as total number of electric lines of force passing normally through a surface in the electric field. ϕ is a scalar quantity. For a closed surface, placed in electric field, the flux is negative when the lines of force enter into the surface and positive when they leave the surface. In order to find the total flux linked with a surface, the surface is divided into infinitesimal small parts and sum of electric flux through all such small elements is find out.

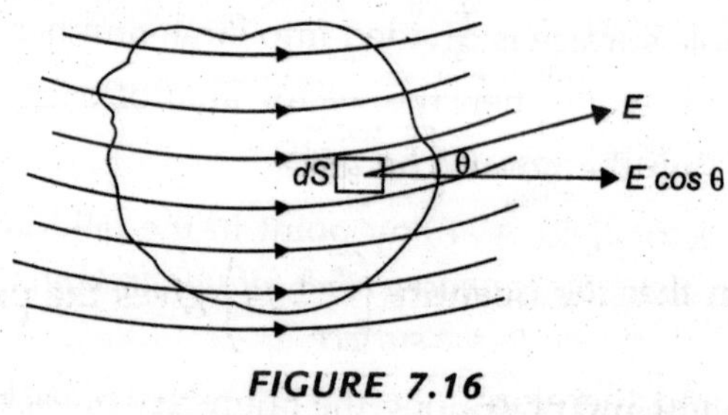

FIGURE 7.16

The flux linked with an element of surface area dS is equal to the product of the component of electric field **E** normal to the surface and area of the element, *i.e.,*

$$d\phi = E \cos\theta\, dS = \mathbf{E}.d\mathbf{S}$$

Therefore, the total electric flux linked through the entire surface S is then given by

$$\phi = \oint_S E \cos \theta \, dS = \oint_S \mathbf{E} \cdot d\mathbf{S}.$$

7.12 GAUSS'S LAW OF ELECTROSTATICS

Statement : "The total normal electric flux over a closed surface in an electric field is equal to $1/\varepsilon_0$ times the total charge enclosed by that surface."

Mathematically it may be expressed as

$$\phi = \iint_S \mathbf{E} \cdot d\mathbf{S} = \oint_S \mathbf{E} \cdot d\mathbf{S} = \frac{1}{\varepsilon_0} \Sigma q.$$

Proof. (*i*) **When the charge lies inside the closed surface:** Let us consider a source producing the field is a point charge + q situated at O inside the closed surface as shown in Figure 7.17.

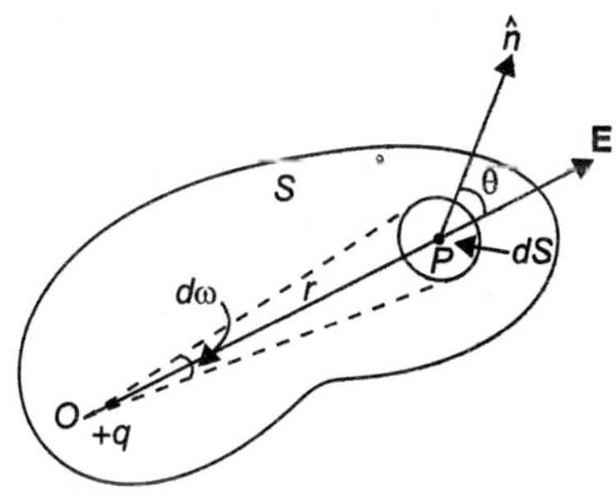

FIGURE **7.17**

Let dS be an infinitesimal element of the surface at point P and $OP = r$. As shown in Figure 7.17, the electric field strength vector $\mathbf{E}$ makes angle θ with the unit vector $\hat{n}$ drawn normal to the surface element dS surrounding point P. The surface integral of the normal component of this electric field over the closed surface is given by $\iint_S \mathbf{E} \cdot \hat{n} \, dS$.

The electric field strength E at the point P is given by

$$\mathbf{E} = \frac{1}{4\pi\varepsilon_0} \frac{q}{r^3} \mathbf{r}$$

$$\therefore \qquad \iint_S \mathbf{E} \cdot \hat{n} \, dS = \frac{1}{4\pi\varepsilon_0} q \iint_S \frac{r \cdot \hat{n}}{r^3} dS \qquad \text{...(1)}$$

Now, it can be seen that the quantity $\left(\frac{\mathbf{r} \cdot \hat{n}}{r} dS \right)$ gives the projection of area dS on the plane perpendicular to $\mathbf{r}$ and therefore.

$$\frac{\text{projected area}}{r^2} = \frac{\mathbf{r} \cdot \hat{n} \, dS}{r^3} = d\omega \qquad \text{...(2)}$$

where $d\omega$ is the solid angle subtended by dS at O.

From equations (1) and (2), we get

$$\iint_S \mathbf{E}.\hat{n}\, dS = \frac{1}{4\pi\varepsilon_0}\, q \iint_S d\omega \qquad ...(3)$$

But $\iint_S d\omega = 4\pi$ = solid angle subtended by entire closed surface at an internal point

$$\therefore \qquad \iint_S \mathbf{E}.\hat{n}\, dS = \frac{1}{4\pi\varepsilon_0}\, q.4\,\pi = \frac{q}{\varepsilon_0} \qquad ...(4)$$

This result is known as Gauss's law for a single point charge enclosed by the surface.

If several point charges $q_1, q_2, q_3, \ldots\ldots. q_n$ be enclosed by the surface S, then total electric field is given by $\Sigma\, q_i\ (= q_1 + q_2 + q_3 + \ldots\ldots. + q_n)$. Each charge subtends a full solid angle 4π and equation (4) becomes

$$\iint_S \mathbf{E}.\hat{n}\, dS = \frac{1}{\varepsilon_0}\sum_{i=1}^{n} q_i \qquad ...(5)$$

(*ii*) **When the charge is situated outside the closed surface:** If the charge q is outside the surface (Figure 7.18), then the surface S can be divided into areas S_1, S_2, S_3 and S_4 each of which subtends the same solid angle at the charge q. But at S_1 and S_3 the directions of the outward drawn normal are away from q, while at S_2 and S_4 they are toward q. Therefore the contributions of two pairs (S_1, S_3) and (S_2, S_4) to the surface integral are equal and opposite. As a result the net surface integral of the normal component of the electric field **E** vanishes, *i.e.*,

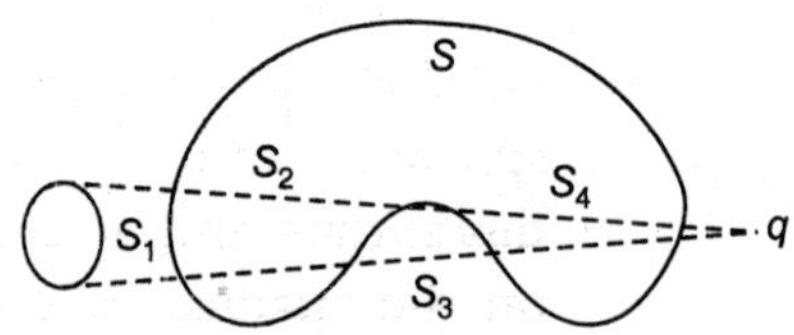

FIGURE 7.18

$$\iint_S \mathbf{E}.\hat{n}\, dS = \iint_S \mathbf{E}.d\mathbf{S} = 0 \qquad ...(6)$$

equations (4) and (6) represents the integral form of the Gauss's law.

7.12.1 Differential Form of Gauss's Law

Let a charge q be distributed over a volume V of the closed surface S and ρ be the charge density; then the charge q may be given as

$$q = \iiint_V \rho\, dV = \int_V \rho\, dV \qquad ...(1)$$

Thus the total flux through the surface S

$$\phi = \oint_S \mathbf{E}.d\mathbf{S} = \frac{1}{\varepsilon_0} q = \frac{1}{\varepsilon_0}\int_V \rho\, dV \qquad ...(2)$$

According to Gauss divergence theorem the surface integral may be converted into volume integral as

$$\oint_S \mathbf{E} \, . \, d\mathbf{S} = \int_V (\text{div.}\,\mathbf{E})\, dV = \int_V (\nabla \, . \, \mathbf{E})\, dV \qquad ...(3)$$

Hence from equation (2), we obtain

$$\int_V \text{div}\,\mathbf{E}\, dV = \frac{1}{\varepsilon_0}\int_V \rho\, dV$$

or
$$\text{div}\,\mathbf{E} = \frac{1}{\varepsilon_0}\rho$$

or
$$\nabla.\mathbf{E} = \frac{1}{\varepsilon_0}\rho \qquad ...(4)$$

As the displacement vector **D** is defined as

$\mathbf{D} = \varepsilon_0\, \mathbf{E}$, we have

$$\nabla.\mathbf{D} = \text{div}\,\mathbf{D} = \rho \qquad ...(5)$$

Equations (4) and (5) are differential form of Gauss's law of electrostatics.

7.12.2 Applications of Gauss's Law of Electrostatics

1. **Derivation of Coulomb's law of electrostatics from Gauss's law:** Consider two point charges q_1 and q_2 separated by a distance 'r'. Draw a Gaussian surface of sphere of radius 'r' with q_1 as centre. **E** must be normal to this surface and must have same magnitude for all points on the surface. **E** and $d\mathbf{S}$ at any point on the drawn Gaussian surface are directed radially outward. The angle between them is zero.

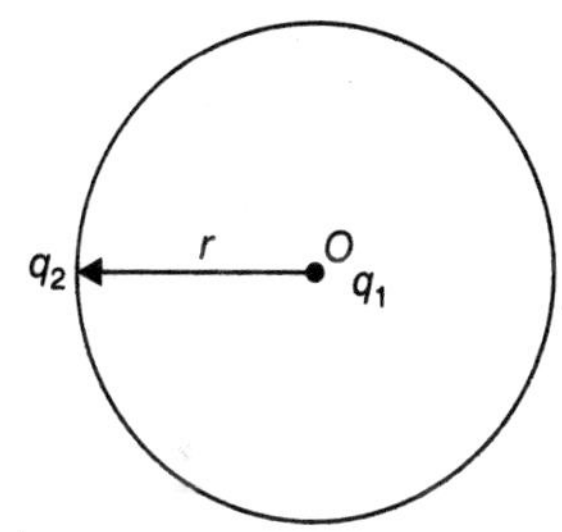

FIGURE 7.19

$\therefore$ $\mathbf{E}.\, d\mathbf{S} = E\, dS \cos\theta = E\, dS \cos 0° = E\, dS$

Hence, by Gauss's theorem

$$\oint \mathbf{E}\,.\,d\mathbf{S} = \frac{q_1}{\varepsilon_0}$$

or
$$E\oint d\mathbf{S} = \frac{q_1}{\varepsilon_0} \qquad \text{[Since } E \text{ is constant]}$$

But
$$\oint d\mathbf{S} = 4\,\pi r^2$$

$\therefore$
$$E.4\pi r^2 = \frac{q_1}{\varepsilon_0} \text{ or } E = \frac{q_1}{4\pi\varepsilon_0 r^2}$$

The force on the charge $q_2 = F = Eq_2 = \dfrac{q_1 q_2}{4\pi\varepsilon_0 r^2}$, which is the Coulomb's law.

2. **Field around a charged straight conductor (Line Charge):** Consider a linear charged conductor. Let the linear charge density be 'λ'. Imagine a Gaussian cylinder of radius 'y' and length 'l', closed at each end by plane caps normal to the axis. By symmetry, all the lines of force go radially outward.

As shown in Figure 7.20, no lines cross the flat ends of the cylinder and hence the $\int E\, dS$ over these end surfaces are zero.

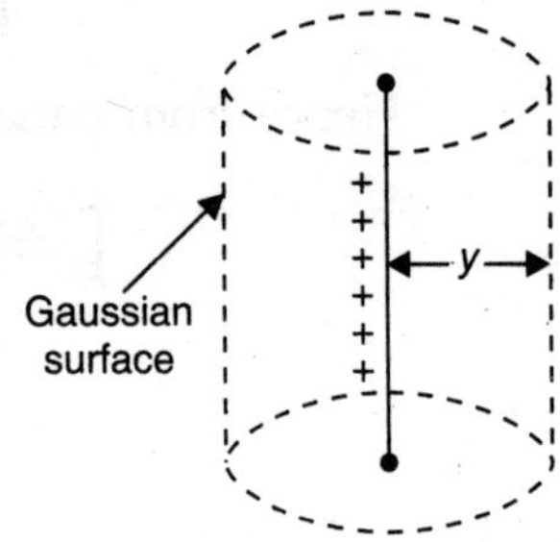

FIGURE 7.20

Over the remaining surface of cylinder, the field is uniform and outward.

∴ Applying Gauss theorem

$$\phi = \frac{q}{\varepsilon_0} = \frac{l\lambda}{\varepsilon_0} \quad \text{...(1)}$$

Net flux passing over the Gaussian surface

$$= \oint E\, dS \cos 0$$

$$= E \oint dS = E\,(2\pi y)\, l \quad \text{...(2)}$$

From equations (1) and (2), $\frac{l\lambda}{\varepsilon_0} = E\,(2\pi y)\, l$ or $\boxed{E = \frac{\lambda}{2\pi\varepsilon_0 y}}$ ...(3)

The direction of **E** is radially outward for a line of the charge.

3. Field due to a uniformly charged hollow cylinder: Let us consider a uniformly charged infinite cylinder of radius '*a*'. Let σ be the surface charge density. Consider a point *P* situated at a distance '*r*' from the axis of the cylinder. Imagine a Gaussian cylinder of radius '*r*' and length '*l*'. The surface density is given by

$$\sigma = \frac{q}{(2\pi a)\, l} \quad \text{or} \quad \boxed{q = \sigma\,(2\pi\, a\, l)}$$

where q = charge contained in the cylinder of length '*l*'.

Now, the charge per unit length of the cylinder $\lambda = \frac{q}{l}$

$$= \frac{\sigma\,(2\pi a l)}{l} = \sigma\,(2\pi a)$$

or $$\sigma = \frac{\lambda}{2\pi a}$$

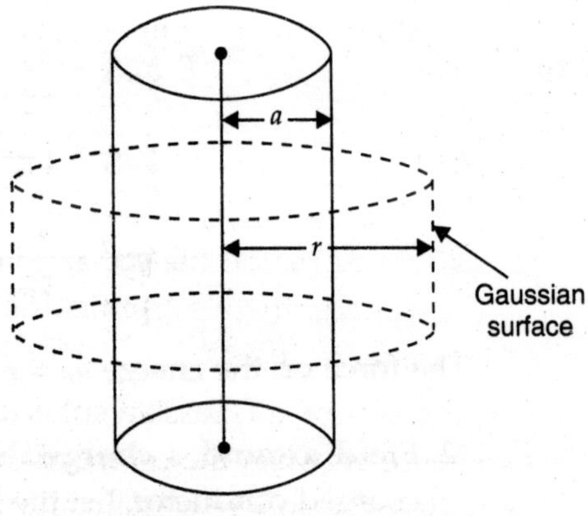

FIGURE 7.21

The total flux over the Gaussian surface

$$= E \oint dS \cos 0$$

$$= E\,(2\pi r\, l) \quad \text{...(1)}$$

But $$\frac{q}{\varepsilon_0} = \frac{\sigma\,(2\pi al)}{\varepsilon_0} = \frac{\left(\dfrac{\lambda}{2\pi a}\right)(2\pi al)}{\varepsilon_0} = \frac{\lambda l}{\varepsilon_0} \quad \ldots(2)$$

From equations (1) and (2),

$$\frac{\lambda l}{\varepsilon_0} = E\,(2\pi\, r\, l)$$

or $$\boxed{E = \frac{\lambda}{2\pi\varepsilon_0 r}}$$

4. Field due to a hollow spherical conductor: Let us consider a hollow spherical conductor of radius R with surface charge density σ. Let q be the net charge of the surface.

$\therefore$ $$q = 4\,\pi\, r^2\,\sigma.$$

(*i*) **Point outside the sphere:** Consider a point P situated outside the sphere at a distance 'r'. Imagine a Caussian spherical surface talking 'r' as radius.

Area of the Gaussian sphere = $4\pi r^2$

By Gauss theorem

Flux over the Gaussian sphere = $E \int dS$

$$= E.\,4\pi r^2 = \frac{q}{\varepsilon_0}$$

or $$\boxed{E = \frac{q}{4\pi\varepsilon_0 r^2}.}$$

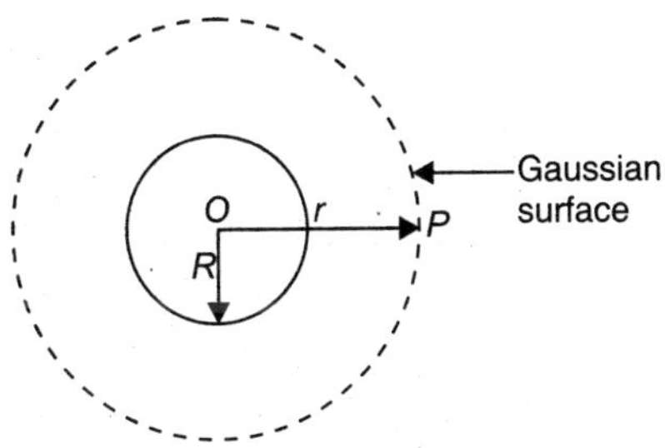

FIGURE 7.22

(*ii*) **Point on the sphere:** When point P is on the sphere $r = R$ then,

$$\boxed{E = \frac{q}{4\pi\varepsilon_0 R^2}.}$$

(*iii*) **Point inside the sphere:** For any point inside the sphere, the electric field is zero, since there is no charge inside.

If we draw a Gaussian spherical surface of radius 'r', the distance between the centre of the sphere and the point P as radius, the sphere contains no charge. Hence the field at any point inside a charged hollow sphere is zero.

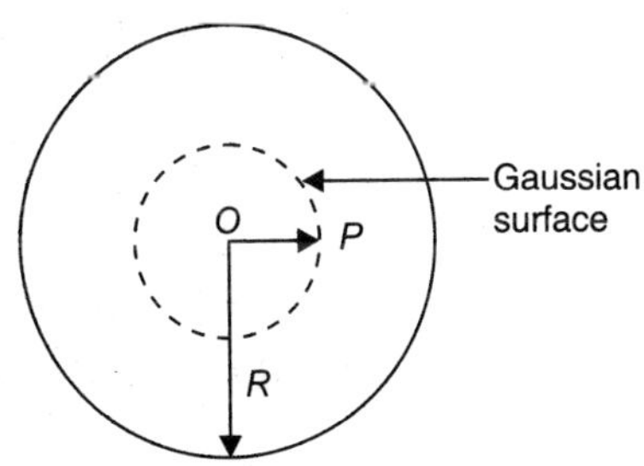

FIGURE 7.23

5. Field due to a solid non-conducting sphere: Let us consider a solid spherical conductor of radius 'R' with volume density 'ρ'. The net charge in the sphere,

$q = \frac{4}{3}\pi R^3 \rho$.

(*i*) **Point outside the sphere:** Consider a point P at a distance 'r' from the centre of the sphere. Imagine a Gaussian spherical surface with 'r' as radius.

Applying Gauss theorem

$$E\,(4\pi r^2) = \frac{q}{\varepsilon_0}$$

or

$$\boxed{E = \frac{q}{4\pi\varepsilon_0 r^2}}$$

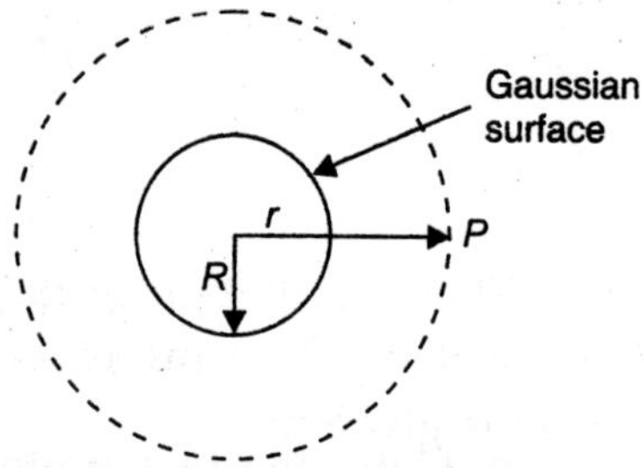

FIGURE 7.24

(*ii*) **Point on the sphere:** When the point P is on the sphere, $r = R$

Then

$$\boxed{E = \frac{q}{4\pi\varepsilon_0 R^2}}.$$

(*iii*) **Point inside the sphere:** To find the field intensity at P inside the solid sphere, consider a Gaussian spherical surface with radius 'r_1'.

The solid sphere consists of two parts :

(*i*) a solid sphere of radius 'r_1'

(*ii*) a shell of thickness '$(R - r_1)$'.

The point P is just on the surface of solid sphere of radius 'r_1'.

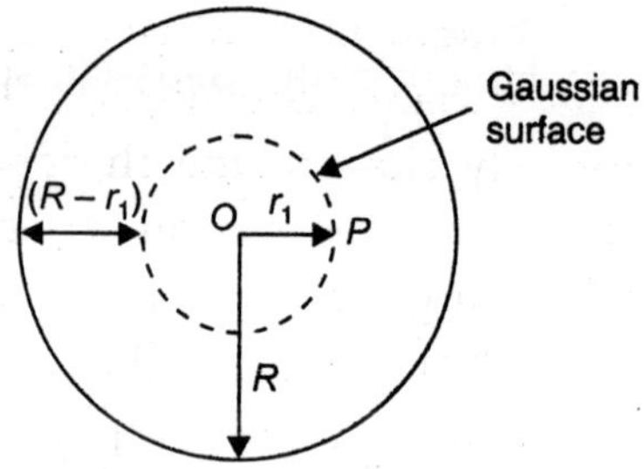

FIGURE 7.25

The field due to the solid sphere is obtained as

$$E\,(4\pi r_1^2) = \frac{q_1}{\varepsilon_0}$$

where

$$q_1 = \frac{4}{3}\pi r_1^3 \rho$$

Then

$$E\,(4\pi r_1^2) = \frac{\left(\frac{4}{3}\pi r_1^3 \rho\right)}{\varepsilon_0}$$

$\Rightarrow$

$$E = \frac{\rho r_1}{3\,\varepsilon_0}$$

But $$q = \frac{4}{3}\pi R^3 \rho,$$

or $$\rho = \frac{q}{\frac{4}{3}\pi R^3}$$

∴ $$E = \frac{\frac{q}{\frac{4}{3}\pi R^3}\cdot r_1}{3\,\varepsilon_0} = \frac{q\,r_1}{4\pi\,\varepsilon_0 R^3}$$

The contribution of the outer shell towards the field is zero, since there is no field inside a charged hollow sphere. Hence the field intensity at a point inside the charged solid sphere is given by

∴ $$\boxed{E = \frac{q\,r_1}{4\pi\,\varepsilon_0 R^3} = \frac{\rho\,r_1}{3\,\varepsilon_0}}$$

i.e., the field is directly proportional to the distance of the point from the centre of the charged sphere.

6. Electric field due to a charged conductor (Coulomb's theorem)

Statement : *The electric field at any point near a charged conductor is $1/\varepsilon_0$ times the surface density of charge on the surface.*

Proof : Let '*AB*' be a large conducting surface and σ be the surface charge density. To find the field at a point *P*, infinitely close to the charged surface, imagine a small cylinder of cross-section '*dS*' drawn with its faces parallel to the charged surface. Let one of the faces pass through *P* and the other through *Q* inside the conductor as shown in Figure 7.26. Normal component of *E*, through the sides of the cylinder, formed by faces enclosing *P* and *Q i.e.*, Gaussian surface is zero.

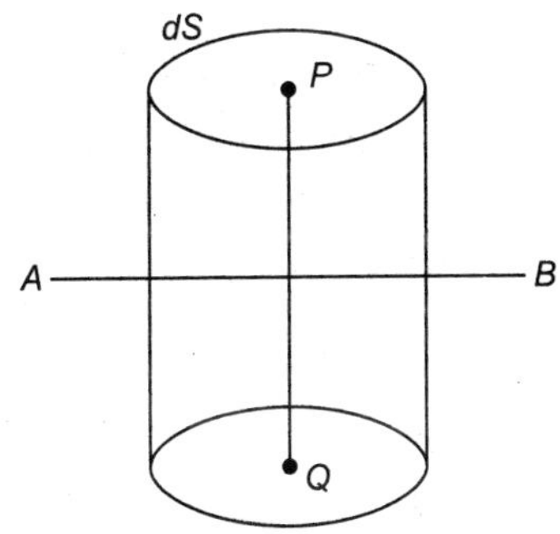

FIGURE 7.26

Since the charge resides on the surface of the conductor, there is no charge on the face enclosing '*Q*' and therefore normal component of *E* through the area *dS* is also zero. Hence the normal component of '*E*' is only through containing '*P*'.

According to Gauss theorem,

$$EdS = \frac{\sigma\,dS}{\varepsilon_0}$$

or $$\boxed{E = \frac{\sigma}{\varepsilon_0}.}$$

7.13 AMPERE'S LAW

The Biot-Savart law in magnetostatics is used to calculate the net magnetic field due to current carrying conductor. It can be applied to show that for an infinity long and straight conductor carrying current i, the magnetic field at any arbitrary point P, situated at distance R as shown in Figure 7.27.

$$B = \frac{\mu_0 i}{2\pi r} \quad \text{...(1)}$$

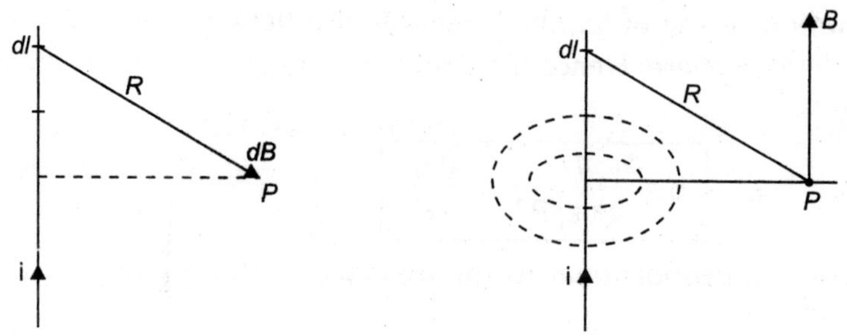

FIGURE 7.27

where μ_0 is a constant called permeability constant ($\mu_0 = 4\pi \times 10^{-7}$ T.m/A) and r is the perpendicular distance between the wire and P. The field lines of **B** are concentric circles around the wire.

If we consider any charge distribution and consider only situation of high symmetry, Ampere's law is preferred over Biot-Savart law to find the magnetic field due to charge distribution.

Ampere's circuital law (or Ampere's law) can be expressed as

$$\oint \mathbf{B}.d\mathbf{l} = \mu_0 i$$

The left hand side of equation (1) represents the dot product $\mathbf{B}.d\mathbf{l}$ integrated around a closed loop (line integral around a path of radius r) and i be the net current encircled by the loop.

Using equation (1), Ampere's law can be derived

$$\oint \mathbf{B}.d\mathbf{l} = \int \frac{\mu_0 i}{2\pi r} dl$$

$$= \frac{\mu_0 i}{2\pi r} \int dl$$

$$= \frac{\mu_0 i}{2\pi r} \times 2\pi r = \mu_0 i$$

We can extends this discussion to show that equation (1) holds good for any nature of the closed path.

Differential Form: In a region of distributed current flow, let **J** be the current density. So the current $i = \int_S \mathbf{J}.d\mathbf{S}$, if $d\mathbf{S}$ is an element of the bounded surface area.

i.e.,
$$\oint \mathbf{B}.d\mathbf{l} = \mu_0 i$$

$$= \mu_0 \int_S \mathbf{J}.d\mathbf{S} \quad ...(1)$$

Using Stoke's theorem

$$\oint \mathbf{B}.d\mathbf{l} = \int_S (\nabla \times \mathbf{B}).d\mathbf{S} \quad ...(2)$$

Equating the R.H.S. expressions of equations (1) and (2)

$$\oint (\nabla \times \mathbf{B}).d\mathbf{S} = \mu_0 \int \mathbf{J}.d\mathbf{S}$$

or
$$\int_S [(\nabla \times \mathbf{B}) - \mu_0 \mathbf{J}] . d\mathbf{S} = 0$$

or
$$\nabla \times \mathbf{B} - \mu_0 \mathbf{J} = 0$$

or
$$\text{curl } \mathbf{B} = \nabla \times \mathbf{B} = \mu_0 \mathbf{J} \quad ...(3)$$

Equation (3) is known as differential form of Ampere's law.

7.14 FARADAY'S LAW OF ELECTROMAGNETIC INDUCTION

If we consider a closed stationary circuit located in a varying magnetic field, the induced electromotive force (e.m.f.) around this circuit is equal to the negative time rate of change of magnetic flux through the circuit

$$e = -\frac{d\phi}{dt} \quad ...(1)$$

where ϕ is the total magnetic flux through the circuit. The negative sign appearing here defines another law, called Lenz's law, which indicates that the direction of current opposes the changes in the flux of B or electromagnetic inertia.

From definition of flux

$$\phi = \int_S \mathbf{B}.d\mathbf{S} \quad ...(2)$$

and since e.m.f. 'e' is work done per unit charge, we have

$$e = \frac{W}{q} = \oint_l (\mathbf{E} + \mathbf{v} \times \mathbf{B}) . d\mathbf{l}$$

In a stationary circuit $\mathbf{v} = 0$ and since $\mathbf{v}$ is parallel to $d\mathbf{l}$ and $\mathbf{v} \times \mathbf{B}$ is perpendicular to $d\mathbf{l}$.

Thus $\mathbf{v} \times \mathbf{B} \,.\, d\mathbf{l} = 0$ ...(3)

Hence $$e = \int_l \mathbf{E} \,.\, d\mathbf{l}$$

and finally from equation (2)

$$\oint \mathbf{E} \,.\, d\mathbf{l} = -\frac{\partial}{\partial t} \int_S \mathbf{B} \,.\, d\mathbf{S}$$

or $$\oint \mathbf{E} \,.\, d\mathbf{l} = -\int \frac{\partial \mathbf{B}}{\partial t} d\mathbf{S} \qquad \text{...(4)}$$

This is Faraday's law written in integral form.

Using Stoke's law, Faraday's law can be mentioned in the differential form.

As $$e = -\frac{\partial \phi}{\partial t}$$

Applying Stoke's law

$$e = \oint \mathbf{E} \,.\, d\mathbf{l}$$

$$= \int_S (\nabla \times \mathbf{E}) . d\mathbf{S}$$

$$\Rightarrow \quad -\frac{\partial \phi}{\partial t} = -\frac{\partial}{\partial t} \int \mathbf{B} \,.\, d\mathbf{S}$$

$$\Rightarrow \quad \int_S \left(\nabla \times \mathbf{E} + \frac{\partial \mathbf{B}}{\partial t} \right) d\mathbf{S} = 0$$

Since this is valid for any surface then

$$\boxed{\nabla \times \mathbf{E} = -\frac{\partial \mathbf{B}}{\partial t}} \qquad \text{...(5)}$$

Faraday's law thus establishes that time varying magnetic fields give rise to electric fields. This further supports the fact that the fields are related to each other and we then must speak of electromagnetic fields rather than electric and magnetic fields.

7.15 DISPLACEMENT CURRENT

The idea of displacement current originated from the study of the discharge of a capacitor. Maxwell asserted that since all electric currents are closed, in the discharge of capacitor, the conduction current is closed by a current known as *displacement current* (arising due to charge in electric displacement in the dielectric) flowing in the dielectric between the capacitor plates.

Let us consider a parallel plate charged capacitor joined together by a conducting wire. The discharge current 'i' flowing through the wire is equal to the rate of charge on plates, *i.e.*,

$$i = \frac{\partial q}{\partial t} \qquad ...(1)$$

But the charge is related to the electric field between the plates by the equation

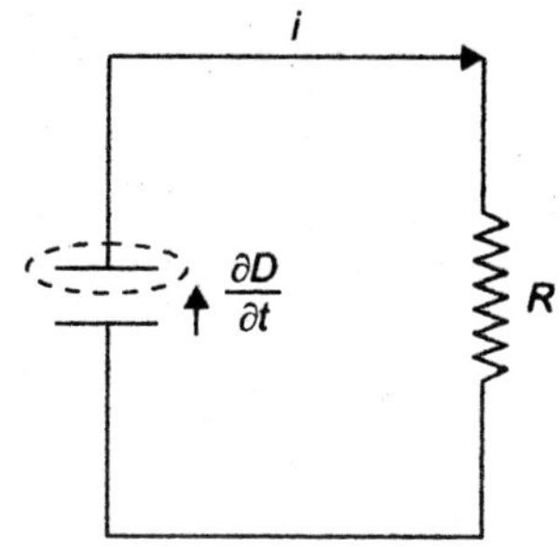

FIGURE 7.28 Discharging of a capacitor

$$q = CV = \frac{\varepsilon_0 A}{d} Ed = \varepsilon_0 EA$$

$$= DA \qquad ...(2) \quad [\because \; D = \varepsilon_0 E]$$

where E = electric field between the plates, A = area of each plate, d = distance between the plates, ε_0 = absolute permittivity of free space.

Therefore from equations (1) and (2), we have

$$i = \frac{\partial q}{\partial t} = \frac{\partial}{\partial t}(DA) = A\frac{\partial D}{\partial t} = \varepsilon_0 A \frac{\partial E}{\partial t}$$

or $$i = A\frac{\partial D}{\partial t} \text{ or } \frac{i}{A} = \frac{\partial D}{\partial t} \qquad ...(3)$$

Hence, $\frac{\partial D}{\partial t}$ can be interpreted as a density of some current which completes the discharge current. Thus, there is a displacement current density denoted by $J_d = \frac{\partial \mathbf{D}}{\partial t}$ (measured by A/m^2) entering the dotted volume (in Figure 7.28) in the region of capacitor plates. The displacement current cannot be detected directly, but its existence is confirmed by electromagnetic theory of light.

Significance of displacement current

1. Displacement current is entirely different from conduction current, Displacement current is a current only in the sense that it produces a magnetic field in the empty space since the conduction current is zero.
2. The magnitude of displacement current is equal to the time rate of change of electric displacement vector **D**. When the field does not vary with time, the displacement current is zero.
3. Displacement current serves to make the total current continuous across discontinuities in the conduction current.
4. The time varying electric field generates a time varying magnetic field because of the term $\frac{\partial \mathbf{D}}{\partial t}$.
5. The term $\frac{\partial \mathbf{D}}{\partial t}$ does not change the fact that the magnetic fields are continuous.

7.16 MAXWELL'S EQUATIONS

There are four fundamental equations of electromagnetism known as *Maxwell's equations* which may be written in differential form as

1. div $\mathbf{D} = \rho$
2. div $\mathbf{B} = 0$
3. curl $\mathbf{E} = -\dfrac{d\mathbf{B}}{dt}$
4. curl $\mathbf{H} = \mathbf{J} + \dfrac{d\mathbf{D}}{dt}$.

In above equations the notations have the following meanings:

J = current density in Amp/met^3

$\mathbf{D}$ = electric displacement vector in Coulomb/met^2

ρ = free charge density in Coul/met^3

$\mathbf{B}$ = magnetic induction in Weber/met^2

$\mathbf{E}$ = electric field intensity in Volt/met or Newton/Coul

$\mathbf{H}$ = magnetic field intensity in Amp/met-turn.

7.16.1 Derivation of Maxwell's Equations

1. Derivation of First Equation

$$\boxed{\text{div } \mathbf{D} = \nabla.\mathbf{D} = \rho}$$

Proof: *"The maxwell first equation is nothing but the differential form of Gauss law of electrostatics."*

Let us consider a surface S bounding a volume V in a dielectric medium. In a dielectric medium total charge consists of free charge. If ρ is the charge density of free charge at a point in a small volume element dV. Then Gauss's law can be expresses as

> "The total normal electrical induction over a closed surface is equal to $\dfrac{1}{\varepsilon_0}$ times of total charge enclosed."

i.e.,
$$\oint_S \mathbf{E}\hat{n}\,d\mathbf{S} = \frac{q}{\varepsilon_0}$$

$$\Rightarrow \quad \oint_S \mathbf{E}\hat{n}\,d\mathbf{S} = \frac{1}{\varepsilon_0}\int_V \rho\,dV$$

where ρ = charge per unit volume

V = volume enclosed by charge.

By Gauss transformation formula

$$\int_V \text{div } \mathbf{E}\, dV = \frac{1}{\varepsilon_0}\int_V \rho\, dV \qquad \left[\because \int_S \vec{\mathbf{A}}\,\hat{n}\, dS = \int_V \text{div } \vec{\mathbf{A}}\, dV\right]$$

$$\Rightarrow \qquad \text{div } \mathbf{E} = \frac{1}{\varepsilon_0}\rho$$

or $$\varepsilon_0 \text{ div } \mathbf{E} = \rho$$

or $$\text{div } \varepsilon_0 \mathbf{E} = \rho \qquad [\because \mathbf{D} = \varepsilon_0 \mathbf{E} + \mathbf{P}]$$

$$\mathbf{P} = 0 \quad \text{then} \quad \mathbf{D} = \varepsilon_0 \mathbf{E}$$

so $$\boxed{\text{div } \mathbf{D} = \rho}$$

2. Derivation of Second Equation

$$\boxed{\text{div } \mathbf{B} = \nabla.\mathbf{B} = 0}$$

"It is nothing but the differential form of Gauss law of magnetostatics."

Since isolated magnetic poles and magnetic currents due to them have no significance. Therefore magnetic lines of force in general are either closed curves or go off to infinity. Consequently the number of magnetic lines of force entering any arbitrary closed surface is exactly the same as leaving it. It means that the flux of magnetic induction B across any closed surface is always zero.

> Gauss law of magnetostatics states that "Total normal magnetic induction over a closed surface is equal to zero."

i.e., $$\int_S \mathbf{B}\,\hat{n}\, dS = 0$$

Applying Gauss transformation formula we get

$$\int_V \text{div } \mathbf{B}\, dV = 0 \qquad \left[\because \int_S \mathbf{A}\,\hat{n} dS = \int_V \text{div } \mathbf{A}\, dV\right]$$

The integrand should vanish for the surface boundary as the volume is arbitrary.

i.e., $$\boxed{\text{div } \mathbf{B} = 0}$$

3. Derivation of Third Law

$$\boxed{\text{curl } \mathbf{E} = -\frac{d\mathbf{B}}{dt}}$$

"It is nothing but the differential form of Faraday's law of electromagnetic induction."

> According to Faraday's law of electromagnetic induction, it is known that e.m.f. induced in a closed loop is defined as negative rate of change of magnetic flux *i.e.,*

$$\boxed{e = -\frac{d\phi}{dt}}$$

where ϕ = magnetic flux

or $$\phi = \int_S \mathbf{B}\hat{n}\,dS$$

$$\phi = \frac{B}{A}$$ where S is any surface having loop as boundary.

So $$\int_l \mathbf{E}\,.\,d\mathbf{l} = -\frac{d\phi}{dt} \qquad ...(1)$$

Putting the value of ϕ in equation (1), we get

$$\int_l \mathbf{E}\,.\,d\mathbf{l} = -\frac{d}{dt}\int_S \mathbf{B}\,.\,\hat{n}\,d\mathbf{S}$$

$$\Rightarrow \quad \int_l \mathbf{E}\,.\,d\mathbf{l} = \int_S -\frac{d\mathbf{B}}{dt}\hat{n}\,d\mathbf{S}$$

Applying Stoke's transformation formula on L.H.S.

$$\int_S \text{curl } \mathbf{E}\hat{n}\,d\mathbf{S} = \int_S -\frac{d\mathbf{B}}{dt}\hat{n}\,d\mathbf{S} \quad \left[\oint_l \mathbf{A}\,d\mathbf{l} = \int_S \text{curl } \mathbf{A}\hat{n}\,dS\right]$$

or $$\int_s \left(\text{curl } \mathbf{E} + \frac{d\mathbf{B}}{dt}\right)\hat{n}\,d\mathbf{S} = 0$$

Further validity of the equation

$$\text{curl } \mathbf{E} + \frac{d\mathbf{B}}{dt} = 0$$

or $$\boxed{\text{curl } \mathbf{E} = -\frac{d\mathbf{B}}{dt}}$$

This is known as Maxwell's third equation.

4. Derivation of Maxwell's Fourth Equation:

"This is nothing but differential form of modified Ampere circuital law."

Ampere's modified circuital law

According to law the work done in carrying a unit magnetic pole once around a closed arbitrary path linked with the current is expressed by

$$\int_l \mathbf{B}\,d\mathbf{l} = \mu_0 \times i$$

i = current enclosed by the path

$$\Rightarrow \quad \int_l \mathbf{B}\,.\,d\mathbf{l} = \mu_0 \int_S \mathbf{J}\,\hat{n}\,dS$$

On applying Stoke's transformation formula in L.H.S.

$$\int_S \text{curl } \mathbf{B}\hat{n}\,dS = \int_S \mu_0 \mathbf{J}\,\hat{n}\,dS \qquad \left\{\oint_l \mathbf{A}.d\mathbf{l} = \int_S \text{curl } \mathbf{A}\,\hat{n}\,dS\right\}$$

$$\Rightarrow \int_S (\text{curl } \mathbf{B} - \mu_0\mathbf{J})\,\hat{n}\,dS = 0$$

For the validity this equation

$$\text{curl } \mathbf{B} - \mu_0\mathbf{J} = 0$$

or $$\text{curl } \mathbf{B} = \mu_0\mathbf{J} \qquad ...(2)$$

It is known as the *fourth equation of Maxwell.*

Taking divergence of both sides

$$\text{div. (curl } \mathbf{B}) = \text{div } (\mu_0\,\mathbf{J})$$

$$0 = \text{div } (\mu_0\,\mathbf{J})$$

$$= \mu_0 \text{ div } \mathbf{J} \quad \{\because \quad \text{div (curl } \mathbf{A}) = 0\}$$

or $$\text{div } \mathbf{J} = 0 \qquad ...(3)$$

Which means that the current is always closed and there are no source and sink. Thus we arrive at contradiction equation (3) is also in conflict with the equation of discontinuity.

But the according to *law of continuity*

$$\text{div } \mathbf{J} = -\frac{d\rho}{dt}$$

So this equation fails and it need of little modification. So Maxwell assume that

$$\text{curl } \mathbf{B} = \mu_0\,\mathbf{J} + \mu_0\mathbf{J}_d \qquad ...(4)$$

Again taking the divergence both sides

$$\text{div (curl } \mathbf{B}) = \mu_0\,(\text{div } \mathbf{J}) + \mu_0\,(\text{div } \mathbf{J}_d)$$

or $$0 = \mu_0\,(\text{div } \mathbf{J}) + \mu_0\,(\text{div } \mathbf{J}_d)$$

By putting $$\text{div } \mathbf{J}_d = \frac{d\rho}{dt} \qquad ...(5)$$

$$\text{div } \mathbf{J}_d = \text{div } \frac{d\mathbf{D}}{dt}$$

$$\boxed{\mathbf{J}_d = \frac{d\mathbf{D}}{dt}}$$ (By Maxwell first equation, div $\mathbf{D} = \rho$ in equation (4))

Putting in equation (4), we get

or $$\text{curl } \mathbf{B} = \mu_0\left(\mathbf{J} + \frac{d\mathbf{D}}{dt}\right)$$

but $$\mathbf{B} = \mu_0 \mathbf{H}$$

$$\text{curl } (\mu_0 \mathbf{H}) = \mu_0 \left(\mathbf{J} + \frac{d\mathbf{D}}{dt} \right)$$

$$\mu_0 \text{ curl } \mathbf{H} = \mu_0 \left(\mathbf{J} + \frac{d\mathbf{D}}{dt} \right)$$

$$\boxed{\text{curl } \mathbf{H} = \mathbf{J} + \frac{d\mathbf{D}}{dt}}$$

7.17 PHYSICAL SIGNIFICANCE OF MAXWELL'S EQUATIONS

(*i*) Let us imagine an arbitrary volume V enclosed by a closed surface S. Now, integrate both sides of the first equation $\nabla.\mathbf{D} = \rho \;\Rightarrow\; \nabla.\mathbf{E} = \frac{\rho}{\varepsilon_0}$ relative to the volume V.

Then, we get

$$\iiint_V \nabla.\mathbf{E}\, dV = \iiint_V \frac{\rho}{\varepsilon_0} dV$$

or $$\boxed{\oiint_S \mathbf{E}.d\mathbf{S} = \frac{q}{\varepsilon_0}} \qquad ...(1)$$

Using Gauss divergence theorem and fact that

$$\iiint_V \rho\, dV = q, \text{ the total charge within } V.$$

Which shows that the *total electric flux over a closed surface S is* $\frac{1}{\varepsilon_0}$ *times the total charge enclosed by the closed surface S*. This is nothing but *Gauss's law in electrostatics*.

(*ii*) Integrate both sides of the second equation $\nabla.\mathbf{B} = 0$ relative to an arbitrary closed volume V enclosed by a closed surface S.

$$\iiint_V \nabla.\mathbf{B}\, dV = 0$$

or $$\boxed{\oiint_S \mathbf{B}.d\mathbf{S} = 0} \qquad ...(2)$$

Which states that the *total magnetic flux over a closed surface is zero*. This confirms the *non-existence of magnetic monopole*.

(*iii*) Now, consider an open surface (two-sided) S with boundary a simple closed curve C. Let us integrate both sides of third equation $\nabla \times \mathbf{E} = -\frac{\partial \mathbf{B}}{\partial t}$ with respect to the open surface S.

$$\iint_S (\nabla . \mathbf{E}) . d\mathbf{S} = -\iint_S \frac{\partial \mathbf{B}}{\partial t} . d\mathbf{S}$$

or

$$\boxed{\oint_C \mathbf{E}.d\mathbf{l} = -\frac{\partial}{\partial t}\left[\iint_S \mathbf{B} . d\mathbf{S}\right]} \quad ...(3) \quad \text{[Using Stoke's theorem]}$$

The left hand side represents electromotive force and right hand side is time rate of change of magnetic flux. Thus, this equation physically signifies that the *electromotive force (e.m.f.) induced in a closed circuit is equal to the time rate of loss of magnetic flux of the surface of the circuit*. This is *Faraday's law of electromagnetic induction in association with Lenz's law.*

(*iv*) If we integrate both sides of the fourth equation $\nabla \times \mathbf{B} = \mu\,[\mathbf{J} + \mathbf{J}_d]$ relative to an open surface S having boundary a simple closed curve C, we get

$$\iint_S (\nabla . \mathbf{B}) . d\mathbf{S} = \iint_S \mu\,(\mathbf{J} + \mathbf{J}_d) . d\mathbf{S}$$

or

$$\boxed{\oint_C \mathbf{B} . d\mathbf{l} = \mu\,[i + i_d]} \quad ...(4) \quad \text{Using Stoke's theorem.}$$

This shows that the *line integral of magnetic induction vector over a closed circuit is equal to μ times the total current (conduction plus displacement current) included in the circuit.* This is simply *"modified Ampere's circuital law"*.

7.18 MAXWELL'S EQUATION IN FREE SPACE AND ISOTROPIC MEDIA: INTEGRAL AND DIFFERENTIAL FORMS

(*i*) **In Free Space:** In absence of any free charges or currents or in the empty space *i.e.*, when q or ρ and i or J are zero. Then Maxwell's equations take the form

Integral Form	**Differential Form**
$\oint \mathbf{E}.d\mathbf{S} = 0$	$\text{div}\,\mathbf{E} = 0$
$\oint \mathbf{B}.d\mathbf{S} = 0$	$\text{div}\,\mathbf{B} = 0$
$\oint \mathbf{E}.d\mathbf{l} = -\int \frac{\partial \mathbf{B}}{\partial t} . d\mathbf{S}$	$\text{curl}\,\mathbf{E} = -\frac{\partial \mathbf{B}}{\partial t}$
$\oint \mathbf{B}.d\mathbf{l} = \mu_0\,\varepsilon_0 \int \frac{\partial \mathbf{E}}{\partial t} . d\mathbf{S}$	$\text{curl}\,\mathbf{B} = \mu_0\,\varepsilon_0 \frac{\partial \mathbf{E}}{\partial t}$

(*ii*) **In Isotropic Medium:** In an isotropic dielectric (or non-conducting isotropic medium)

$$\mathbf{D} = \varepsilon\mathbf{E},\ \mathbf{B} = \mu\mathbf{H},\ \mathbf{J} = \sigma\mathbf{E} = 0 \text{ and } \rho = 0$$

Then Maxwell's equations take the form

$$\oint \mathbf{E}.d\mathbf{S} = 0 \qquad \text{div } \mathbf{E} = 0$$

$$\oint \mathbf{B}.d\mathbf{S} = 0 \qquad \text{div } \mathbf{B} = 0 \ \Rightarrow\ \text{div } \mathbf{H} = 0$$

$$\oint \mathbf{E}.dl = -\int \frac{\partial \vec{B}}{\partial t}.d\vec{S} \qquad \text{curl } \mathbf{E} = -\mu \frac{\partial \mathbf{H}}{\partial t}$$

$$\oint \mathbf{H}.dl = \varepsilon \int \frac{\partial \vec{E}}{\partial t}.d\vec{S} \qquad \text{curl } \mathbf{H} = \varepsilon \frac{\partial \mathbf{E}}{\partial t}$$

7.19 PROPAGATION OF ELECTROMAGNETIC WAVES IN FREE SPACE

For free space $\rho = 0$

$$\mathbf{J} = 0$$

$$\mathbf{B} = \mu_0 \mathbf{H}$$

$$\mathbf{D} = \varepsilon_0 \mathbf{E}$$

Maxwell's equation for free space.

I. $\text{div } \mathbf{D} = \rho$

Putting $\mathbf{D} = \varepsilon_0 \mathbf{E}$

and $\rho = 0$

in above equation we getting

$$\text{div } (\varepsilon_0 \mathbf{E}) = 0$$

or $$\varepsilon_0 \text{ div } \mathbf{E} = 0$$

or $$\boxed{\text{div } \mathbf{E} = 0} \qquad ...(1)$$

II. $\text{div } \mathbf{B} = 0$

Putting $\mathbf{B} = \mu_0 \mathbf{H}$

or $$\text{div } \mu_0 \mathbf{H} = 0$$

or $$\mu_0 \text{ div } H = 0$$

or $$\boxed{\text{div } \mathbf{H} = 0} \qquad ...(2)$$

III. $$\text{curl } \mathbf{E} = -\frac{d\mathbf{B}}{dt}$$

Putting $B = \mu_0 \mathbf{H}$ in above relation

$$\text{curl } \mathbf{E} = -\frac{d(\mu_0 \mathbf{H})}{dt}$$

or
$$\boxed{\text{curl } \mathbf{E} = -\mu_0 \frac{d\mathbf{H}}{dt}} \qquad ...(3)$$

IV. $$\text{curl } \mathbf{H} = \mathbf{J} + \frac{d\mathbf{D}}{dt}$$

Putting $\mathbf{D} = \varepsilon_0 \mathbf{E}, \mathbf{J} = 0$

we get

$$\text{curl } \mathbf{H} = 0 + \frac{d(\varepsilon_0 \mathbf{E})}{dt}$$

or
$$\boxed{\text{curl } \mathbf{H} = \varepsilon_0 \frac{d\mathbf{E}}{dt}} \qquad ...(4)$$

If we take curl of equation (3), we get

$$\text{curl (curl } \mathbf{E}) = -\mu_0 \frac{d}{dt} (\text{curl } \mathbf{H}) \qquad ...(5)$$

From equation (4), we get

$$[\text{Gradient (div } \mathbf{E}) - \nabla^2 \mathbf{E}] = -\mu_0 \varepsilon_0 \frac{d^2\mathbf{E}}{dt^2} \qquad ...(6)$$

from equation (1) $\text{div } \mathbf{E} = 0$

So the equation (4) becomes

$$\nabla^2 \mathbf{E} = \mu_0 \varepsilon_0 \frac{d^2\mathbf{E}}{dt^2} \qquad ...(7)$$

Suppose $E = E_x + E_y + E_z$

then L.H.S. is

$$\frac{\partial^2 E_x}{\partial x^2} + \frac{\partial^2 E_y}{\partial y^2} + \frac{\partial^2 E_z}{\partial z^2} = \mu_0 \varepsilon_0 \frac{d^2 E}{dt^2} \qquad ...(8)$$

The wave equation is

$$\frac{d^2y}{dx^2} = \frac{1}{v^2}\left(\frac{d^2y}{dt^2}\right) \qquad ...(9)$$

where v = velocity of wave

x = path of propagation wave

On comprising the equation (8) and (9), we get

$$v^2 = \frac{1}{\mu_0\varepsilon_0} \text{ and } v = \frac{1}{\sqrt{\mu_0\varepsilon_0}}$$

Since a time varying electric field travels in free space with the velocity

$$v = \frac{1}{\sqrt{\mu_0\varepsilon_0}}$$

for magnetic field

$$\mu_0 = 4\pi \times 10^{-7} \text{ Henry/farad}$$
$$\varepsilon_0 = 8.8542 \times 10^{-12} \text{ Farad/met}$$
$$v = 2.9999 \times 10^8 \text{ met/sec.}$$

Similarly taking curl of equation (4), we getting

$$\text{curl (curl } \mathbf{H}) = \varepsilon_0 \cdot \frac{d}{dt} \text{ curl } \mathbf{E} \text{ [From equation (3)]}$$

$$= -\mu_0\,\varepsilon_0 \frac{\partial^2 \mathbf{H}}{\partial t^2}$$

$$(\text{grad (div } \mathbf{H}) - \nabla^2\mathbf{H}) = -\varepsilon_0\,\mu_0 \frac{d^2\mathbf{H}}{dt^2}$$

$$H = H_x + H_y + H_z \qquad [\because \text{ div } H = 0 \text{ from equation (2)}]$$

$$\nabla^2 \mathbf{H} = \mu_0\,\varepsilon_0 \frac{d^2\mathbf{H}}{dt^2}$$

$$\frac{\partial^2 H_x}{\partial x^2} + \frac{\partial^2 H_y}{\partial y^2} + \frac{\partial^2 H_z}{\partial z^2} = \mu_0\,\varepsilon_0 \frac{d^2 H}{dt^2} \qquad ...(10)$$

The wave equation is

$$\frac{d^2y}{dx^2} = -\frac{1}{v^2}\frac{d^2y}{dt^2} \qquad ...(11)$$

On comprising equations (10) and (11), and taking

$$\mu_0 = 4\pi \times 10^{-7} \text{ Henry/met}$$

$$\varepsilon_0 = 8.8542 \times 10^{-12} \text{ Farad/met.},$$

we get,

$$\frac{1}{v^2} = \mu_0 \varepsilon_0$$

$$\boxed{v = \frac{1}{\sqrt{\mu_0 \varepsilon_0}}} \qquad ...(12)$$

$$v = \frac{1}{\sqrt{\mu_0 \varepsilon_0}} = 2.9999 \times 10^8 \text{ met/sec.}$$

7.20 POYNTING THEOREM — (ENERGY OF ELECTROMAGNETIC FIELD)

Energy is transported through space by means of electromagnetic waves, let the material inside the surface S be isotropic homogeneous and characterised by permeability μ and permittivity ε and conductivity σ.

Maxwell's IIIrd and IVth equations are as follows

$$\text{curl } \mathbf{E} = -\frac{\partial \mathbf{B}}{\partial t} \qquad ...(1)$$

and

$$\text{curl } \mathbf{H} = \mathbf{J} + \frac{\partial \mathbf{D}}{\partial t} \qquad ...(2)$$

Taking scalar product of both sides of equation (1) and equation (2) respectively with **H** and **E**

$$\mathbf{H} \text{ curl } \mathbf{E} = -\mathbf{H}\frac{\partial \mathbf{B}}{dt} \qquad ...(3)$$

and

$$\mathbf{E} \text{ curl } \mathbf{H} = \mathbf{E}\left(\mathbf{J} + \frac{\partial \mathbf{D}}{\partial t}\right) \qquad ...(4)$$

Now, subtracting equation (3) from equation (4), we get

$$(\mathbf{E} \text{ curl } \mathbf{H} - \mathbf{H} \text{ curl } \mathbf{E}) = \mathbf{E}\left(\mathbf{J} + \frac{\partial \mathbf{D}}{\partial t}\right) + \mathbf{H}\frac{\partial \mathbf{B}}{\partial t}$$

$$(\mathbf{E} \text{ curl } \mathbf{H} - \mathbf{H} \text{ curl } \mathbf{E}) = \mathbf{E}.\mathbf{J} + \mathbf{E}.\frac{\partial \mathbf{D}}{\partial t} + \mathbf{H}.\frac{\partial \mathbf{B}}{\partial t} \qquad ...(5)$$

Now, using vector identity

$$\text{div}\ (\mathbf{E} \times \mathbf{H}) = \mathbf{H}\ \text{curl}\ \mathbf{E} - \mathbf{E}\ \text{curl}\ \mathbf{E}$$

putting above value in L.H.S. equation (5) may be expressed as

$$-\ \text{div}\ (\mathbf{E} \times \mathbf{H}) = \mathbf{J.E} + \mathbf{E}.\frac{\partial \mathbf{D}}{\partial t} + \mathbf{H}.\frac{\partial \mathbf{B}}{\partial t} \qquad ...(6)$$

Now, if the medium is linear so that the relations

$$\mathbf{B} = \mu\mathbf{H} \text{ and } \mathbf{D} = \varepsilon\ \mathbf{E}$$

apply, then we may write

$$\mathbf{E}.\frac{\partial \mathbf{D}}{\partial t} = \mathbf{E}\frac{\partial}{\partial t}(\varepsilon\ \mathbf{E}) = \frac{1}{2}\varepsilon\frac{\partial}{\partial t}E^2 = \frac{\partial}{\partial t}\left(\frac{1}{2}\mathbf{E}.\mathbf{D}\right)$$

and

$$\mathbf{H}.\frac{\partial \mathbf{B}}{\partial t} = \mathbf{H}\frac{\partial}{\partial t}(\mu\ \mathbf{H}) = \frac{1}{2}\mu\frac{\partial}{\partial t}H^2 = \frac{\partial}{\partial t}\left(\frac{1}{2}\mathbf{H}.\mathbf{B}\right)$$

$$-\ \text{div}\ (\mathbf{E} \times \mathbf{H}) = \mathbf{J.E} + \left[\mathbf{E}\frac{\partial}{\partial t}(\varepsilon\mathbf{E}) + \mathbf{H}\frac{\partial}{\partial t}(\mu\mathbf{H})\right]$$

$$= \mathbf{J.E} + \frac{1}{2}\left[\frac{\partial}{\partial t}(E\varepsilon E) + \frac{\partial}{\partial t}(H\mu H)\right]$$

$$= \mathbf{J.E} + \frac{1}{2}\left[\frac{\partial}{\partial t}(\varepsilon E^2) + \frac{\partial}{\partial t}(\mu H^2)\right]$$

or

$$\text{div}\ (\mathbf{E} \times \mathbf{H}) + \mathbf{J.E} + \frac{1}{2}\left[\frac{\partial}{\partial t}(E\ \varepsilon\ E) + \frac{\partial}{\partial t}(H\mu H)\right] = 0$$

or

$$\text{div}\ (\mathbf{E} \times \mathbf{H}) + \mathbf{J.E} + \frac{1}{2}\left[\frac{\partial}{\partial t}(E.D) + \frac{\partial}{\partial t}(B.H)\right] = 0$$

or

$$\text{div}\ (\mathbf{E} \times \mathbf{H}) + \mathbf{J.E} + \frac{1}{2}\frac{\partial}{\partial t}[(\mathbf{E.D}) + (\mathbf{B.H})] = 0$$

Now, integrating this equation over the volume V bounded by the surface S, we get

$$\int_V \text{div}\ (\mathbf{E} \times \mathbf{H})\ dV + \int_V \mathbf{J}.\mathbf{E}\ dV + \int_V \frac{1}{2}\frac{d}{dt}[\mathbf{B}.\mathbf{D} + \mathbf{B}.\text{H}]\ dV = 0 \qquad ...(7)$$

$$\Rightarrow \qquad -\int_V \text{div}\ (\vec{E} \times \vec{H})\ dV = \int_V \vec{J}.\vec{E}\ dV + \int_V \frac{1}{2}\frac{d}{dt}[\vec{E}.\vec{D} + \vec{B}.\vec{H}]\ dV \qquad ...(8)$$

By applying Gauss transformation formula in L.H.S.

$$-\int_S (\mathbf{E}\times\mathbf{H})\,\hat{n}\,dS = \int_V \mathbf{J}.\mathbf{E}\,dV + \int_V \frac{1}{2}\frac{d}{dt}[\mathbf{E}.\mathbf{D}+\mathbf{B}.\mathbf{H}]\,dV \qquad ...(9)$$

$$-\int_V \mathbf{J}.\mathbf{E}\,dV = \int_V \frac{1}{2}\frac{d}{dt}[\mathbf{E}.\mathbf{D}+\mathbf{B}.\mathbf{H}]\,dV + \int_S (\mathbf{E}\times\mathbf{H})\,\hat{n}\,d\mathbf{S} \qquad ...(10)$$

The above equation (10) represents the poynting theorem.

7.20.1 Physical Significance or Interpretation of terms of Poynting Theorem

(*a*) **Interpretation of term $\int_V \mathbf{J}.\mathbf{E}\,\mathbf{dV}$:** The current distribution represented by the vector **J** can be considered as made up of various changes q_j moving with velocity v_j,

$$\int_V \mathbf{J}.\mathbf{E}\,dV = \int_V (\mathbf{J}.dV)\,E$$

$$= \int_V (\mathbf{J}.d\mathbf{S}\,d\mathbf{l})\,E \qquad [\vec{J}.d\vec{S} = I]$$

$$= \int_V I\,dl\,E \qquad [\because\ \mathbf{J}.d\mathbf{S} = \mathbf{I}]$$

$$= \int_V \frac{dq}{dt}\,dl\,E$$

$$\because \quad \boxed{I = \frac{dq}{dt} \Rightarrow \mathbf{J}.d\mathbf{S}}$$

$$\Rightarrow \quad \mathbf{J}.d\mathbf{S} = \int dq\,v_j\,E$$

$$\Rightarrow \quad \mathbf{J}.d\mathbf{S} = \sum_j q_j(v_j\,E_j) \qquad ...(1)$$

where E_j denotes the electric field at the position of charge q_j. According to Lorentz force equation, the force experienced by point charge is given by

$$F = q\,(\mathbf{E} + \mathbf{v}\times\mathbf{B})$$

We know that workdone is equal to force x displacement.

$$dW = \mathbf{F}.d\mathbf{l}$$

$$= q\,(E + \mathbf{v}\times\mathbf{B})\,d\mathbf{l}$$

$$= q.E\,d\mathbf{l} + q.(\mathbf{v}\times\mathbf{B})\,d\mathbf{l}$$

$$= q\mathbf{E}\,d\mathbf{l} + d\mathbf{l}\left(\frac{d\mathbf{l}}{dt}\times\mathbf{B}\right)q$$

$$= q\mathbf{E}\,d\mathbf{l} + \left(d\mathbf{l}\times\frac{d\mathbf{l}}{dt}\right).\mathbf{B}\,q$$

$$dW = q\,E\,d\mathbf{l} + 0 \qquad [\because\ d\mathbf{l}\times d\mathbf{l} = 0]$$

or

$$\frac{dW}{dt} = q\,E\,\frac{d\mathbf{l}}{dt}$$

$$\frac{dW}{dt} = q\,\mathbf{E.v}$$

taking all the charges into consideration, we get

$$\frac{dW}{dt} = \sum_j q_j(E_j\,v_j) \qquad ...(2)$$

From equations (1) and (2)

This shows that this term represents the rate at which the work is done by the field on the charges.

(*b*) **Interpretation of term**

$$\int_V \frac{1}{2}\frac{\partial}{\partial t}[\mathbf{E}\,.\,\mathbf{D}+\mathbf{H}\,.\,\mathbf{B}]\,dV$$

or

$$\int_V \frac{1}{2}\frac{d}{dt}[\varepsilon\,E^2+\mu\,H^2]\,dV$$

The above equation represents the *time rate of increase in energy stored in the electric and magnetic field* respectively in volume.

(*c*) **Interpretation of the term**

$$\int_S (\mathbf{E}\times\mathbf{H})\,\hat{n}\,dS$$

Since L.H.S. shows the *rate of change of workdone* or power so this term must shows the *power flow into volume V* across the surface *S* or the power flow out of volume *V* across the surface *S*

$$\int_S (\mathbf{E}\times\mathbf{H})\,\hat{n}\,dS = \int_S \mathbf{P}\,\hat{n}\,d\mathbf{S}$$

where $\mathbf{P} = \mathbf{E}\times\mathbf{H}$

where $\mathbf{P} = \mathbf{E}\times\mathbf{H}$ is known as **Poynting vector** and hence the meaning of power density associated with electromagnetic field at a point

Dimension of $P = [\mathrm{M\,T^{-3}}]$

unit of P = Watt/meter^2.

7.21 LAW OF CONSERVATION OF ENERGY FOR E.M. FIELD

We know that by equations

$$\mathbf{J}.\mathbf{E} + \frac{1}{2}\left[\frac{d}{dt}(\mathbf{E}.\mathbf{D}) + \frac{d}{dt}(\mathbf{H}.\mathbf{B})\right] + \text{div}\,(\mathbf{E}\times\mathbf{H}) = 0$$

For a non conductive media $\mathbf{J} = 0$. we get

$$\frac{1}{2}\left[\frac{d}{dt}(\mathbf{E}.\mathbf{D}) + \frac{d}{dt}(\mathbf{H}.\mathbf{B})\right] + \text{div}\,(\mathbf{E}\times\mathbf{H}) = 0$$

$$\Rightarrow \quad \frac{1}{2}\left[\frac{d}{dt}(\mathbf{E}.\mathbf{D}) + \frac{d}{dt}(\mathbf{H}.\mathbf{B})\right] + \text{div}\,\mathbf{P} = 0 \qquad [\because\ P = E\times H]$$

$$\Rightarrow \quad \frac{d}{dt}\left[\frac{1}{2}(\varepsilon E^2) + \frac{1}{2}(\mu H^2)\right] + \text{div}\,\mathbf{P} = 0$$

$$\boxed{\text{Energy density} = \frac{1}{2}(\varepsilon E^2) + \frac{1}{2}(\mu H^2) = u}$$

$$\frac{d}{dt}.u + \text{div}\,\mathbf{P} = 0$$

$$\Rightarrow \quad \text{div}\,\mathbf{P} + \frac{du}{dt} = 0 \qquad \text{...(1)}$$

This equation is same as equation of the continuity

$$\text{div}\,\mathbf{J} + \frac{d\rho}{dt} = 0$$

Since equation of continuity represents the law of conservation of charge. So by analogy the previous equation must represent the law of conversation of energy for electromagnetic field.

7.22 FUNDAMENTAL CONCEPTS OF ELECTROMAGNETISM

1. $$\boxed{E = \frac{1}{4\pi\varepsilon_0}\frac{q}{r^2}}$$

where E = Electric potential energy, energy of electric field or intensity of electric field

q = charge place at distance r from infinitesimal test charge

r = distance between point charge and infinitesimal charge

ε_0 = permittivity of free space

$= 8.854 \times 10^{-12}$ Coul^2/Newton-met^2

2. $$\mathbf{D} = \varepsilon_0 \mathbf{E} + \mathbf{P}$$

where D = displacement vector

$\mathbf{P}$ = polarisation vector (electric dipole moment per unit volume)

$$P = \frac{n(q\,r)}{V}$$

where n = number of term

V = volume

$q\,r$ = electric dipole moment

3. $$\phi = \mathbf{B.A}$$

where ϕ = magnetic flux

B = magnetic intensity

A = area

$$B = \frac{\phi}{A} = \text{Weber/met}^2$$

4. $\mathbf{F} = i\,B\,l \sin\theta$ is Lorentz force.

where $\mathbf{F}$ = The magnetic force on the conductor

i = current in magnetic field

l = whole length of the conductor

B = uniform field of conduction

from above equation

$$\mathbf{B} = \frac{\mathbf{F}}{il} = \frac{\text{Newton}}{\text{Amp} \times \text{met}} = \text{Tesla}$$

1 Tesla = 10^4 Gauss

5. **Biot-Savart's Law**

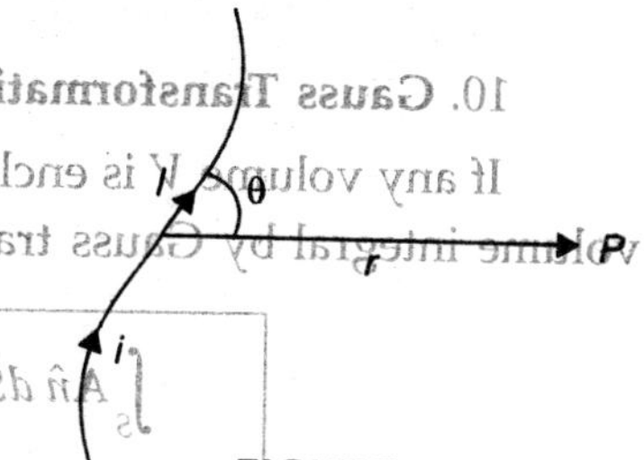

FIGURE 7.29

$$\mathbf{B} = \frac{\mu_0}{4\pi}\frac{i\,\Delta l \sin\theta}{r^2}$$

where μ_0 = permeability of free space

6. $$\mathbf{B} = \mu_0 \mathbf{H}$$

where **H** = magnetising force

7. $$\mathbf{B} = \mu_0 (\mathbf{M} + \mathbf{H})$$

where **M** = magnetic intensity

H = intensity of magnetisation or magnetic moment per unit volume

8. $\mathbf{A} = i\mathbf{A}_x + j\mathbf{A}_y + k\mathbf{A}_z$

$\mathbf{B} = i\mathbf{B}_x + j\mathbf{B}_y + k\mathbf{B}_z$

(*i*) $\mathbf{A}.\mathbf{B} = A_x B_x + A_y B_y + A_z B_z$

(*ii*) $$\mathbf{A} \times \mathbf{B} = \begin{vmatrix} i & j & k \\ A_x & A_y & A_z \\ B_x & B_y & B_z \end{vmatrix}$$

(*iii*) $\mathbf{A}.(\mathbf{B} \times \mathbf{C}) = \mathbf{C}.(\mathbf{A} \times \mathbf{B}) - \mathbf{B}.(\mathbf{A} \times \mathbf{C})$

(*iv*) $\mathbf{A} \times (\mathbf{B} \times \mathbf{C}) = (\mathbf{A}.\mathbf{C}).\mathbf{B} - (\mathbf{A}.\mathbf{B}).\mathbf{C}$

9. ∇ = del operator

(*i*) $$\nabla = i\frac{\partial}{\partial x} + j\frac{\partial}{\partial y} + k\frac{\partial}{\partial z}$$

(*ii*) $$\nabla\phi = \text{gradient of } \phi = \text{grad } \phi\; i\frac{\partial \phi}{\partial x} + j\frac{\partial \phi}{\partial y} + k\frac{\partial \phi}{\partial z}$$

(*iii*) $$\nabla.\mathbf{A} = \text{divergence of } \mathbf{A} = \text{div } \mathbf{A} = \frac{\partial A_x}{\partial x} + \frac{\partial A_y}{\partial y} + \frac{\partial A_z}{\partial z}$$

(*iv*) $$\nabla \times \mathbf{A} = \text{curl of } \mathbf{A} = \text{curl } \mathbf{A} = \begin{vmatrix} \hat{i} & \hat{j} & \hat{k} \\ \frac{\partial}{\partial x} & \frac{\partial}{\partial y} & \frac{\partial}{\partial z} \\ A_x & A_y & A_z \end{vmatrix}$$

10. Gauss Transformation Formula

If any volume V is enclosed by the surface S then surface integral can be changed into volume integral by Gauss transformation formula in the following manner

$$\int_S \mathbf{A}\hat{n}\, dS = \int_V div\, \mathbf{A}\, dV$$

N.B. (*i*) curl (grad ϕ) = 0

and (*ii*) div (curl **A**) = 0

11. Stoke's Transformation Formula

By this transformation we change line integral into surface integral

$$\oint_l \mathbf{A} \cdot d\mathbf{l} = \int_S \text{curl}\, \mathbf{A}\, \hat{n}\, dS$$

12. The Equation of Continuity

$$I = -\frac{dq}{dt} \quad \Rightarrow \quad \int_S \mathbf{J}\hat{n}\, dS$$

J = Current density

$$= \frac{I}{A}$$

$$\int_S \mathbf{J}\hat{n}\, dS = -\frac{d}{dt}\int_V \rho\, dV = -\frac{dq}{dt}$$

Applying Gauss transformation

ρ = charge per unit volume or charge density

$$\int_v \text{div}\, \mathbf{J}\, dV = \int_v \frac{d\rho}{dt}\, dV$$

$$\Rightarrow \quad \int_v \text{div}\, \mathbf{J}\, dV = -\int \frac{d\rho}{dt}\, dV$$

$$\Rightarrow \quad \text{div}\, \mathbf{J} = -\frac{d\rho}{dt}$$

This is the continuity equation.

7.23 NUMERICAL PROBLEMS BASED ON ELECTROMAGNETISM

Problem 1. *Find the gradient, divergence and curl of* $|\mathbf{r}|$, *where* $\mathbf{r}$ *is the position vector.*

Hint: $\mathbf{r} = i\,x + j\,y + k\,z \quad |\mathbf{r}| = (x^2 + y^2 + z^2)^{1/2}$

Then grad $|\mathbf{r}|$ $\quad \nabla|\mathbf{r}| = \nabla(x^2 + y^2 + z^2)^{1/2} = \boxed{\hat{r}}$

where $\nabla = i\dfrac{\partial}{\partial x} + j\dfrac{\partial}{\partial y} + k\dfrac{\partial}{\partial z}$.

$$\text{divergence of } \mathbf{r} = \nabla.\mathbf{r} = \left(i\frac{\partial}{\partial x} + j\frac{\partial}{\partial y} + k\frac{\partial}{\partial z}\right).(i\,x + j\,y + k\,z) = \boxed{3}$$

$$\text{curl of } \mathbf{r} = \nabla \times \mathbf{r} = \boxed{0}$$

Problem 2. *If* $\phi = 5xy - 3y^2 z^3$, *find* $\nabla\phi$ *at the point* $(-1, 2, -1)$.

Hint: $$\nabla\phi = \left(i\frac{\partial}{\partial x} + j\frac{\partial}{\partial y} + k\frac{\partial}{\partial z}\right)(5xy - 3y^2 z^3)$$

$\Rightarrow$ *at* $(-1, 2, -1)$ $\nabla\phi = \boxed{10i + 7j - 36k.}$

Problem 3. *If* $\mathbf{A} = 3xyz^2\, i + 2xy^3\, j - x^2yz\, k$, *find* $\nabla.\mathbf{A}$ *at a point* $(1, -1, 1)$.

Hint: $$\nabla.\mathbf{A} = \left(i\frac{\partial}{\partial x} + j\frac{\partial}{\partial y} + \hat{k}\frac{\partial}{\partial z}\right).\,(3xyz^2\, i + 2xy^3\, j - x^2yz\, k) \text{ at } (1, -1, 1) = \boxed{4 \text{ units}}$$

Problem 4. *Prove that* $\nabla^2\left(\dfrac{1}{r}\right) = 0$.

Hint: Here $\quad \mathbf{r} = ix + jy + kz$, then $r = |\mathbf{r}| = \sqrt{(x^2 + y^2 + z^2)}$

Now $$\nabla^2\left(\frac{1}{r}\right) = \nabla.\left\{\nabla\left(\frac{1}{r}\right)\right\}.$$

Problem 5. *Find the work done in moving a particle along the curve* $y = x^2$ *by a force* $\vec{F} = i\,(2xy) - j\,(y^3)$ *from* $(0, 0)$ *to* $(2, 4)$.

Hint: $$W = \int \vec{F}\,.\,d\vec{r} = \int F_1\,dx + F_2\,dy$$

Here $\quad F_1 = (2xy),\ F_2 = y^3$

$$W = \int (2xy\,dx + y^3\,dy)$$

Now $\quad y = x^2 \quad \therefore \quad dy = 2x\,dx$

$$W = \int (2x.x^2\,dx - x^6.2x\,dx) = \int_0^2 (2x^3 - 2x^7)\,dx = \boxed{-56 \text{ unit}}$$

Problem 6. *Evaluate the line integral of function $\vec{F} = i\,(6x) + j\,(4y)$ between the points (0, 0) and (2, 2) in x-y plane.*

Hint: $\int \mathbf{F}.d\mathbf{l} = \int_M^N (F_x dx + F_y\, dy + F_z dz)$; x-y plane mean $z = 0$

$$= \int_0^2 6x\,dx + \int_0^2 4y\,dy = \boxed{20.}$$

Problem 7. *Obtain the values of the following:*

(i) curl grad ϕ *(ii) grad div* **A** *and*

(iii) div curl **A**.

Hint: (*i*) $\boxed{0}$

(*ii*) $i\dfrac{\partial}{\partial x}\left(\dfrac{\partial A_x}{\partial x} + \dfrac{\partial A_y}{\partial y} + \dfrac{\partial A_z}{\partial z}\right) + j\dfrac{\partial}{\partial y}\left(\dfrac{\partial A_x}{\partial x} + \dfrac{\partial A_y}{\partial y} + \dfrac{\partial A_z}{\partial z}\right) + k\dfrac{\partial}{\partial z}\left(\dfrac{\partial A_x}{\partial x} + \dfrac{\partial A_y}{\partial y} + \dfrac{\partial A_z}{\partial z}\right)$

(*iii*) $\boxed{0}$.

Problem 8. *Prove that $\iint_S \text{curl}\,\mathbf{F}\,.\,d\mathbf{S} = 0$ for any closed surface.*

Hint: Gauss's theorem $\iint_S \text{curl}\,\mathbf{F}\,.\,d\mathbf{S} = \iiint_V \nabla.(\text{curl}\,\mathbf{F})\,dV$

$$= \iiint_V \text{div.curl}\,\mathbf{F}\,dV$$

div curl **F** = 0

Problem 9. *Using Stoke's theorem, prove that*

$$\oint_C \mathbf{r}.d\mathbf{l} = 0, \text{ where } \mathbf{r} \text{ is position vector.}$$

Hint: $\oint \mathbf{A}.d\mathbf{l} = \iint \text{curl}\,\mathbf{A}\,.d\mathbf{S}$, here curl $\mathbf{r} = 0$

Hence $\int \mathbf{r}.d\mathbf{l} = 0.$

Problem 10. *Electric field associated with a charge body is given by* $\mathbf{E} = 4i + j + 7k$ *and the surface is given by* $\mathbf{S} = 8i + 3j$. *Calculate the electric flux coming out the surface.*

Hint: $\phi = \mathbf{E.S} = (4i + j + 7k).(8i + 3j) = \boxed{35 \text{ unit.}}$

Problem 11. *Show that* **F** *= $(2xy + z^3)\, i + x^2\, j + 3xz\, k$ is a conservative field.*

Hint: $\nabla \times \mathbf{F} = \boxed{0}$, Hence **F** is conservative.

Problem 12. *Examine whether the field given by* **E** *= $a\,(xy^2\, i + y^3\, j)$ is conservative. (here i and j are unit vectors), a is constant.*

Hint: $\nabla \times \mathbf{E} = -2\, ax\, y\hat{k} \quad \therefore \quad \nabla \times \mathbf{E} \neq 0$

Hence field is not conservative.

Problem 13. *If* **E** *= 3i + 4j + 8k. Calculate the electric flux through a surface area 100 units lying in x-y plane.*

Hint: $\phi = \mathbf{E.S} = (3i + 4j + 8k).\, 100\, k$

$= 800\ \text{Nm}^2\text{C}^{-1}$

Problem 14. *The electric field* **E** *in a certain space is given by*

$$E_x = (ax + by + c),\ E_y = 0,\ E_z = 0.$$

Use Gauss's theorem, to evaluate the charge enclosed in a cube with one of its sides of length L along the x-axis, the two faces of the cube are perpendicular to the x-axis lying in the planes x = l and x = l + L.

Hint: According to Gauss theorem of divergence

$$\iint \mathbf{E}\,.\,d\mathbf{S} = \iiint_V div\, \mathbf{E}\, dV$$

$$= \iiint_V \left(\frac{\partial E_x}{\partial x} + \frac{\partial E_y}{\partial y} + \frac{\partial E_z}{\partial z} \right) dV$$

$$= \iiint_V a\, dV = a\, V = a\, L^3 \qquad ...(i)$$

From the Gauss's law in electrostatics

$$\iint_S \mathbf{E}\,.\,d\mathbf{S} = \frac{q}{\varepsilon_0} \qquad ...(ii)$$

From equations (*i*) and (*ii*), $a\, L^3 = \frac{q}{\varepsilon_0} \Rightarrow \boxed{q = \varepsilon_0\, a\, L^3.}$

Problem 15. *A long wire carries a current of 5 mA. Find the integral of* **B** *around the path of radius 10 cm, enclosing the wire.* $\frac{\mu_0}{4\,\pi} = 10^{-7}$ *S.I. units.*

Hint: $\oint \mathbf{B}.d\mathbf{l} = \mu_0\, i = \boxed{62.9 \times 10^{-10}}$ Tesla meter.

Problem 16. *A ring of radius 10 cm and resistance 2 Ohm is placed in a magnetic field of 0.5 Weber/m² which is perpendicular to the plane of the ring. If the magnetic field is decreasing at the rate of 0.1 Weber/m² per second, find the values of the induced e.m.f. and the induced current in the ring.*

Hint: The time rate of change of magnetic flux is

$$\frac{d\phi}{dt} = \frac{d}{dt}(BA) = A\frac{dB}{dt}$$

$$A = \pi r^2 = 3.14 \times (10 \times 10^{-2})^2 = 3.14 \times 10^{-2}\,\text{m}^2, \quad \frac{dB}{dt} = \frac{0.1}{1}\ \text{Weber/m}^2\,\text{sec}$$

$$\frac{d\phi}{dt} = 3.14 \times 10^{-3}\ \text{Weber/sec.}$$

According Faraday's law $e = \frac{d\phi}{dt} = 3.14 \times 10^{-3}$ Volts

If R is resistance of the loop, then induced current

$$i = \frac{e}{R} = \frac{3.14}{2} = 1.57\ \text{mA.}$$

Problem 17. *Express displacement current, appeared inside a parallel plate capacitor, in terms of electric field (E) and electric potential.*

Hint: Frame Gauss's law

$$\oint_S \mathbf{E}.d\mathbf{S} = \frac{q}{\varepsilon_0}$$

Assuming the Gaussian surface enclosing the charges of in the plate, then above equation becomes $EA = \frac{q}{\varepsilon_0}$, where A = area of each plate of the capacitor $\Rightarrow$ $q = \varepsilon_0 EA$

$$\Rightarrow \quad \frac{dq}{dt} = \varepsilon_0 A\frac{dE}{dt} \Rightarrow i_d = \varepsilon_0 A\frac{dE}{dt} \text{ where } i_d = \text{displacement current}$$

We know that $\quad q = CV \Rightarrow \frac{dq}{dt} = C\frac{dV}{dt} \Rightarrow i_d = C\frac{dV}{dt}$

Alternatively: $\quad i_d = \varepsilon_0 A\frac{dE}{dt}. \Rightarrow i_d = \frac{\varepsilon_0 A}{d}.\frac{d}{dt}(Ed)$; d = spacing of plates

$$\Rightarrow \quad i_d = \frac{\varepsilon_0 A}{d}.\frac{dV}{dt} \qquad (\because V = E.d)$$

$$\Rightarrow \quad i_d = C\frac{dV}{dt} \qquad \left(\because C = \frac{\varepsilon_0 A}{d}\right)$$

Problem 18. *An infinite long wire is stretched horizontally 4 m above the surface of the earth. It has a charge of 1 μ C/cm. Calculate electric field at a point on earth vertically below the wire.*

$$\textbf{Hint : } E = \frac{\lambda}{2\pi\varepsilon_0 r} = \frac{2\lambda}{4\pi\varepsilon_0 r} = \frac{9\times10^9\times2\times10^{-4}}{4} = 4.5 \times 10^5 \text{ N/Coul.}$$

Problem 19. *Assuming that all the energy from a 1000 W lamp is radiated uniformly, calculate the average values of the intensities of electric and magnetic fields of radiation at a distance of 2 m from the lamp.*

Hint: Energy of lamp = 1000 W; Area illuminated = $4\pi r^2 = 4\pi (2)^2 = 16\pi$ met^2

$$\therefore \quad \text{Energy radiated per unit area per second} = \frac{1000}{16\pi}$$

$$\text{Hence from Poynting theorem } |\mathbf{S}| = |\mathbf{E}\times\mathbf{H}| = EH = \frac{1000}{16\pi} \qquad ...(i)$$

and
$$\frac{E}{H} = \sqrt{\frac{\mu_0}{\varepsilon_0}} = 376.72 \qquad ...(ii)$$

$$\text{Eq. } (i) \times \text{Eq. } (ii) \Rightarrow E^2 = \frac{1000}{16\pi} \times 376.2 \Rightarrow E = 48.87 \text{ V/m.}$$

EXERCISES

1. Define gradient of a scalar point function. Give examples.
2. Define divergence of a vector point function with example. Give its physical significance.
3. Define curl of a vector point function with example. Give its physical significance.
4. Write a short note on (*i*) Equipotential surfaces, (*ii*) Displacement current.
5. What are the line, surface and volume integrals?
6. State and prove Stoke's theorem.
7. State and prove Gauss's divergence theorem.
8. State and prove Gauss theorem. Derive Coulomb's law of electrostatics from Gauss's theorem.
9. Obtain an expression of the field intensity at any point due to a line charge by Gauss's theorem.
10. Obtain an expression for the field intensity at any point due to infinite cylindrical charge by Gauss law.
11. State Ampere's law in magnetostatics in integral form and from that deduce its differential form.
12. State Ampere's circuital law.
13. Distinguish conduction current and displacement current.

14. What is the significance of displacement current?
15. Obtain the Maxwell's equations of electromagnetism from fundamental laws of electricity and magnetism.
16. Write the Maxwell's equations in integral and differential forms. Discuss their physical significance.
17. Derive an expression for the speed of light in free space.
18. Explain the concept of displacement current and show that it lead to the modification of Ampere's law. Show that in free space the Maxwell's equations lead to the relation

$$\nabla^2 E = \varepsilon_0 \mu_0 \left(\frac{\partial^2 E}{\partial t^2}\right).$$

19. Define Poynting vector. Derive an expression for it and explain its physical significance for e.m. waves in free space.

SUGGESTION FOR FURTHER READING

1. David J. Grifiths, Introduction to Electrodynamics, Prentice-Hall of India Pvt. Ltd., New Delhi, 2007.
2. Matthew N.O Sadiku, Elements of Electrodynamics, Oxford University Press, New Delhi, 2008.
3. Uma Mukherji, Electromagnetic Field Theory and Wave Propagation, Narosa Publishing House, New Delhi, 2008.
4. S.K. Gupta, Electromagnetic Field Theory, Krishna Prakashan Media (P) Ltd., Meerut, 2005.
5. B.S. Guru and H.R. Hiziroglu, Electromagnetic Field Theory Fundamentals, Cambridge University Press, New Delhi, 2007.
6. R. Gowri, Electromagnetic Fields and Waves, S.K. Kataria and Sons, Delhi, 2009.
7. Devraj Singh, Engineering Physics, Vol. 2, Dhanpat Rai and Co. (P) Ltd., New Delhi, 2009.
8. Devraj Singh, Fundamentals of Engineering Physics, Vol. 1, Dhanpat Rai and Co. (P) Ltd., New Delhi, 2009.
9. Parul Dawar, Electromagnetic Field Theory, S.K. Kataria and Sons, New Delhi, 2009.
10. D.C. Tayal, Electricity and Magnetism, Himalayan Publishing House, Mumbai, 2007.

Chapter 8

SPECIAL THEORY OF RELATIVITY

8.1 INTRODUCTION

With the acceptance of the electromagnetic theory of light only two general problems remained. One of these was that of luminiferous ether, a hypothetical medium suggested as the carrier of light waves, just as air or water carries sound waves. The ether was assumed to have some very unusual properties, *e.g.*, being massless but having high elasticity. A number of experiments performed to give evidence of the ether, most notably by A.A Michelson in 1881 and by Michelson and E.W. Morley in 1887, failed to support ether hypothesis. With the publication of special theory of relativity in 1905 Albert Einstein, the ether was shown to be unnecessary to the electromagnetic theory. Much before Einstein it was considered that the velocity of light, like other material's particle velocity, should depend on the relative motion between the observer and the source of light. However, experiments failed to disapprove the constancy of velocity of light to generate a new branch of physics. According to classical mechanics space and time are absolute and they are not interconnected.

This absolute nature is broken by relativity which generally starts that both space and time are a function of observer's velocity as well as observables velocity. At high velocities both space and time combine to form space-time continuum.

The theory which deals with the relativity of motion and rest is called the theory of relativity. Einstein was the pioneer in this field and did a lot which helps us to decode the mysteries of the universe.

It has two parts :

Special Theory of Relativity. It deals with objects and system which are either moving at a constant speed with respect to one another or at rest.

General Theory of Relativity. It deals with objects and system which are speeding up or slowing down with respect to one another.

Here we are only dealing with Special Theory of Relativity only.

8.2 FRAMES OF REFERENCE

We have seen that motion of the body has no meaning unless it is described with respect to some well-defined co-ordinate system. The co-ordinate system is known as the "**frame of reference**". Thus a frame of reference is a coordinate system relative to which the position and motion of the body is described. The reference frame may be at rest or under motion. Its choice is totally at our own convenience. Suppose we want to study about anything from Earth. Thus we are considering *frame of reference* as the Earth. However, Earth is not a stationary object. Similarly, a passenger sitting in a train observes outside trees moving backwards, which is not the fact. Actually in the reference system attached to the earth the train is moving forward.

There are two types of *Frames of Reference :*

8.2.1 Inertial (unaccelerated)

This obeys *Newton's laws of inertia* and other laws of *Newtonian mechanics*. In this frame, a body not acted upon by external force, is at rest, or moves with a constant velocity. Actually for different types of frames of references the laws of mechanics becomes different. We can approximate a reference frame which moves with constant velocity relative to the distant stars as a best approximation of inertial frame.

8.2.2 Non-Inertial (accelerated)

This doesn't obey *Newton's laws of inertia* and other laws of *Newtonian mechanics*. In this frame, a body not acted upon by external force, is accelerated. A coordinate system fixed in the Earth is a non-inertial frame if the object motion is observed from the outside of the Earth as it itself rotates about its axis and also the sun.

Ether: Newton's concept of stationary frame of reference gave the idea of the *'ether'* frame in 19th century. As we measure the speed of any moving object with respect to any stationary object, the velocity of light also requires a stationary frame for its velocity measurement. Hence it seemed a fairly logical postulate to think for the ether medium having unusual properties like zero density and 100% transmittivity. These properties are essential for its undetectibility. It was assumed to fill all the space by ether and called as a *universal frame of reference*, relative to which any motion can be defined.

8.3 GALILEAN TRANSFORMATION

In this system the frames were taken as inertial one, which obey Newton's laws. Also any reference frame moving with uniform velocity with respect to another inertial frame is also an inertial frame. Our experience shows that any experiment performed in a uniformly moving vehicle will be the same if we perform the same experiment in any stationary laboratory. It implies the universality of every inertial frame.

Galileo's principle of relativity tells that laws of mechanics are the same in all inertial frames of reference. The transformation from one inertial frame of reference to another is called the *Galilean Transformation.*

Consider two frames of reference S and S' as in Figure 8.1.

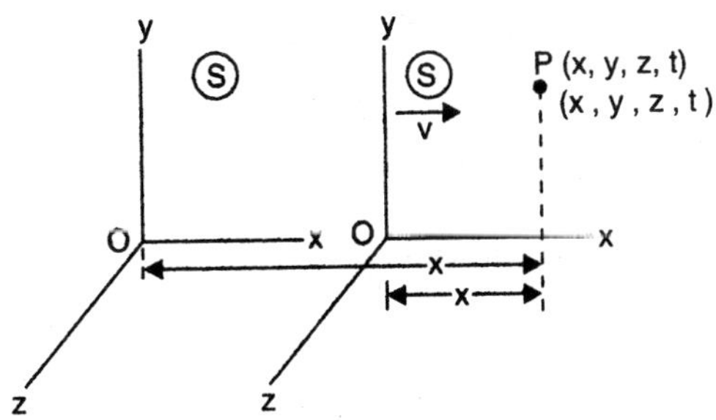

FIGURE 8.1 **Stationary and moving frames of reference**

The Galilean transformation establishes a relationship between the coordinates (x, y, z, t) of an event at S and the coordinates (x', y', z', t') of the same event at S'. We assume S' moves in such a way that x and x' are in one straight line and have some positive directions. For convenience, we are restricting the motion of the moving frame in x-direction only. Also O and O' coincide at $t = 0$ and S' moves with a uniform velocity 'v' with respect to S.

I. Coordinate Transformation

From Figure 8.1 above,

$$x = x' + vt \quad \Rightarrow \quad x' = x - vt$$

also,
$$y' = y;\ z' = z;\ t' = t$$

So the Galilean transformations for S and S' frame are

$$\boxed{\begin{aligned} x' &= x - vt \\ y' &= y \\ z' &= z \\ t' &= t \end{aligned}}$$

II. Velocity Transformation

Consider the movement of a moving body. Let the body be at the position P_1 at t_1 and P_2 at t_2. Now at P_1 the coordinates in system S are (x_1, y_1, z_1, t_1) and P_2 the coordinates in system S' are (x_2, y_2, z_2, t_2)

Hence, for S' system, using coordinate transformation as above,

$$x_1' = x_1 - vt_1;\ y_1' = y_1;\ z_1' = z_1$$

and,
$$x_2' = x_2 - vt_2;\ y_2' = y_2;\ z_2' = z_2$$

So,
$$x_2' - x_1' = (x_2 - x_1) - v(t_2 - t_1)$$

$$\Rightarrow \quad \frac{x_2' - x_1'}{t_2 - t_1} = \frac{x_2 - x_1}{t_2 - t_1} - v.$$

Let,
$$u_x' = \frac{x_2' - x_1'}{t_2 - t_1};\ u_x = \frac{x_2 - x_1}{t_2 - t_1}.$$

So the Galilean transformation for velocity is,

$$u_x' = u_x - v$$

$$u_y' = u_y$$

$$u_z' = u_z$$

III. Acceleration Transformation

We now consider the body in S moves along x-axis with acceleration 'a' and a' with respect to S'. So at some moment of time,

$$u = u' + v \quad \ldots(1)$$

After time Δt, the velocity in both systems increased and the new velocities are

$$u + \Delta u = (u' + \Delta u') + v$$

Using equation (1) we have,

$$[(u' + v) + \Delta u] = (u' + \Delta u') + v$$

$$\Delta u = \Delta u'$$

$$\Rightarrow \quad \frac{\Delta u}{\Delta t} = \frac{\Delta u'}{\Delta t}$$

$$\Rightarrow \quad a = a'$$

This is Galilean transformation for acceleration

If F and F' are the forces in S-frame and S' frame respectively, then, $F = ma = ma' = F'$.

It shows that the acceleration and forces do not change from one inertial frame to another inertial frame. Hence force and acceleration are invariant under Galilean transformation.

8.4 MICHELSON-MORLEY EXPERIMENT

The requirement to measure the speed of light necessitates the need of *reference frame* or coordinate frame with respect to which we can measure the velocity. It was assumed initially that *ether* is present in the space through which light travels. This ether is perfectly transparent to light and the material body may pass through it without any resistance. The orbital velocity of earth is 30 km/s with respect to ether. If ether exists, the spinning and orbital motion of the earth should be through it. An observer on the earth would sense an 'ether wind' whose velocity is 'v' relative to the earth. Hence, as ether was assumed to be at rest, it was considered possible to measure absolute velocity of light moving through it. The experimental set up used for the verification of ether medium by Michelson and Morley is described below :

Michelson and Morley mounted the interferometer on a massive stone slab for stability and floated the apparatus in mercury so that we can rotate the apparatus smoothly about a central pin. The light path was increased by making it travel through mirrors on the slab to reflect the beam back and forth through eight round trips. Light from a monochromatic source S after being made parallel by a collimating lens L falls on semi-silvered glass plate G inclined at 45° to thc beam. It is then divided into two parts. First part, ray 1 travels towards mirror M_1 and second part ray '2' travels towards mirror M_2. Then they reflected back along their original paths meet again at the semi-silvered face of G and enter a short focus telescope T. These two ray produce interference fringes in the field of view of T. A desired path difference can be introduced between the two reflected rays by moving M_1. The optical distances of M_1 and M_2 from G are made equal by using a compensating glass plate G'. If the apparatus is at rest, the two reflected rays would take equal time to return the glass plate G. But actually the whole apparatus is moving along with the Earth.

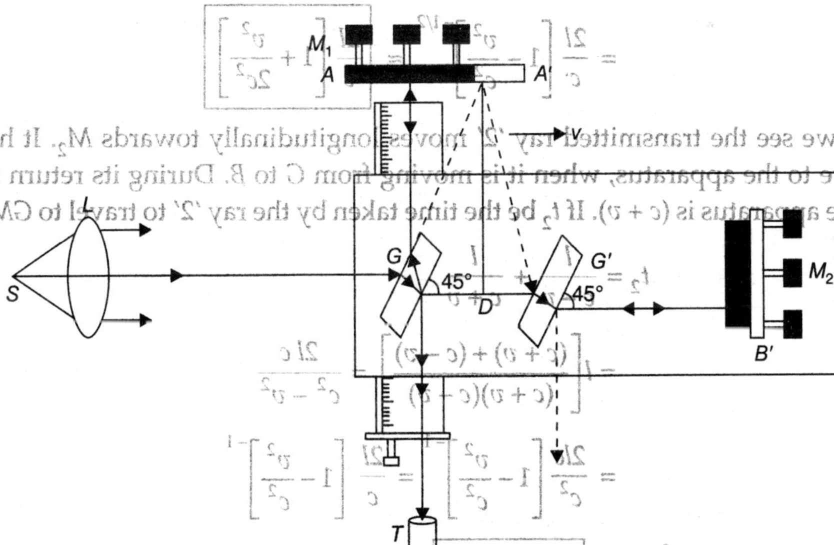

FIGURE 8.2 Michelson-Morley Experiment

Let us suppose the direction of motion of earth is in the direction of the initial beam due to the motion of the earth the optical paths traversed by both the rays are not the same. The reflections at M_1 and M_2 do not take place at A and B, but at A' and B' respectively. Thus the time taken by the two rays to travel to the mirrors and back to G will be different in this case.

Theory: Let M_1 and M_2 be at distances 'l' from G. Also let 'c' and 'v' are velocities of light and the apparatus or Earth respectively. Now the reflected ray '1' from G strikes M_1 at A' and not at A due to Earth's notion. The total path of the ray '1' from G to A' and back to G' will be GAG'.

By laws of reflection,

$$GA' = A'G'$$

$$GA'G' = GA' + A'G' = 2GA'$$

However, $$(GA')^2 = (GD)^2 + (A'D)^2$$

$$= (AA')^2 + (A'D)^2 \qquad ...(1)$$

If 't' be the time taken by the ray to move from G to A' then,

$$GA' = ct;\ AA = vt \text{ and } A'D = l$$

So from equation (1)

$$c^2t^2 = v^2t^2 + l^2 \quad \Rightarrow \quad t = \frac{l}{\sqrt{c^2 - v^2}}$$

If t_1 be the time taken by the ray '1' to travel $GA'G'$, then,

$$t_1 = 2t = \frac{2l}{\sqrt{c^2 - v^2}}$$

$$= \frac{2l}{c}\left[1 - \frac{v^2}{c^2}\right]^{-1/2} \approx \boxed{\frac{2l}{c}\left[1 + \frac{v^2}{2c^2}\right]} \qquad ...(2)$$

Now, we see the transmitted ray '2' moves longitudinally towards M_2. It has velocity $(c - v)$ relative to the apparatus, when it is moving from G to B. During its return its velocity relative to the apparatus is $(c + v)$. If t_2 be the time taken by the ray '2' to travel to GM_2'G', then,

$$t_2 = \frac{l}{c - v} + \frac{l}{c + v}$$

$$= l\left[\frac{(c + v) + (c - v)}{(c + v)(c - v)}\right] = \frac{2l\,c}{c^2 - v^2}$$

$$= \frac{2lc}{c^2}\left[1 - \frac{v^2}{c^2}\right]^{-1} = \frac{2l}{c}\left[1 - \frac{v^2}{c^2}\right]^{-1}$$

$$= \boxed{\frac{2l}{c}\left[1 + \frac{v^2}{c^2}\right]} \qquad ...(3)$$

Thus from equations (2) and (3), the difference in times to travel of longitudinal and transverse journey is

$$\Delta t = t_2 - t_1$$

$$= \frac{2l}{c}\left[1 + \frac{v^2}{c^2} - 1 - \frac{v^2}{2c^2}\right] = \boxed{\frac{lv^2}{c^3}} \qquad ...(4)$$

The number of fringes shift between the two rays is given by

$$N = \frac{\text{Path difference}}{\text{Wavelength}} = \frac{c\Delta t}{\lambda} = \boxed{\frac{lv^2}{\lambda c^2}} \qquad ...(5)$$

This shift will be observed if Earth is made at rest. However, this is not possible. Hence Michelson and Morley performed the experiment in two steps, firstly, by setting up the apparatus as shown in Figure 8.2 and secondly, by turning the apparatus through 90°.

When the apparatus was turned through 90°, the positions of M_1 and M_2 are changed. Now the path difference is in opposite direction, *i.e.*, the path difference is $-\frac{lv^2}{\lambda c^2}$ fringes.

The resultant path difference now becomes

$$\frac{lv^2}{\lambda c^2} - \left(-\frac{lv^2}{\lambda c^2}\right) = \frac{2lv^2}{\lambda c^2} \text{ fringes}$$

So a fringe shift of $\frac{2lv^2}{\lambda c^2}$ is expected.

Using $l = 10$ m, $\lambda = 5000$ Å, $v = 3 \times 10^6$ cm/s and

$c = 3 \times 10^{10}$ cm/s. We get a fringe shift, n as $\boxed{n = \frac{2lv^2}{\lambda c^2}} = 0.4$ fringe.

Thus a fringe shift of less than half of fringe was expected, but no change was observed actually.

Though, Michelson and Morley could observe a shift of about 0.01 fringe only using the same experiment.

They repeated the experiment at different locations in Europe and USA in day and night and at different seasons, but could not detect any measurable shift. So, *it was a complete Negative Result.*

"This result suggests that it is impossible to measure the speed of earth relative to the fixed frame of reference, i.e., ether. Further no existence of the ether was detected. It was also concluded that the speed of light in vaccum is same in all inertial frames".

8.4.1 Explanation of the Negative Result of Michelson-Morley Experiment

The negative result of Michelson-Morley experiment was explained by considering following assumptions.

(*i*) **Constancy of speed of light.** If we consider the speed of light as constant then there would be no path difference between reflected and transmitted rays and so no fringe shift will be observed.

(*ii*) **Ether drag hypothesis.** If we consider that material bodies are dragged with the ether medium then there won't be any relative motion between earth and ether. So no path difference should be observed and there will be no fringe shift.

(*iii*) **Lorentz Fitzgerald contraction hypothesis.** In 1892, *Lorentz -Fitzgerald* proposed a hypothesis that length of all material bodies moving with velocity 'v' contract in the direction of motion by a factor of $\sqrt{\left(1-\frac{v^2}{c^2}\right)}$. This hypothesis retains the concept of ether frame.

If l_o is the length of the moving object at rest with respect to ether frame then length of body moving with velocity 'v' can be written as

$$l = l_o\sqrt{\left(1-\frac{v^2}{c^2}\right)}$$

According to this hypothesis t_2 from equation. (3) can be modified as

$$t_2 = \frac{2l}{c}\left(1-\frac{v^2}{c^2}\right)^{1/2}\left(1+\frac{v^2}{c^2}\right)$$

$$= \frac{2l}{c}\left(1-\frac{v^2}{2c^2}\right)\left(1+\frac{v^2}{c^2}\right)$$

$$= \frac{2l}{c}\left(1+\frac{v^2}{c^2}-\frac{v^2}{2c^2}-\frac{v^4}{2c^4}\right)$$

$$= \frac{2l}{c}\left(1+\frac{v^2}{2c^2}\right) \text{ (neglecting higher powers of } v/c)$$

This is equal to time 't_1' as given by equation. (2).

Thus there is no path difference and hence no fringe shift is observed in experiment.

8.5 POSTULATES OF SPECIAL THEORY OF RELATIVITY

The negative result of Michelson-Morley experiment gave rise to postulates of Special Theory of Relativity by Einstein. Which are given below :

First Postulate: The fundamental laws of physics have the same form for all inertial frames. No preferred inertial system exists. This is the principle of relativity.

Second Postulate: The speed of light in free space has the same value in all inertial frames of reference. This is the principle of constancy of the speed of light.

8.6 LORENTZ TRANSFORMATION

Lorentz tried to replace Galilean transformation to give results in agreement with Michelson and Morley experiment and second postulate of the special theory of relativity and developed his transformation equations in 1890.

I. Coordinate transformation

Here we consider two inertial systems S and S'. We assume S' moves in such a way that x and x' are one straight line and have same positive directions. Also O and O' coincides at $t = 0$ and S' moves with a velocity 'v' with respect to S. For our convenience, we take the axis of two coordinate systems coincide at $t = t' = 0$. (Refer to Figure 8.1)

Let a pulse of light be generated at $t = 0$ at the origin which grows in the space.

Now we consider the situation when the pulse reaches at point P. Let (x, y, z, t) and (x', y', z', t') are the position and time coordinates of P measured by observers at O and O' in frames S and S' respectively. When the pulse is observed from S, we have the velocity of light as distance divided by the time taken; hence,

$$c = \frac{\sqrt{(x^2 + y^2 + z^2)}}{t}$$

$$\Rightarrow \quad x^2 + y^2 + z^2 - c^2t^2 = 0 \quad \text{...(1)}$$

Similarly, when the pulse in observed from S';

$$(x')^2 + (y')^2 + (z')^2 - c^2(t')^2 = 0 \quad \text{...(2)}$$

whereas

$$y = y' \,;\, z = z' \quad \text{...(3)}$$

Using equation (3) into equations (1) and (2) we have,

$$x^2 + y^2 + z^2 - c^2t^2 = (x')^2 + (y')^2 + (z')^2 - c^2(t')^2$$

$$\Rightarrow \quad x^2 + y^2 + z^2 - c^2t^2 = (x')^2 + y^2 + z^2 - c^2(t')^2$$

$$\Rightarrow \quad x^2 - c^2t^2 = (x')^2 - c^2(t')^2 \quad \text{...(4)}$$

Let the transformation between x and x' is represented by

$$x' = \lambda(x - vt) \quad ...(5)$$

where λ is independent of x and t. Suppose the system S in moving relative to S' along x-direction (because the motion is relative) then,

$$x = \lambda' (x' + vt') \quad ...(6)$$

where λ' being independent of x' and t'

From equations (5) and (6) we have,

$$x = \lambda' (\lambda(x - vt) + vt')$$

$$\Rightarrow \quad vt' = \frac{x}{\lambda'} - \lambda(x - vt)$$

$$= \lambda\left[\frac{x}{\lambda\lambda'} - x + vt\right]$$

$$= \lambda\left[vt - x\left(1 - \frac{1}{\lambda\lambda'}\right)\right]$$

$$t' = \lambda\left[t - \frac{x}{v}\left(1 - \frac{1}{\lambda\lambda'}\right)\right] \quad ...(7)$$

Now using equations (7) and (5) into equation (4) we have,

$$x^2 - c^2t^2 = [\lambda(x - vt)]^2 - c^2\left[\lambda\left\{t - \frac{x}{v}\left(1 - \frac{1}{\lambda\lambda'}\right)\right\}\right]^2$$

$$\Rightarrow \quad x^2 - c^2t^2 - \lambda^2 [x^2 - 2xvt + v^2t^2] + c^2\lambda^2\left[t^2 - \frac{2xt}{v}\left(1 - \frac{1}{\lambda\lambda'}\right) + \left(\frac{x}{v}\right)^2 + \left(1 - \frac{1}{\lambda\lambda'}\right)^2\right] = 0$$

This is an identity and hence coefficients of various powers of x and t must vanish separately. So, equating the coefficient of t^2 from both sides to zero, we have,

$$-c^2 - \lambda^2v^2 + c^2\lambda^2 = 0 \quad ...(8)$$

$$\Rightarrow \quad c^2 = \lambda^2(c^2 - v^2)$$

$$\Rightarrow \quad \lambda^2 = \frac{c^2}{c^2 - v^2} \quad ...(9)$$

$$\Rightarrow \quad \lambda = \frac{1}{\sqrt{\left(1 - \frac{v^2}{c^2}\right)}} \quad ...(10)$$

Now equating the coefficient of 'xt' from both sides to zero, we have,

$$2v\lambda^2 + c^2\lambda^2\left[-\frac{2}{v}\left(1 - \frac{1}{\lambda\lambda'}\right)\right] = 0$$

$$\Rightarrow \qquad 2v\lambda^2 = \frac{2c^2\lambda^2}{v}\left(1 - \frac{1}{\lambda\lambda'}\right)$$

$$\Rightarrow \qquad v^2\lambda\lambda' = c^2[\lambda\lambda' - 1]$$

$$\Rightarrow \qquad \lambda\lambda'(c^2 - v^2) = c^2 \qquad ...(11)$$

Comparing equations (8) and (11) we have,

$$\lambda\lambda'(c^2 - v^2) = c^2 = \lambda^2\,[c^2 - v^2]$$

$$\lambda\lambda' = \lambda^2 \quad \Rightarrow \qquad \lambda = \lambda'$$

So using equation (10) $\lambda = \lambda' = \dfrac{c}{\sqrt{c^2 - v^2}} = \dfrac{1}{\sqrt{1 - v^2/c^2}}$

Substituting equation (9) into equation (7) we have,

$$t' = \frac{1}{\sqrt{1 - v^2/c^2}}\left[t - \frac{x}{v}\left(1 - \frac{1}{\lambda^2}\right)\right]$$

$$= \frac{1}{\sqrt{1 - v^2/c^2}}\left[t - \frac{x}{v}\left(1 - \frac{c^2 - v^2}{c^2}\right)\right]$$

$$= \frac{1}{\sqrt{1 - v^2/c^2}}\left[t - \frac{x}{v}\frac{v^2}{c^2}\right]$$

$$t' = \frac{t - vx/c^2}{\sqrt{1 - v^2/c^2}} \qquad ...(12)$$

Substituting equation (9) into equation (5) we have,

$$x' = \frac{x - vt}{\sqrt{1 - v^2/c^2}} \qquad ...(13)$$

Hence,

Lorentz transformation equations	*Inverse Lorentz transformation equation*
$x' = \dfrac{x - vt}{\sqrt{\left(1 - \dfrac{v^2}{c^2}\right)}}$	$x = \dfrac{x' + vt}{\sqrt{\left(1 - \dfrac{v^2}{c^2}\right)}}$
$y' = y$	$y = y'$
$z' = z$	$z = z'$
$t' = \dfrac{\left(t - \dfrac{vx}{c^2}\right)}{\sqrt{\left(1 - \dfrac{v^2}{c^2}\right)}}$	$t = \dfrac{t' + \dfrac{vx'}{c^2}}{\sqrt{\left(1 - \dfrac{v^2}{c^2}\right)}}$

II. Velocity Transformation

Let the velocities of the particle in S and S' frames are $u(u_x, u_y, u_z)$ and $u'(u_x', u_y', u'_z)$ respectively

So, $$u_x = \frac{dx}{dt} \text{ and } u_x' = \frac{dx'}{dt'}$$

However from equation (13)

$$x' = \lambda\,(x - vt)$$

where $\lambda = \dfrac{1}{\sqrt{\left(1-\dfrac{v^2}{c^2}\right)}}$; $y = y'$; $z = z'$

On differentiating,

$$dx' = \lambda(dx - vdt);\ dy' = dy\ ;\ dz' = dz$$

Also from equation (12) $\quad t' = \lambda\left(t - \dfrac{xv}{c^2}\right)$

On differentiating,

$$dt' = \lambda\left(dt - \frac{v}{c^2}\,dx\right)$$

Hence,
$$\frac{dx'}{dt'} = \frac{\lambda(dx - vdt)}{\lambda\left(dt - \dfrac{v}{c^2}\,dx\right)} = \frac{dx - vdt}{dt - v/c^2\,dx} \qquad ...(14)$$

$$\frac{dy'}{dt'} = \frac{dy}{\lambda\left(dt - \dfrac{v}{c^2}\,dx\right)} \qquad ...(15)$$

$$\frac{dz'}{dt'} = \frac{dz}{\lambda\left[dt - \dfrac{v}{c^2}\,dx\right]} \qquad ...(16)$$

Using equation (14) we have,

$$\frac{dx'}{dt'} = \frac{\dfrac{dx}{dt} - v}{1 - \dfrac{v}{c^2}\dfrac{dx}{dt}}$$ (By dividing numerator and denominator by dt)

$$\Rightarrow \qquad u'_x = \frac{u_x - v}{1 - \frac{v}{c^2} u_x} \qquad ...(17)$$

Similarly,
$$u'_y = \frac{uy}{\lambda\left[1 - \frac{v}{c^2} u_x\right]} \qquad ...(18)$$

and
$$u'_z = \frac{u_z}{\lambda\left[1 - \frac{v}{c^2} u_x\right]} \qquad ...(19)$$

Also from equation (17)

$$u'_x - \frac{v}{c^2} u_x u'_x = u_x - v$$

$$\Rightarrow \qquad u_x = \frac{u'_x + v}{1 + \frac{v}{c^2} u'_x} \qquad ...(20)$$

From equation (18)

$$\lambda u'_y = \left[1 - \frac{v}{c^2} u_x\right] = u_y$$

$$\Rightarrow \qquad \lambda u'_y = \left[1 - \frac{v}{c^2}\left\{\frac{u'_x + v}{1 + (v/c^2) u'_x}\right\}\right] = \lambda u'_y\left[1 - \frac{(v/c^2)u'_x + v^2/c^2}{1 + v/c^2\, u_x}\right]$$

$$\Rightarrow \qquad u_y = \lambda u'_y\left[\frac{1 + (v/c^2)u'_x + v/c^2\, u'_x - v^2/c^2}{1 + v/c^2\, u'_x}\right]$$

$$= \lambda u'_y\left[\frac{1 - v^2/c^2}{1 + v/c^2\, u'_x}\right]$$

$$= \lambda u'_y\left[\frac{1/\lambda^2}{1 + v/c^2\, u'_x}\right] = u'_y\left[\frac{1/\lambda}{1 + v/c^2\, u_x}\right]$$

$$u_y = \frac{u'_y\sqrt{1 - v^2/c^2}}{1 + v/c^2\, u'_x} \qquad ...(21)$$

Similarly
$$u_z = \frac{u'_z\sqrt{1 - v^2/c^2}}{1 + v/c^2\, u_x} \qquad ...(22)$$

(*a*) For Non-relativistic Case

here, $v << c$ and $\lambda \approx 1$

So $$u_x = u'_x + v_1, u_y = u'_y \; ; u_z = u'_z$$

(*b*) When the motion is along x-axis only

here, $$u_x = u'_x \; ; u_y = 0, u_z = 0$$

$$\Rightarrow \qquad u_x^1 = \frac{u_x - v}{1 - \frac{v}{c^2} u_x}$$

(*c*) For relativistic case

when, $u_x = c$ then

$$u'_x = \frac{c - v}{1 - \frac{v}{c^2} c} = \frac{c - v}{(c - v)/c} = c$$

So, $$u = u' = c$$

It proves the constancy of the speed of light in any inertial reference frame.

(*d*) For $v > c$

Here the left hand side becomes negative and it has no physical significance.

8.7 LENGTH CONTRACTION

We consider a rod parallel to x-axis in frame S at rest.

The length of the rod in frame S (Stationary frame) is

$$l = x_2 - x_1$$

where x'_1 and x'_2 are the abcissae of the ends of the rod and in moving system S' it is measured as

$$l' = x'_2 - x'_1$$

where x'_1 and x'_2 are the abcissae of the ends of the same rod.

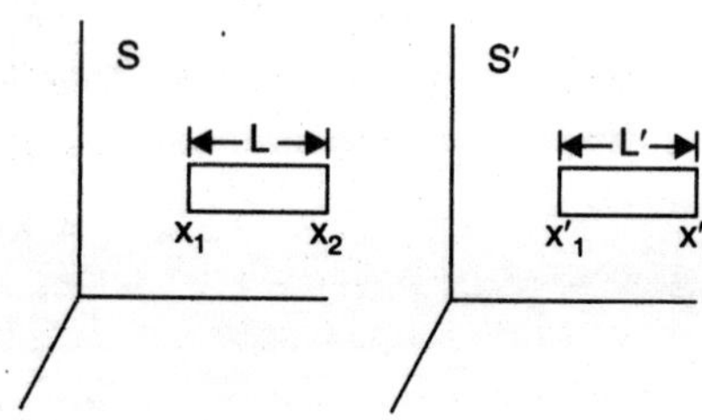

FIGURE 8.3

Using Lorentz transformation we have,

$$x = \frac{x' + vt'}{\sqrt{1 - \frac{v^2}{c^2}}}$$

So,
$$x_1 = \frac{x_1' + vt'}{\sqrt{1 - \frac{v^2}{c^2}}}$$

$$x_2 = \frac{x_2' + vt'}{\sqrt{1 - \frac{v^2}{c^2}}}$$

Hence,
$$l = x_2 - x_1$$

$$= \frac{x_2' + vt'}{\sqrt{1 - \frac{v^2}{c^2}}} - \frac{x_1' + vt'}{\sqrt{1 - \frac{v^2}{c^2}}}$$

$$= \frac{x_2' - x_1'}{\sqrt{1 - \frac{v^2}{c^2}}} = \frac{L'}{\sqrt{1 - \frac{v^2}{c^2}}}$$

$\Rightarrow$
$$l' = l\sqrt{1 - \frac{v^2}{c^2}}$$

This length is also known as *proper length.* Here $\left[1 - \frac{v^2}{c^2}\right] < 1$. Hence the rod seems to be contracted from the measured value of length in moving frame when observed from a stationary frame. This is called as *Length Contraction.*

The body's length is thus measured to be the maximum when it is at rest relative to the observer. As soon as it starts moving with a velocity 'v' relative to the observer, its measured length is contracted in the direction of its motion, whereas its dimensions perpendicular to the direction of motion remains unaffected. *Fitzgerald and Lorentz* proposed this idea of length contraction long before *Einstein's special Theory of relativity.* So, this is also known as **Lorentz-Fitzgerald contraction.** In Present particle physics experiments in high energy accelerators the colliding particles (nucleons, proton 5, etc.) have achieved a very high velocity approaching 'c', the length contraction is regularly observed.

8.8 TIME DILATION

We consider here a rail car (*i.e.,* a moving frame) moving with velocity 'v'. A mirror is fixed to its roof inside the car and an observer is at rest inside the car. He starts a light pulse at an instant vertically towards the mirror and the light signal reflects immediately back to the observer with a velocity 'c'.

The time interval is measured as,

$$\Delta t' = \frac{2h}{c} \qquad ...(1)$$

Here, 'h' is the vertical height of the mirror from the floor of the rail car as shown in Figure. 8.4. Also an observer on the ground outside the rail car (*i.e.*, a stationary frame) observes that both the mirror and the light source are moving with a velocity 'v' which is now at the point C.

If 'Δt' is the time taken for the light pulse to reach at C from A, then from Figure. 8.4 we have,

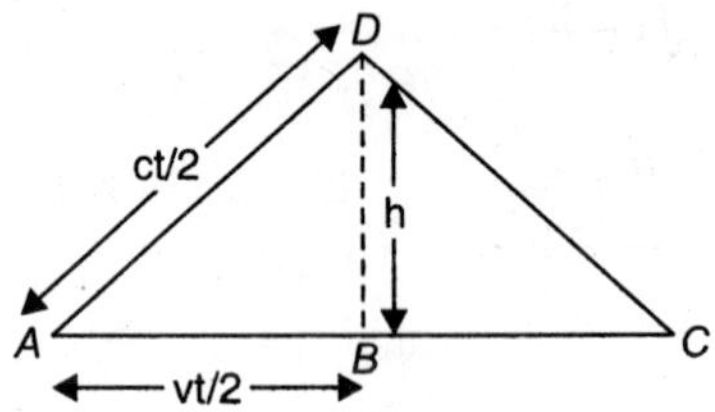

FIGURE 8.4 **Representation of moving light pulse and mirror from outside the rail**

$$h^2 + \left(v\frac{\Delta t}{2}\right)^2 = \left(c\frac{\Delta t}{2}\right)^2$$

$$\Rightarrow \qquad \Delta t = \frac{2h/c}{\sqrt{1-\frac{v^2}{c^2}}} = \frac{\Delta t'}{\sqrt{1-\frac{v^2}{c^2}}} \quad \text{(using equation 1)}$$

Hence,
$$\Delta t = \Delta t\sqrt{1-\frac{v^2}{c^2}} \qquad ...(2)$$

So, time measured in the moving frame becomes lengthened or dilated. Hence, from the point of view of the observer at the stationary frame the events will be found happening at a slower rate in the moving frame. This is known as *Time Dilation effect*. The effect has been experimentally verified by the decay of Muons in 1941 and also in 1971 in a jet plane travel.

8.9 ADDITION OF VELOCITIES

Consider two frame S and S', where S' is moving towards positive x-axis relative to S with velocity v. Suppose a particle moves with velocity component u_x, u_y, u_z relative to frame S. We have to find velocity component u_x', u_y', u_z' of the particle relative to S'.

According to Lorentz's transformation

$$x' = \frac{x - vt}{\sqrt{1 - v^2/c^2}} \quad \text{...(1)}$$

and

$$t' = \frac{t - \frac{vx}{c^2}}{\sqrt{1 - v^2/c^2}} \quad \text{...(2)}$$

Differentiate equations (1) and (2), we get

$$dx' = \frac{dx - vdt}{\sqrt{1 - v^2/c^2}} \quad \text{...(3)}$$

and

$$dt' = \frac{dt - \frac{vdx}{c^2}}{\sqrt{1 - v^2/c^2}} \quad \text{...(4)}$$

As, $u'_x = \dfrac{dx'}{dt'}$

$$\therefore \quad u'_x = \frac{dx - vdt}{dt - \frac{vdx}{c^2}} = \frac{\frac{dx}{dt} - v}{1 - \frac{v}{c^2}\frac{dx}{dt}}$$

or

$$u'_x = \frac{u_x - v}{1 - \frac{u_x v}{c^2}}$$

Since $y' = y$ ...(5)

and $dy' = dy$ ...(6)

From equations (4) and (6), we get

$$\frac{dy'}{dt'} = \frac{dy}{\left(dt - \frac{vdx}{c^2}\right) \Big/ \left(1 - \frac{v^2}{c^2}\right)^{1/2}} = \frac{\frac{dy}{dt}\sqrt{1 - v^2/c^2}}{1 - \frac{v}{c^2}\frac{dx}{dt}}$$

$$u'_y = \frac{u_y\sqrt{1 - v^2/c^2}}{1 - \frac{v . u_x}{c^2}} \quad \text{...(7)}$$

Similarly, $z' = z$ ...(8)

and $dz' = dz$ ...(9)

From equations, (4) and (8), we have

$$u'_z = \frac{u_z\sqrt{1-v^2/c^2}}{1-\dfrac{v.u_x}{c^2}}$$

The inverse velocity transformations can be obtained by replacing v with (– v) and interchanging primed and unprimed co-ordinates. Thus, the inverse velocities transformations are :

$$u_x = \frac{u'_x + v}{1+\dfrac{u'_x v}{c^2}},\ u_y = \frac{u'_y\sqrt{\left(1-\dfrac{v^2}{c^2}\right)}}{\dfrac{1+u'_x v}{c^2}} \text{ and } u_z = \frac{u'_z\sqrt{\left(1-\dfrac{v^2}{c^2}\right)}}{1+\dfrac{u'_x v}{c^2}}$$

Note.

1. If particle is moving in frame S' with velocity of light.

$$u'_x = c$$

From
$$u_x = \frac{u'_x + v}{1+\dfrac{u'_x v}{c^2}}$$

We have
$$u_x = \frac{c + v}{1+\dfrac{cv}{c^2}} \qquad (\because\ c = u'_x)$$

which means relativistic combination of velocities is consistent.

2. If $u'_x = c = v$, then from equation

$$u_x = \frac{u'_x + v}{1+\dfrac{u'_x v}{c^2}}, \text{ we have}$$

$$u_x = \frac{c + c}{1+\dfrac{c^2}{c^2}} = c$$

Thus addition of velocity of light to velocity of light gives *velocity of light*.

8.10 VARIATION OF MASS WITH VELOCITY

According to Newtonian mechanics the mass of a body does not change with velocity. However, conservation laws, especially here the law of conservation of momentum, hold for any inertial system. Hence, in order to maintain the momentum conserved in any isolated system, mass of the body must be related to its velocity. So according to Einstein, the mass of the body in motion is different from the mass of the body at rest. We consider two inertial frames S and S' as in Figure 8.5.

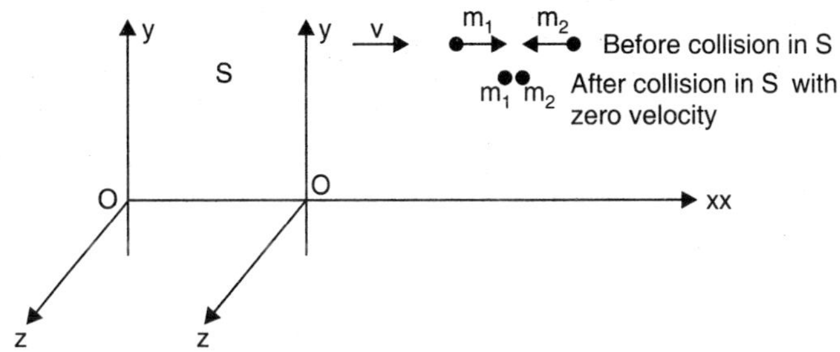

Figure 8.5 Collision between masses viewed from stationary and moving frames of reference

We now consider the collision of two bodies in S' and view it from the S. Let the two particles of masses m_1 and m_2 are travelling with velocity u' and $-u'$ parallel to x-axis in S'. The two bodies collide and after collision they coalesced into one body.

In System S : Before Collision: Mass of bodies are m_1 and m_2. Let the their velocities are u_1 and u_2 respectively.

In System S : After Collision: Mass of the coalesced body is $(m_1 + m_2)$ and the velocity is v.

Using law of addition of velocities,

$$u_1 = \frac{u' + v}{1 + \frac{u'v'}{c^2}} \text{ and } u_2 = \frac{-u' + v}{1 - \frac{u'v}{c^2}} \qquad ...(1)$$

Applying the *principle of conservation of momentum* of the system before and after the collision, we have,

$$m_1u_1 + m_2u_2 = (m_1 + m_2)\, v$$

$$\Rightarrow \quad m_1\left[\frac{u' + v}{1 + \frac{u'v}{c^2}}\right] + m_2\left[\frac{-u' + v}{1 - \frac{u'v}{c^2}}\right] = (m_1 + m_2)v$$

$$\Rightarrow \quad m_1\left[\frac{u' + v}{1 + \frac{u'v}{c^2}} - v\right] = m_2\left[v - \frac{-u' + v}{1 - \frac{u'v}{c^2}}\right]$$

$$\Rightarrow \quad \frac{m_1}{m_2} = \left[\frac{1 + \frac{u'v}{c^2}}{1 - \frac{u'v}{c^2}}\right] \qquad ...(2)$$

Now, using equations (1) and (2), we have

$$\Rightarrow \qquad \frac{m_1}{m_2} = \left[\frac{\sqrt{1-(u_2/c)^2}}{\sqrt{1-(u_1/c)^2}}\right] \qquad ...(3)$$

Let the body of mass m_2 is moving with zero velocity in S before collision, *i.e.*, $u_2 = 0$, hence, using equation (3), we have,

$$\frac{m_1}{m_2} = \left[\frac{1}{\sqrt{1-(u_1/c)^2}}\right] \qquad ...(4)$$

Using common notation as $m_1 = m$, $m_2 = m_0$, $u_1 = v$, we have by using equation (4).

$$\boxed{m = \frac{m_0}{\sqrt{1-\frac{v^2}{c^2}}}} \qquad ...(5)$$

This is the relativistic formula for variation of mass with velocity, where m_o is the rest mass and m is the relativistic mass of the body. There are a large numbers of experimental observations of this enhancement of mass of particles in high energy physics.

I. When $v << c$

$$\therefore \qquad v^2 << c^2, \frac{v^2}{c^2} \text{ is negligible as compared to 1} \Rightarrow \qquad m = m_0$$

When velocity of the moving particle is much smaller as compared to velocity of light, relativistic mass equals the rest mass.

II. When $v \approx c$

$$\therefore \qquad v^2 \approx c^2, \frac{v^2}{c^2} \approx 1 \Rightarrow \left[1-\frac{v^2}{c^2}\right] < 1 \Rightarrow m > m_0$$

When velocity of the moving particle is comparable to velocity of light, relativistic mass of the body appears to be greater than the rest mass.

III. When $v = c$

$$\therefore \qquad v^2 = c^2, \frac{v^2}{c^2} = 1 \Rightarrow m = \infty$$

When velocity of the moving particle is exactly equal to velocity of light, relativistic mass of the body appears to be infinite and this is an impractical concept.

IV. When $v > c$

$$\therefore \qquad v^2 >> c^2, \frac{v^2}{c^2} > 0 \Rightarrow m = \text{Imaginary}$$

When velocity of the moving particle is greater as compared to velocity of light, relativistic mass becomes imaginary and this is an impractical concept.

8.11 EQUIVALENCE OF MASS AND ENERGY

According to classical mechanics, force F is given by the time derivative of the product of mass (m) and velocity (v), *i.e.*, momentum of the body, and is given by

$$F = \frac{d}{dt}(mv)$$

However, according to the theory of relativity, both m and v are variable.

So, $$F = \frac{d}{dt}(mv) = m\frac{dv}{dt} + v\frac{dm}{dt} \quad ...(1)$$

When a particle is displaced through a distance 'dx' on application of a force F, then the increase in kinetic energy is given by,

$$dk = F \times dx$$

$$= \left[m\frac{dv}{dt} + v\frac{dm}{dt}\right] \times \frac{dx}{dt} \times dt$$

$$= \left[m\frac{dv}{dt} + v\frac{dm}{dt}\right] \times v \times dt$$

$\Rightarrow$ $$dk = v^2\, dm + mv\, dv \quad ...(2)$$

Now, we know that

$$m = \frac{m_0}{\sqrt{1-\left(\frac{v}{c}\right)^2}}$$

$\Rightarrow$ $$m^2 = \frac{m_0^2}{\left(\frac{c^2 - v^2}{c^2}\right)} \Rightarrow m^2 = \frac{m_0^2\, c^2}{c^2 - v^2}$$

$\Rightarrow$ $$m^2c^2 - m^2v^2 = m_0{}^2\, c^2$$

On differentiation both sides, we get,

$$2m \times dm \times c^2 - [2m \times dm \times v^2 + 2v \times dv \times m^2] = 0$$

$\Rightarrow$ $$2\, m \times dm \times c^2 = 2m\, [dm \times v^2 + v \times m \times dv]$$

$\Rightarrow$ $$c^2\, dm = v^2\, dm + vm\, dv \quad ...(3)$$

By comparing equations (2) and (3), we have,

$$dK = c^2\, dm \qquad ...(4)$$

Now, we can consider that the body initially at rest and on application of force it acquires a velocity v. Also the mass of the body increases from m_0 to m. Thereby, the total kinetic energy acquired by the body is given by.

$$\int dk = \int_{m_0}^{m} c^2\, dm$$

$$\Rightarrow \qquad k = c^2 (m - m_0) \qquad ...(5)$$

This is the increase in kinetic energy due to the increase in mass.

Now, the total energy of a moving body is the total of kinetic energy due to motion and the energy at rest.

So, $$E = k + m_0 c^2 = c^2 (m - m_0) + m_0 c^2 = m c^2$$

Hence, $$\boxed{E = m c^2}$$

This equation gives the universal equivalence between mass and energy. This is another major contribution of Einstein. It signifies the manifestation of energy into the matter and vice-versa. Relativity shows that mass is not any independent quantity, however, it is a form of energy. Its experimental evidences are well known in nuclear fission and fusion reactions.

The conversion of mass and energy and the reverse are also illustrated in nature itself. When a particle collides with an anti-particle there is a mutual annihilation and the total mass is converted into radiant energy. Thus, the conservation of mass takes place into energy. Other way, when a radiant energy comes near a charged nucleus, particle and anti-particle are created. Thus the energy is converted into mass.

8.12 NUMERICAL PROBLEMS BASED ON SPECIAL THEORY OF RELATIVITY

Problem 1. *The mean life of a meson is* 2×10^{-6} *s. Calculate the mean life of a meson moving with a velocity 0.9c?*

Hint: $$t = \frac{t_0}{\sqrt{1 - \frac{v^2}{c^2}}}$$ [**Ans.** = 3.33×10^{-6} sec]

Problem 2. *Calculate the rest energy of electron and proton in eV. Given* $m_e = 9.11 \times 10^{-31}$ *kg and* $m_p = 1.673 \times 10^{-21}$ *kg.*

Hint: $$E = m c^2$$ [**Ans.** 941×10^6 eV]

Problem 3. *At what speed will be mass of a body is 2.25 times its rest mass?*

Hint: $$m = \frac{m_0}{\sqrt{1 - \frac{v^2}{c^2}}}$$ [**Ans.** 2.68×10^8 m/s]

Problem 4. *The binding energy of hydrogen is 13.6 eV. Find the loss of mass in the formation of one atom of hydrogen.*

Hint: $E = m c^2$ **[Ans.** 2.42×10^{-35} kg]

Problem 5. *The rest mass of an electron is 9.1×10^{-31} kg. What will be its mass if it was moving with 4/5th the speed of light?*

Hint: $$m = \frac{m_0}{\sqrt{1 - \frac{v^2}{c^2}}}$$ **[Ans.** 15.16×10^{-31} kg]

Problem 6. *A rocket is chasing enemy's space-ship. An observer on the earth observes the speed of rocket to be 2.5×10^8 m/sec and that of space-ship 2×10^8 m/sec. Calculate (i) the velocity of enemy's ship as seen by rocket (ii) the velocity of rocket as seen by enemy.*

Hint: $$u_x = \frac{u_x' + v}{1 + \frac{u_x' v}{c^2}}$$ **[Ans.** 1.14×10^8 m/s]

Problem 7. *The rest mass of an electron is 9.1×10^{-31} kg. What will be its mass if it was moving with 3/5th the speed of light.*

Hint: $$m = \frac{m_0}{\sqrt{1 - \frac{v^2}{c^2}}},$$

$$m = \frac{9.1 \times 10^{-31}}{\sqrt{1 - (0.6)^2}} = 11.38 \times 10^{-31} \text{ kg.}$$

EXERCISES

1. What do you understand by inertial and non inertial frames of references? Give examples.
2. What do you understand by concept of frame of reference ?
3. Explain, Earth is not an inertial frame.
4. What are Galilean transformation equations? Derive them.
5. State the conclusion of Michelson-Morley experiment.
6. Draw neat and clean diagram of Michelson-Morley experiment set up and discuss its working.
7. What are the postulate of special theory of relativity?
8. What are Lorentz's transformation? Derive them.
9. Derive an expression for variation of mass with velocity.

10. What do you understand by famous universal relation of mass energy equivalence?
11. Write short notes
    (*i*) Relativistics addition of velocities
    (*ii*) Mass energy equivalence
    (*iii*) Time dilation
    (*iv*) Length contraction.
12. A body of mass m is moving with velocity v. Show that $m = \frac{m_0}{\sqrt{\left(1-\frac{v^2}{c^2}\right)}}$ where m_0 is rest mass of body and c is speed of light.
13. What is proper length? Explain how Lorentz's transformation account for the phenomenon of length contraction.
14. Describe the Michelson-Morley experiment and explain the physical significance of negative result.
15. Deduce an expression for the variation of mass with velocity and depict it graphically. Also prove that no material particle can have a velocity equal or greater than the velocity of light.
[*Amity University, Jan. 2007 (5 Marks)*]

SUGGESTION FOR FURTHER READING

1. D. Bohm, The Special Theory of Relativity, Benjamin, New York, 1965.
2. M. Born, Einstein's Theory of Relativity, Dover, New York, 1962.
3. J.H. Smith, Introduction to Special Relativity, Benjamin, New York, 1965.
4. Devraj Singh, Engineering Physics, Vol. 1, Dhanpat Rai & Co. (P) Ltd., New Delhi, 2008.
5. Devraj Singh, Fundamentals of Engineering Physics, Vol. 1, Dhanpat Rai & Co. (P) Ltd., New Delhi, 2009.
6. R. Murugeshan and K. Sivaprasath, Modern Physics, S. Chand & Co., Delhi, 2008.
7. S.L. Kakani and Subhra Kakani, Modern Physics, Viva Books Pvt. Ltd. Delhi, 2009.

Chapter 9

WAVE MECHANICS

9.1 INTRODUCTION

The de Broglie concept of matter waves brought the break through in the development of quantum theory in 1925. In classical mechanics, Newton's law describe the motion of particles. The future history of particle is completely determined by its initial position and momentum together with the forces that act upon it. But the concept of matter waves introduced the uncertainty in the position and momentum of a particle. Moreover, according to de Broglie theory, a material particle is associated with the wave. So the mathematical reformation using a wave function associated with matter wave is needed. Such a mathematical formation known as *quantum mechanics* was developed in 1926 by Schrödinger. Schrödinger described the amplitude of matter waves by a complex quantity $\psi(x, y, z, t)$ known as *wave function* or the state of the system. It describes the particular dynamical system under observation.

9.2 DE BROGLIE WAVES OR MATTER WAVES

The wave associated with the matter in motion is called *matter wave*. The idea of matter wave was first proposed by a French scientist Louis de Broglie in 1924. The Basis of this idea is the dual nature of light. According to dual nature of light, sometime light energy behaves like a wave motion as in reflection, refraction, interference, diffraction and polarisation. On the other hand, sometimes light behaves like a particle as in photoelectric effect, compton effect, Bohr theory, and absorption and emission of light. This idea of dual nature was generalised by de Broglie for the matter, as the matter and energy, both are related by an equation $E = mc^2$.

According to the dual nature the matter sometimes behaves like a wave motion, while in other time it has a particle nature.

The energy of the light radiation in wave nature is given by

$$E = mc^2 \quad \text{...(1)}$$

where c = velocity of light.

The energy of the light radiation in particle form (photon nature)

$$E = h\nu \quad \text{...(2)}$$

where ν = frequency of wave, h = Planck's constant.

Since the two equations (1) and (2) represent the energy of light radiation,

$$\therefore \quad E = mc^2 = h\nu$$

$$\Rightarrow \quad mc = \frac{h\nu}{c}$$

$$\Rightarrow \quad mc = \frac{h}{\lambda}, \text{ where } \nu = c\lambda^{-1}$$

λ = wavelength of radiation

or
$$\lambda = \frac{h}{mc}$$

$$\Rightarrow \quad \boxed{\lambda = \frac{h}{p}} \quad \text{...(3)}$$

where $p = mc$ = momentum of light radiation. The equation (3), which is valid for a radiation de Broglie suggested that it must be true for the matter in motion. Hence if a particle of mass is moving with velocity v, then

$$\lambda = \frac{h}{p} = \frac{h}{mv}$$

This relation $\lambda = \frac{h}{p} = \frac{h}{mv}$ is known as *de Broglie equation*.

This equation shows that the material particle of a mass m, moving with a velocity v, is always associated with a wave, whose wavelength is equal to the ratio of the Planck's constant to the momentum of the particle

$$\lambda = \frac{h}{p} = \frac{h}{mv}$$

This equation shows that the de Broglie wavelength is inversely proportional to the momentum. Larger is the momentum, smaller will be the wavelength, while smaller momentum will produce larger de Broglie wavelength.

9.2.1 de Broglie Wavelength of an Electron

Let us consider an electron is attracted by a certain potential V. The motion of electron will perform some work and will appear as the kinetic energy of the electron, or the decrease in potential energy will appear as the kinetic energy of the electron

$$eV = \frac{1}{2} mv^2$$

where V and v are the potential and velocity of the electron respectively and m is the mass of the electron

or
$$v = \sqrt{\frac{2eV}{m}}$$

∴ de Broglie equation becomes

$$\lambda = \frac{h}{mv} = \frac{h}{m\sqrt{\frac{2eV}{m}}}$$

$$\lambda = \frac{h}{\sqrt{2\,m\,e\,V}}$$

For an electron, m, e and h are constant

$$\lambda \propto \frac{l}{\sqrt{V}}$$

This shows that for different potential, we get de Broglie wavelength

For an electron $m = 9.1 \times 10^{-31}$ kg, $h = 6.626 \times 10^{-34}$ J.s

$e = 1.6 \times 10^{-19}$ C

The de Broglie wavelength of an electron

$$\lambda - \frac{6.626 \times 10^{-34}}{\sqrt{(2 \times 9.1 \times 10^{-31} \times 1.6 \times 10^{-19}\,\text{V})}} = \frac{12.27 \times 10^{-10}}{\sqrt{V}}\ \text{m}$$

$$= \sqrt{\frac{150}{V}} \times 10^{-10}\,\text{m}$$

$$\Rightarrow \quad \boxed{\lambda = \sqrt{\frac{150}{V}}\ \text{Å}}$$

9.2.2 Relativistic Mass in de Broglie Wavelength of an Electron

When the accelerating potential V is very high, the kinetic energy of the electron or velocity of the electron is very high and comparable to the velocity of the light. In this case the relativistic kinetic energy of the electron is used.

The relativistic kinetic energy is given by

$$E_k = mc^2 - m_0c^2$$

$$\Rightarrow \quad eV = \frac{m_0c^2}{\sqrt{\left(1-\frac{v^2}{c^2}\right)}} - m_0c^2$$

$$\Rightarrow \quad \frac{m_0c^2}{\sqrt{\left(1-\frac{v^2}{c^2}\right)}} = eV + m_0c^2$$

or

$$\sqrt{\left(1-\frac{v^2}{c^2}\right)} = \frac{m_0c^2}{eV + m_0c^2} \qquad \text{...(4)}$$

On squaring both sides, we get

$$1 - \frac{v^2}{c^2} = \left[\frac{m_0c^2}{eV + m_0c^2}\right]^2$$

$$\Rightarrow \quad \frac{v^2}{c^2} = 1 - \left[\frac{m_0c^2}{eV + m_0c^2}\right]^2$$

$$v^2 = c^2\left[1 - \left\{\frac{m_0c^2}{eV + m_0c^2}\right\}^2\right] \qquad \text{...(5)}$$

$$\Rightarrow \quad v = c\left[1 - \left\{\frac{m_0c^2}{eV + m_0c^2}\right\}^2\right]^{1/2}$$

$$= c\left[\frac{(eV)^2 + m_0^2c^4 + 2eVm_0c^2 - m_0^2c^4}{(eV + m_0c^2)^2}\right]^{1/2}$$

$$= c\left[\frac{eV\,(eV + 2m_0c^2)}{(eV + m_0c^2)^2}\right]^{1/2}$$

∴ The de Broglie wavelength of an electron

$$\lambda = \frac{h}{mv} = \frac{h}{m_0 v}\sqrt{\left(1-\frac{v^2}{c^2}\right)}$$

$$\Rightarrow \qquad \lambda = \frac{h}{m_0 v}\sqrt{\left(1-\frac{v^2}{c^2}\right)}$$

Putting the value of $\sqrt{\left(1-\frac{v^2}{c^2}\right)}$ and v

$$\lambda = \frac{h}{m_0}\frac{m_0 c^2}{(eV + m_0 c^2)}\cdot\frac{(eV + m_0 c^2)}{c[eV\{eV + 2m_0 c^2]^{1/2}}$$

or

$$\boxed{\lambda = \frac{hc}{eV(eV + 2m_0 c^2)^{1/2}}}.$$

9.3 EXPERIMENTAL VERIFICATION OF WAVE–PARTICLE DUALISM: DAVISSON AND GERMER EXPERIMENT

The schematic diagram of the experimental arrangement is given in Figure 9.1. A collimated beam of electrons are scattered by a nickel target T capable of rotating about a vertical axis passing through the point of incidence. The scattered beam is received by Faraday's cylinder F moving over a circular scale. Thus, the intensity of scattered beam and angle θ between the incident and scattered beam can be measured. The experiment was performed by varying the acceleration voltage of the electron and the angle θ as shown in Figure 9.2.

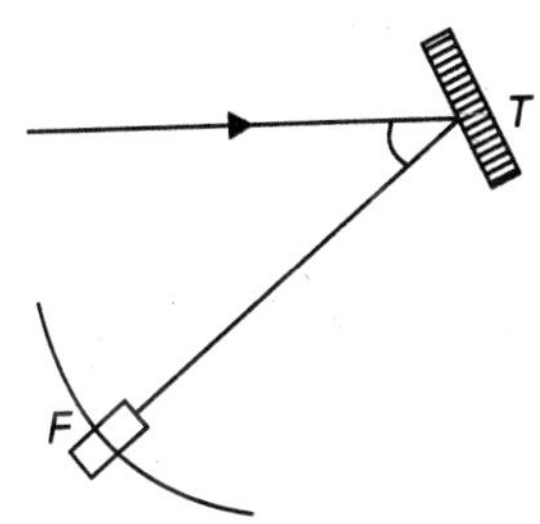

FIGURE 9.1

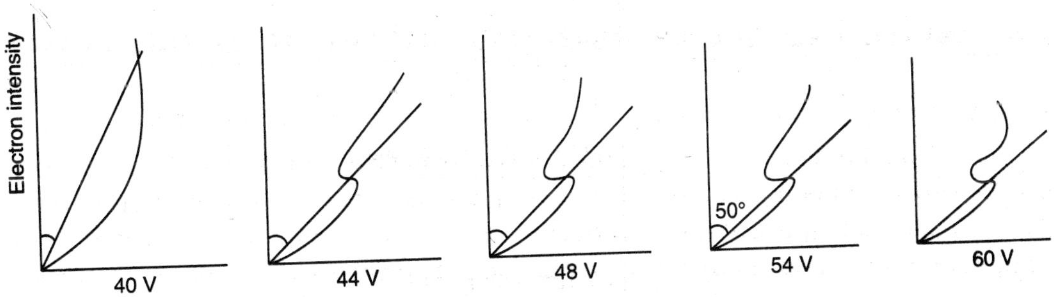

FIGURE 9.2 Graph showing the variation of electron intensity with θ for different values of accelerating voltage

In the experiment the variation of the intensity of the scattered electron beam for different values of θ are recorded for a particular value of the accelerating voltage. The experiment is repeated with different values of this voltage. Then a polar graph is drawn for each value of the voltage showing the intensity of the scattered electrons at different direction. At 40 V the graph is smooth. With increasing voltage the graph develops a peak at 50° and the peak becomes most pronounced for a voltage 54 V.

Bragg has demonstrated that when a beam of X-rays are scattered by a crystal, the maxima occur along the directions given by

$$2d \sin \theta = n\lambda \qquad \text{...(1)}$$

where d is the spacing between the crystal plane, n is the order of the maximum and θ is the angle between the crystal plane and the incident X-ray beam; called the *glancing angle* (Figure 9.3).

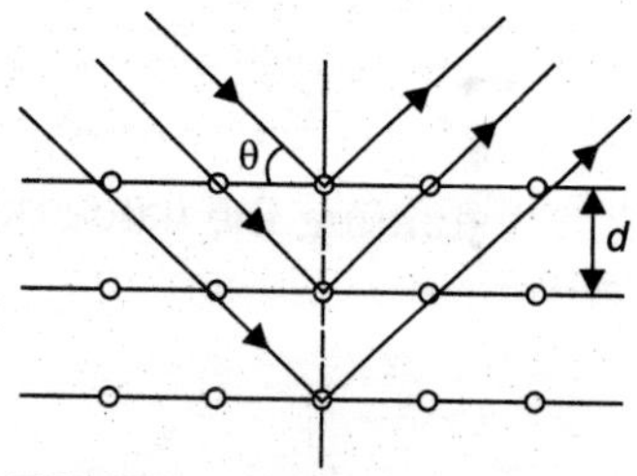

FIGURE 9.3 **Bragg's condition for diffraction**

For this experiment $d = 0.91$ Å, $n = 1$ and $\theta = 90 - \dfrac{50}{2} = 65°$.

Substituting these values in equation (1), we get wavelength

$$\lambda = 2 \times 0.91 \times \sin 65° = 1.65 \text{ Å}.$$

In the Davisson and Germer experiment the maximum intensity occurs for an accelerating voltage 54 volts.

Thus, the de Broglie wavelength of electron may be obtained as

$$\lambda = \frac{h}{p} = \frac{h}{\sqrt{2meV}}$$

$$= \frac{6.625 \times 10^{-34}}{\sqrt{2 \times 9.1 \times 10^{-31} \times 1.6 \times 10^{-19} \times 54}} = 1.66 \text{ Å}.$$

The agreement between the two values of wavelength is striking leading us to believe that the electrons have undergone diffraction through the crystal.

9.4 PHASE AND GROUP VELOCITIES OF THE DE BROGLIE WAVE

The group of wave need not have the same velocity as the individual one.

The amplitude of the de Broglie wave is associated with the moving body represents the probability of finding a body at a particular time and space. The wave equation $y = A \cos (\omega t - kx)$ does not represent the de Broglie wave. The de Broglie wave is represented by the combination of several such waves. Thus, the de Broglie waves can be obtained by the superposition of the several waves. Hence the mathematical expression of the de Broglie waves is obtained by the superposition of number of waves or group of waves or wave group. The diagramatic representation of the de Broglie wave may be shown in the form of *wave packet* or *wave group*.

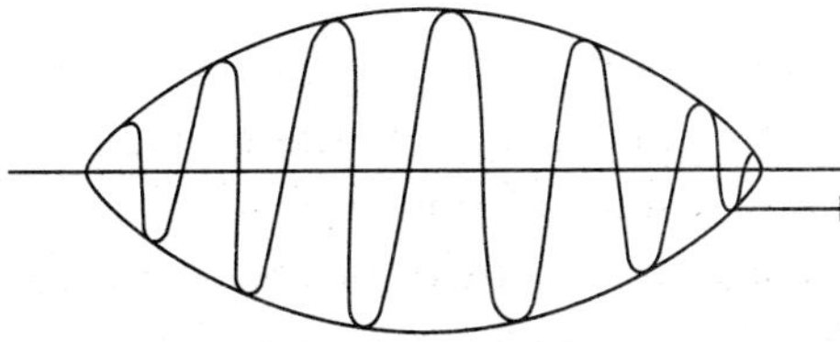

FIGURE 9.4

Let us find the velocity of the wave packet or de Broglie wave. We assume that the wave packet (or group) is formed due to combination of the two waves having equal amplitude and different their angular frequency by $d\omega$ and wave velocity dk. The waves are represented mathematically as

$$y_1 = A\cos(\omega t - kx)$$
$$y_2 = A\cos[(\omega + d\omega)t - (k + dk)x]$$

where $k = \dfrac{2\pi}{\lambda}$, $\omega = 2\pi\nu$.

According to the superposition principle, the resultant displacement Y at any time t at any point x is the sum of two displacements

$$Y = y_1 + y_2$$
$$= A[\cos(\omega t - kx) + \cos\{(\omega + d\omega)t - (k + dk)x\}]$$
$$= 2A\cos\frac{[(2\omega + d\omega)t - (2k + dk)x]}{2}\cos\frac{[d\omega\, t - dk\, x]}{2}$$

$\because$ $d\omega$ and dk are very small as compared to ω and k

$\therefore$ $2\omega + d\omega \cong 2\omega$ and $2k + dk \cong 2k$

Hence $$Y = 2A\cos(\omega t - kx)\cos\left(\frac{d\omega\, t}{2} - \frac{dk\, x}{2}\right) \qquad ...(1)$$

This equation represents a wave of angular frequency ω and wave number k moving in the same direction superimposed by a modulated wave of angular frequency $d\omega$ and wave vector dk. Thus, the superposition of the two waves results a new waves and successively they form a new wave.

Equation (1) describes the *de Broglie wave* or *wave packet* or *wave group*.

Form Equation (1), two types of velocities are defined as :

The **phase velocity** of the de Broglie waves is defined by

$$v_p = \frac{\omega}{k} \qquad ...(2)$$

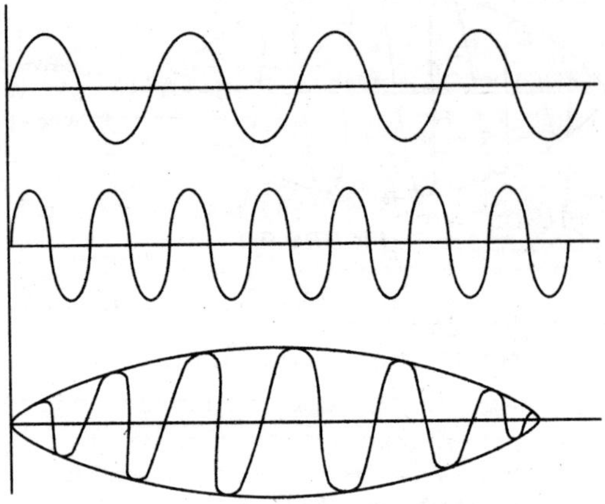

FIGURE 9.5

and the **group velocity** of the de Broglie wave is defined as

$$v_g = \frac{d\omega}{dk} \qquad ...(3)$$

The group velocity of the de Broglie wave depends upon the manner in which the phase velocity of the medium varies or constant.

The phase velocity is defined by

$$v_p = \frac{\omega}{k} = \frac{2\pi\nu}{2\pi/\lambda} = \nu\lambda$$

$$v_p = \nu\lambda = v \qquad ...(4)$$

where v = velocity of wave.

This shows that the *phase velocity* of the de Broglie wave is same as the *wave velocity*. In other words, the *phase velocity* is also called *wave velocity*.

Let us consider a de Broglie wave associated with a moving particle of a rest mass m_0 and velocity v. The angular frequency of the de Broglie wave is given by

$$\omega = 2\pi\nu$$

$$\Rightarrow \qquad \omega = 2\pi\frac{E}{h}$$

$$\Rightarrow \qquad \boxed{\omega = \frac{2\pi c^2}{h}\frac{m_o}{\sqrt{\left(1-\frac{v^2}{c^2}\right)}}} \qquad ...(5)$$

Similarly the wave vector k is given by

$$k = \frac{2\pi}{\lambda}$$

Using de Broglie relation $\lambda = \dfrac{h}{p}$

$$\therefore \qquad k = \frac{2\pi p}{h}$$

$$\Rightarrow \qquad k = \frac{2\pi mv}{h}$$

$$\Rightarrow \qquad k = \frac{2\pi v}{h} \frac{m_o}{\sqrt{\left(1 - \frac{v^2}{c^2}\right)}} \qquad \text{...(6)}$$

$\therefore$ The phase velocity of the de Broglie wave

$$v_p = \frac{\omega}{k}$$

Substituting the value of ω and k from equations (5) and (6), we get

$$v_p = \frac{2\pi c^2}{h} \frac{m_o}{(1 - v^2/c^2)^{1/2}} \cdot \frac{h(1 - v^2/c^2)^{1/2}}{2\pi v m_0}$$

$$\Rightarrow \qquad v_p = \frac{c^2}{v} \Rightarrow v_p = \left(\frac{c}{v}\right)c$$

$$\because \qquad \frac{c}{v} >> 1, \text{ for material particle}$$

$$\therefore \qquad \boxed{v_p > c}$$

This shows the phase velocity of the de Broglie wave is greater than the velocity of light.

The group velocity of the de Broglie wave is defined as

$$v_g = \frac{d\omega}{dk}.$$

9.4.1 Relation between the Phase Velocity and Group Velocity of the de Broglie Wave

The wave velocity is given by $v = \nu\lambda$

$$= 2\pi\nu\frac{\lambda}{2\pi}$$

$$\Rightarrow \qquad v = \frac{\omega}{k}$$

$$\Rightarrow \qquad v = \frac{\omega}{k} = v_p$$

where $\omega = 2\pi\nu$ = angular frequency and $k = \dfrac{2\pi}{\lambda}$ = wave number.

This shows that the phase velocity of the de Broglie wave is equal to the wave velocity and hence phase velocity is also known as wave velocity.

Phase velocity is defined by

$$v_p = \frac{\omega}{k} \quad \Rightarrow \quad \omega = kv_p \qquad ...(1)$$

The group velocity of a de Broglie wave is defined by

$$v_g = \frac{d\omega}{dk} \qquad ...(2)$$

Putting the value of $\omega = kv_p$ in equation (2), we get

$$v_g = \frac{d}{dk}(kv_p)$$

$$\Rightarrow \qquad v_g = v_p + k\frac{dv_p}{dk}$$

Putting the value of k

$$v_g = v_p + \frac{2\pi}{\lambda}\frac{dv_p}{d\left(\frac{2\pi}{\lambda}\right)}$$

$$\Rightarrow \qquad v_g = v_p + \frac{1}{\lambda}\frac{dv_p}{d\left(\frac{1}{\lambda}\right)}$$

Now $\qquad d\left(\frac{1}{\lambda}\right) = -\frac{1}{\lambda^2}\,d\lambda$

$$\therefore \qquad \boxed{v_g = v_p - \lambda\frac{dv_p}{d\lambda}}$$

This is a relation between the phase velocity and group velocity for a dispersive medium.

For *normal dispersion* the quantity $\frac{dv_p}{d\lambda}$ is a positive quantity. Therefore for normal dispersion, group velocity is less than the phase velocity. For *anomalous dispersion*, the quantity $\frac{dv_p}{d\lambda}$ is a negative quantity, the group velocity is greater than the phase velocity.

For *non-dispersive medium* $v_p = \frac{\omega}{k}$ = constant $\therefore$ $\frac{dv_p}{d\lambda} = 0$. Hence $v_g = v_p$. Hence for nondispersive medium the group velocity is equal to the phase velocity. For electromagnetic waves in vacuum, the speed of light (c) is constant. Therefore group velocity v_g and the phase velocity v_n of the light radiations are same.

9.5 HEISENBERG UNCERTAINTY PRINCIPLE

As per the de Broglie's hypothesis, it can be said that, a particle is considered as a group, or packet of waves. The constituent waves are of nearly equal wavelengths, whose values are centered around the de Broglie wavelength $\lambda = h/p$. The wave packet is formed due to the superposition of waves. The amplitude of wave packet is zero everywhere in space, except only over a small region in which the particle is located. Essentially, the wave packet's extent is limited over this small region of space, but travels with a velocity equal to its group velocity, which is equal to the particle velocity.

Because that the wave packet is spread in space, it implies that the position of the particle cannot be pin pointed or specified accurately. It can only be said that, the particle is somewhere within the region over which the wave packet is spread.

Let the motion of the particle be along X-axis. If the wave packet extends over a finite distance Δx, along the axis of motion, then the two points at which the amplitude of the wave group becomes zero successively, will be separated by a distance Δx (Figure 9.6).

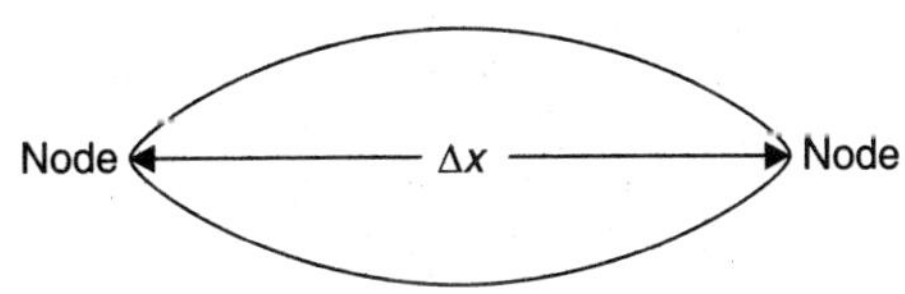

FIGURE 9.6 **Spatial extension of wave packet**

Two points where the amplitude becomes zero are the *nodes*. Any measurement made to find the position of the particle will have a minimum error equal to the distance between the two nodes. This is inevitable due to the wave nature of matter.

The amplitude of the wave packet is given by

$$2\,A\cos\left[\left(\frac{d\omega}{2}\right)t-\left(\frac{dk}{2}\right)x\right]$$

At nodes, the amplitude is zero.

i.e., $$2A\cos\left[\frac{d\omega}{2}t-\frac{dk}{2}x\right]=0$$

$\because$ $$2A\neq 0,\ \text{Hence}\ \cos\left[\frac{d\omega}{2}t-\frac{dk}{2}x\right]=0$$

or $$\frac{d\omega}{2}t-\frac{dk}{2}x=\frac{\pi}{2},\frac{3\pi}{2},\frac{5\pi}{2},\ldots\ldots\ \frac{(2n+1)\pi}{2}.$$

If x_1 and x_2 be the two positions at which the nodes are formed, then, for the first node

$$\left(\frac{d\omega}{2}\right)t-\frac{dk}{2}x_1=(2n+1)\frac{\pi}{2} \qquad \ldots(1)$$

Since, there is a change of a phase π, between the successive nodes, the equation for second node is

$$\frac{d\omega}{2}t-\frac{dk}{2}x_2=(2n+1)\frac{\pi}{2}+\pi \qquad \ldots(2)$$

Subtracting equation (1) from equation (2), we have

$$\frac{dk}{2}(x_1-x_2)=\pi$$

$\because$ $$x_1-x_2=dx$$

$$\frac{dk}{2}(dx)=\pi$$

or $$dx=\frac{2\pi}{dk}\ \Rightarrow\ \boxed{dx=\frac{2\pi}{dk}} \qquad \ldots(3)$$

which is the fundamental error in the measurement of the position of particle.

But, we know that $k=\dfrac{2\pi}{\lambda}=\dfrac{2\pi p_x}{h}$

(by using de Broglie equation $\lambda=h/p_x$),

where, p_x is the momentum along x-axis and h is the Planck's constant

or $$\Delta k = \frac{2\pi \Delta p_x}{h} \qquad ...(4)$$

where Δp_x represents the fundamental error in the measurement of the momentum of the particle.

Substituting for Δk in equation (3) from equation (4), we have

$$\Delta x = \frac{h}{\Delta p_x}$$

or $$\Delta x \Delta p_x = h \qquad ...(5)$$

But as per Fourier analysis, the width Δx of a single wave group which represents the superposition of waves, whose propagation constants vary in the range Δk, obey the relation.

$$\Delta x = \frac{1}{\Delta k} \qquad ...(6)$$

Whereas, equation (3) is obeyed, when the superposition is due to only two wave trains. Hence for general superposition, equation (6) is valid.

∴ Substituting for Δk from equation (6) in equation (4), we have

$$\frac{1}{\Delta x} = \frac{2\pi(\Delta p_x)}{h}$$

$$\Delta x \, \Delta p_x \approx \frac{h}{2\pi} \qquad ...(7)$$

Since, the fundamental errors involved in the measurement of x and p_x could as best be equal to, or greater than, but will never be less than Δx and Δp_x respectively, the equation is more rationally expressed as

$$\boxed{\Delta x \, \Delta p_x \geq \frac{h}{2\pi}.}$$

9.5.1 Statement of Heisenberg's Uncertainty Principle

> In any simultaneous determination of the position and momentum of the particle, the product of the corresponding uncertainties inherently present in the measurement is equal to, or greater than $(h/2\pi)$.

The Heisenberg uncertainty principle could also be expressed in terms of the uncertainties involved in the measurement of physical variables-pair, such as energy and time, angular displacement (θ) and angular momentum (L). Both the pairs are conjugate in the sense that, their respective products have the dimensions of Planck's constant.

Thus, we have the uncertainty relations

$$\Delta E\, \Delta t \geq \frac{h}{2\pi}$$

and
$$\Delta L\, \Delta \theta \geq \frac{h}{2\pi}$$

where, the notation Δ is association with the respective variables, indicates the minimum uncertainty involved in the measurement of the corresponding variables.

9.5.2 Physical Significance of Heisenberg's Uncertainty Principle

In classical mechanics, by the very definition itself, a particle occupies a definite place in space, and possesses a definite momentum, knowing the particle's position and momentum at any given instant of time, it is possible to evaluate its position and momentum at any later point of time, and the trajectory of the particle could be continuously traced. But, such a notion breaks down in atomic scales, wherein a new mechanics, known as the *wave mechanics*, or *quantum mechanics* takes over, for which one of the fundamental principle is the *Heisenberg's uncertainty principle.*

As we realise from the principle, it is impossible to determine simultaneously both the position and momentum of a particle accurately. Any efforts made to make the measurement of the position of the particle such as an electron very accurately, results in a large uncertainty in the measurement of momentum and vice versa. In effect, irrespective of the level of sophistification arranged for the measurements, the measured values cannot have accuracies exceeding what is permitted under the condition

$$\Delta x\, \Delta p_x \geq (h/2\pi)$$

where, Δx and Δp_x are the uncertainties in the measured values of the position and momentum of the particle.

The consequences arising from the above conditions, *i.e.*, the uncertainty principle are as follows :

An increase in accuracy in the measurement of position means a decrease in the value of Δx.

As per the above condition, a decrement in Δx must simultaneously result in a proportionate increment in the value of Δp_x, since the right hand side of the identity is a constant. The physical significance of the above arguments is that, one should not think of the exact position, or, an accurate value of momentum of a particle. Instead, one should think of the probability of finding the particle at a certain position, or of the probable value for the momentum of the particle. The estimation of such probabilities are made by means of certain mathematical functions, named *probability density functions* in quantum mechanics.

9.5.3 Impossibility of Existence of Electrons in the Atomic Nucleus

We know the theory of relativity that, the energy E is expressed as

$$E = mc^2 = \frac{m_0c^2}{\sqrt{(1 - v^2/c^2)}}$$

or

$$E^2 = \frac{m_0^2c^4}{1 - \frac{v^2}{c^2}} = \frac{m_0^2c^6}{(c^2 - v^2)} \qquad \text{...(1)}$$

Also, we know that, momentum p is given by

$$p = mv = \frac{m_0v}{\sqrt{\left(1 - \frac{v^2}{c^2}\right)}}$$

or

$$p^2 = \frac{m_0^2v^2}{(1 - v^2/c^2)} = \frac{m_0^2v^2c^2}{(c^2 - v^2)}$$

Multiplying by c^2, we have

$$p^2c^2 = \frac{m_0^2v^2c^2}{(c^2 - v^2)} \qquad \text{...(2)}$$

Subtracting equation (2) by equation (1), we have

$$E^2 - p^2c^2 = \frac{m_0^2c^4(c^2 - v^2)}{(c^2 - v^2)}$$

or

$$E^2 - p^2c^2 = m_0^2c^4$$

$$\Rightarrow \qquad \boxed{E = p^2c^2 + m_0^2c^4} \qquad \text{...(3)}$$

Heisenberg's uncertainty principle states that

$$\Delta p_x \, \Delta x \geq \frac{h}{2\pi} \qquad \text{...(4)}$$

We know the diameter of the nucleus is of the order of 10^{-14} m. If an electron is expected to exist inside the nucleus, then the uncertainty in its position Δx must not exceed the size of the nucleus.

i.e., $\Delta x \leq 10^{-14}$ m

In such an event, the uncertainty in its momentum as per equation (4) becomes

$$\Delta p_x \geq \frac{h}{2\pi\, \Delta x}$$

i.e.,
$$\Delta p_x \geq \frac{6.63 \times 10^{-34}}{2\pi \times 10^{-14}}$$

i.e.,
$$\Delta p_x \geq 1.055 \times 10^{-20} \text{ Ns} \qquad \text{...(5)}$$

which is the uncertainty in the momentum of the electron, in which case, the momentum of the electron must at least be equal to the uncertainty in the momentum, *i.e.*, 1.055×10^{-20}.

$\therefore$
$$p_x \geq 1.055 \times 10^{-20} \text{ Ns} \qquad \text{...(6)}$$

Now, from equation (3)

$$E^2 = p^2c^2 + m_0^2 c^4 = c^2(p^2 + m_0^2 c^2)$$

we know, the rest mass of the electron

$$m_0 = 9.11 \times 10^{-31} \text{ kg}$$

Now, by making use of the condition in equation (6), we can say that in order that the electron may exist within the nucleus, its energy E must be such that

$$E^2 \geq (3 \times 10^8)^2\left[(1.055 \times 10^{-20})^2 + (9.11 \times 10^{-31})^2 (3 \times 10^8)^2\right]$$

i.e.,
$$E^2 \geq (3 \times 10^8)^2\left[1.113 \times 10^{-40} + 7.4692 \times 10^{-44}\right]$$

Since the second term in the bracket is smaller by more than 3 orders of magnitude compared to the first term, it is neglected.

Taking the square roots on both sides, and simplifying, we get the condition

$$E \geq 3.1649 \times 10^{-12} \text{ J}$$

Expressing in eV, we get

$$E \geq 20 \text{ MeV}$$

This means to say that, in order that an electron may exist inside the nucleus, its energy must be greater than or equal to 20 MeV. But, the experimental investigations on β-decay reveal that, the kinetic energy of the β-particles (which are same as electrons) is of the order of 3 and 4 MeV.

This clearly indicates that, electrons cannot exist within the nucleus.

9.6 WAVE FUNCTION AND ITS PHYSICAL INTERPRETATION

The wave function $\psi(\mathbf{r}, t)$ explains the behaviour of the particle at a given position **r** at a given time. The magnitude of wave function is large in regions where probability of finding the particle is high and its magnitude is small in regions where the probability of finding the particle is low. Therefore, ψ *may be taken as a measure of the probability of the particle around a particular position.*

It is already assumed that the wave associated with a particle in motion may be represented by a complex variable quantity $\psi(x, y, z, t)$ is known as the *wave function*. Since it is complex quantity, it may be expressed as

$$\psi(x, y, z, t) = a + ib \qquad ...(1)$$

where a and b are real functions of variables (x, y, z, t) and $i = \sqrt{(-1)}$.

The complex conjugate of ψ is written as ψ^* can be obtained by changing i to $-i$. Therefore

$$\psi^*(x, y, z, t) = a - ib \qquad ...(2)$$

Multiplying equation (1) by equation (2), we get

$$\psi(x, y, z, t)\, \psi^*(x, y, z, t) = a^2 - i^2b^2 \qquad (\because\ i^2 = -1)$$

$$\Rightarrow \quad \psi(x, y, z, t)\, \psi^*(x, y, z, t) = a^2 + b^2 \qquad ...(3)$$

We may write the product on left hand side of equation (3)

$$P = |\psi(x, y, z, t)|^2$$

$$\therefore \quad |\psi(x, y, z, t)|^2 = a^2 + b^2 \qquad ...(4)$$

Therefore, the product of ψ and ψ^* is real and positive if $\psi \neq 0$. Its positive square root is denoted by $|\psi(x, y, z, t)|$. It represents the modulus of ψ.

The quantity $|\psi(x, y, z, t)|^2$ is called the *probability density i.e.*, probability per unit volume.

9.6.1 Max Born Interpretation of ψ

Max Born used equation (4) to give the interpretation of $\psi(x, y, z, t)$, which is as follows :

Consider a small element of volume dV defined by the coordinates $(x, x + dx)$, $(y, y + dy)$ and $(z, z + dz)$ or $(r, r + dr)$.

Then the probability of finding the particle existing within the element of volume dV is given by

$$PdV = \psi(x, y, z, t)\, \psi^*(x, y, z, t)\, dV$$

$$= |\psi(x, y, z, t)|^2 dV$$

For the motion of a particle in one-dimension the quantity

$$Pdx = \psi(x, t)\, \psi^*(x, t) = |\psi(x, t)|^2 dx$$

is the probability that the particle will be found over a small distance dx at position x at time t. $|\psi(x,t)|^2$ is called the probability per unit distance.

$$\because \quad \psi(x, y, z, t) = u(x, y, z)\, e^{-i\omega t}$$

and
$$\psi^*(x, y, z, t) = u^*(x, y, z)\, e^{+i\omega t}$$

$$PdV = u(x, y, z).\, u^*(x, y, z)\, dV$$

$$= |u(x, y, z)|^2\, dV$$

Therefore, the wave function, which satisfies the time independent Schrödinger equation, the probability is always independent of time.

9.6.2 Limitations of ψ

The solutions of ψ obtained from the Schrödinger equation are subject to following limitations :

(*i*) ψ must be finite for all values x, y, z.

(*ii*) ψ must be single-valued function *i.e.*, for each set of values of x, y, z ; ψ must have one value only.

(*iii*) ψ must be continuous in all regions except in those regions where the potential energy $V(x, y, z) = \infty$.

9.6.3 Normalisation of Wave Function

The solution of three dimension time independent Schrödinger wave equation for a particle gives, an expression for the wave function of particle, which is function of (x, y, z) and contains an infinite constant. The value of the constant along with its sign may be determined as follows :

Since $|\psi(x, y, z)|^2\, dV$ is the probability that the particle will be found in a volume element dV surrounding the point at position (x, y, z), the total probability that the particle will be somewhere in space must be equal to 1.

Thus, we have

$$|\psi(x, y, z)|^2\, dV = 1$$

where ψ is a function of space coordinates (x, y, z).

From this normalising condition, we can find the value of the constant and its sign. A wave function which satisfies the above condition is said to be normalised (to unity).

The normalising condition for the wave function for the motion of a particle in one dimension is $|\psi(x)|^2\, dx = 1$.

9.7 OPERATORS

An operator is a rule by means of which a given function is changed into another function.

The measurable quantities like energy, momentum, position etc. are called observable. Each observable has a definite operator associated with it.

Operators Associated with Dynamical Variables

9.7.1 Energy Operator

The Schrödinger equation

$$H\psi = E\psi$$

informs, that the operator associates with energy E is Hamiltonian H, *i.e.*, E_{op} or $E = H$

The time independent form of E_{op} is

$$E_{op} = H = -\frac{\hbar^2}{2m}\nabla^2 + V$$

The time dependent Schrödinger equation is

$$H\psi = i\hbar\frac{\partial\psi}{\partial t} \quad i.e., \ (H)\psi = \left(i\hbar\frac{\partial}{\partial t}\right)\psi$$

i.e., time dependent value of Hamiltonian is

$$H = i\hbar\frac{\partial}{\partial t}$$

Thus we have

$$E_{op} = H = -\frac{\hbar^2}{2m}\nabla^2 + V = i\hbar\frac{\partial}{\partial t}. \quad \Rightarrow \quad \boxed{E_{op} = i\hbar\frac{\partial}{\partial t}}$$

9.7.2 Momentum Operator

As Hamiltonian $\qquad H$ = kinetic energy + potential energy

$$= \frac{p^2}{2m} + V$$

Also

$$H = -\frac{\hbar^2}{2m}\nabla^2 + V$$

$$\therefore \quad \frac{p^2}{2m} + V = -\frac{\hbar^2}{2m}\nabla^2 + V \ i.e., \ p^2 = -\hbar^2\nabla^2$$

Thus, the operator associated with momentum p_{op} is

$$p_{op}^2 = -\hbar^2\nabla^2 = \frac{\hbar^2}{i^2}\nabla^2$$

$$p_{op} \ \text{or} \ \hat{p} = \frac{\hbar}{i}\nabla$$

If p_x, p_y, p_z are components of momentum p then

$$(i\,p_x + j\,p_y + k\,p_z) = \frac{\hbar}{i}\left(i\frac{\partial}{\partial x} + j\frac{\partial}{\partial y} + k\frac{\partial}{\partial z}\right)$$

$\therefore$ Therefore operators associated with momentum components are

$$(p_x)_{op} \text{ or } \boxed{\hat{p}_x = \frac{\hbar}{i}\frac{\partial}{\partial x},\ \hat{p}_y = \frac{\hbar}{i}\frac{\partial}{\partial y} \text{ and } \hat{p}_z = \frac{\hbar}{i}\frac{\partial}{\partial z}}.$$

9.7.3 Kinetic Energy Operator

As $$H = \textit{kinetic energy} + \textit{potential energy} = T_{op} + V_{op}$$

Also $$H = -\frac{\hbar^2}{2m}\nabla^2 + V$$

$\therefore$ Kinetic energy operator T_{op} or $\hat{T} = -\dfrac{\hbar^2}{2m}\nabla^2.$

9.7.4 Velocity Operator

$$(KE)_{op} = \frac{1}{2} m v_{op}^2 = -\frac{\hbar^2}{2m}\nabla^2$$

$\therefore$ Operator associated with velocity v is given by

$$v_{op}^2 = -\frac{\hbar^2}{2m}\nabla^2 = \frac{\hbar^2}{i^2}\frac{\nabla^2}{m^2}$$

$$v_{op} \text{ or } \boxed{\hat{v} = \frac{\hbar}{im}\nabla}.$$

9.7.5 Potential Energy Operator

The operator associaied with potential energy

$$\boxed{V_{op} = V}$$

9.7.6 Position Operator

The operator associated with position

$$\boxed{r_{op} = r}.$$

9.8 EXPECTATION VALUES OF DYNAMICAL QUANTITIES

According to Born the wave function ψ has probabilistic interpretation, therefore it is essential to calculate the average or expectation value of any dynamical quantity defined by the wave function. In physics such dynamical quantities are space-coordinates, momentum and energy of the system.

The average or expectation value of a dynamical quantity is the mathematical expectation for the result of a single measurement.

Or

It may be defined as the average of the result of a large number of measurements on independent systems.

The expectation value of any observable f, for normalised wave function may be written as

$$<f> = \int P(r, t) f_{op} \, d\tau \qquad ...(1)$$

$$= \int \psi^* (r, t) f_{op} \psi(r, t) \, d\tau \qquad ...(2)$$

where f_{op} is operator associated with observable f.

The expression value of energy is written as

$$< E > = \int \psi^* p_{op} \psi \, d\tau = \int \psi^* \left(\frac{h}{i}\nabla\right) \psi \, d\tau \qquad ...(3)$$

The expectation value of momentum, is written as

$$< p > = \int \psi^* p_{op} \psi \, d\tau = \int \psi^* \left(\frac{\hbar}{i}\vec{\nabla}\right) \psi \, d\tau$$

$$= - i\hbar \int \psi^* \nabla \psi \, d\tau \qquad ...(4)$$

This equation is equivalent to three component equations defining the expectation values of x, y and z components of momentum *i.e.*,

$$< p_x > = - i\hbar \int \psi^* \frac{\partial \psi}{\partial x} \, d\tau$$

$$< p_y > = - i\hbar \int \psi^* \frac{\partial \psi}{\partial y} \, d\tau \qquad ...(5)$$

$$< p_z > = - i\hbar \int \psi^* \frac{\partial \psi}{\partial z} \, d\tau.$$

9.9 THE SCHRÖDINGER WAVE EQUATION

So far, we have made a lot of progress concerning the properties and interpretation of the wave function, but as yet we have had very little to say about how the wavefunction may be derived in a general situation, that is to say, we do not have on hand a "wave equation" for a wave function. *There is no true derivation of this equation*, but its form can be motivated by physical and mathematical arguments at a wide variety of levels of sophistication. Here, we will offer a simple derivation based on what we have learned so far about the wave function.

The Schrödinger equation has *two 'forms'*, **one** in which time explicitly appears, and so describes how the wave function of a particle will evolve in time. In general, the wave function behaves like a wave, and so the equation is often referred to as **time dependent Schrödinger wave equation.** The **other** is the equation in which the time dependence has been 'removed' and hence is known as the **time independent Schrödinger equation** and is found to describe, amongst other things, what the allowed energies are of the particle.

These are not two separate, independent equations—the time independent equation can be derived readily from the time dependent equation.

9.9.1 Derivation of the Schrödinger Wave Equation

(*a*) **The time Dependent Schrödinger Wave Equation:** Let us consider a microparticle. Let ψ be the wave function associated with the motion of this microparticle ψ function represents the wave field of the particle. It is similar to **E** and **B** used to describe the electromagnetic waves and to transverse displacement for waves on a string.

For one-dimensional case, the classical wave equation has the following form.

$$\frac{\partial^2 \xi}{\partial x^2} = \frac{1}{v^2}\frac{\partial^2 \xi}{\partial t^2} \quad \text{...(1)}$$

The solution of above equation is the familiar plane wave

$$\xi(x, t) = A\, e^{i(kx - \omega t)} \quad \text{...(2)}$$

where $\omega = vk$ and v is phase velocity.

For microparticle, ω and k may be replaced with E and p using Einstein and de Broglie relations

$$\left.\begin{aligned} \omega &= \frac{E}{\hbar} \\ \text{and} \quad k &= \frac{p}{\hbar} \end{aligned}\right\} \quad \begin{aligned} \omega &= 2\pi\nu = \frac{2\pi E}{h} = \frac{E}{\hbar} \\ k &= \frac{2\pi}{\lambda} = \frac{2\pi p}{h} = \frac{p}{\hbar} \end{aligned} \quad \text{...(3)}$$

Substituting equation (3) in equation (2) and replacing $\xi(x, t)$ with wave function $\psi(x, t)$

$$\psi(x, t) = A\, e^{-i(Et - px)/\hbar} \quad \text{...(4)}$$

Differentiating equation (4) with respect to t

$$\frac{\partial \psi}{\partial t} = -\frac{i}{h} E\psi \quad \text{...(5)}$$

Differentiating equation (4) with respect to x

$$\frac{\partial \psi}{\partial x} = \frac{i}{\hbar} p\psi \quad \text{...(6)}$$

Again we differentiating equation (6) with respect to x

$$\frac{\partial^2 \psi}{\partial x^2} = -\frac{p^2}{\hbar^2}\psi \qquad ...(7)$$

The kinetic energy and momentum of a free particle are related by the expression

$$\text{K.E.} = \frac{p^2}{2m}$$

The total energy of the particle

Total energy = Kinetic energy + Potential energy

i.e.,

$$E = \frac{p^2}{2m} + V \qquad ...(8)$$

Multiplying equation (8) by ψ, we get

$$E\psi = \frac{p^2}{2m}\psi + V\psi \qquad ...(9)$$

Substituting $E\psi$ and $p^2\psi$ using equations (5) and (7) in equations (9)

$$-\frac{\hbar}{i}\frac{\partial \psi}{\partial t} = -\frac{\hbar^2}{2m}\frac{\partial^2 \psi}{\partial x^2} + V\psi$$

$$\Rightarrow \quad \boxed{-\frac{\hbar^2}{2m}\frac{\partial^2 \psi}{\partial x^2} + V\psi = i\hbar\frac{\partial \psi}{\partial t}} \qquad ...(10)$$

This is *one dimensional time dependent Schrödinger's* wave equation.

(*b*) **The Time Independent Schrödinger Equation:** In many cases the potential energy V of a particle does not depend on time, it varies only with position of the particle and field is said to be *stationary*. In such cases Schrödinger's time independent wave equation is used.

In this case the wave function is written as product of $\psi(x)$ and $\psi(t)$

$$\psi(x, t) = \psi(x)\,\psi(t) \qquad ...(1)$$

Substituting this in Schrödinger's time dependent wave equation

$$-\frac{\hbar^2}{2m}\psi(t)\frac{\partial^2 \psi(x)}{\partial x^2} + V(x)\psi(x)\psi(t) = i\hbar\,\psi(x)\frac{\partial \psi(t)}{\partial t}$$

and dividing by $\psi(x, t) = \psi(x)\psi(t)$, we get

$$-\frac{\hbar^2}{2m}\frac{1}{\psi(x)}\frac{\partial^2 \psi(x)}{\partial x^2} + V(x) = i\hbar\frac{1}{\psi(t)}\frac{\partial \psi(t)}{\partial t} \qquad ...(2)$$

The right hand side of equation (2) is a function of t only and the left hand side a function of x only. Therefore equation (2) must be valid for any x and t, it can be so only if the two sides of equation (2) are equal to a constant. Setting this constant equal to energy E, we get

$$-\frac{\hbar^2}{2m}\frac{1}{\psi(x)}\frac{\partial^2\psi(x)}{\partial x^2} + V(x) = E$$

or

$$\boxed{\frac{-\hbar^2}{2m}\frac{\partial^2\psi}{\partial x^2} + V\psi = E\psi} \quad ...(3)$$

which is *time independent Schrödinger's wave equation.*

In three dimensions, the time independent Schrödinger equation is written as

$$\frac{-\hbar^2}{2m}\nabla^2\psi + V(x)\psi = E\psi \quad ...(4)$$

where $\nabla^2 = \frac{\partial^2}{\partial x^2} + \frac{\partial^2}{\partial y^2} + \frac{\partial^2}{\partial z^2}$.

Equation (3) is frequently written in the form

$$\boxed{H\psi = E\psi} \quad ...(5)$$

where H is Hamiltonian operator

$$H = -\frac{\hbar^2}{2m}\nabla^2 + V. \quad ...(6)$$

9.9.2 Applications of the Schrödinger Equation

(*i*) **Motion of a Free Particle:** Consider an electron moving freely in space in the positive x-direction and not acted upon by any force. Since no force is acting on electron and its potential energy is zero.

We know that Schrödinger's time independent equation

$$-\frac{\hbar^2}{2m}\frac{\partial^2\psi}{\partial x^2} + V\psi = E\psi \quad ...(1)$$

$$\frac{\partial^2\psi}{\partial x^2} + \frac{8\pi^2 m}{h^2}E\psi = 0 \qquad \left[\text{as } \hbar = \frac{h}{2\pi} \text{ and } V = 0\right] \quad ...(2)$$

If $$k^2 = \frac{8\pi^2 mE}{h^2} \quad ...(3)$$

The above equation reduces to

$$\frac{\partial^2 \psi}{\partial x^2} + k^2\psi = 0 \qquad ...(4)$$

The general solution of equation (4) is

$$\psi(x) = A\, e^{ikx} + B\, e^{-ikx} \qquad ...(5)$$

where A and B are constants. As it is assumed that the waves propagating only in the positive x-direction, we can write

$$\psi(x, t) = A\, e^{ikx}\, e^{-i\omega t}$$

Since the electron is moving freely, there are no boundary conditions and hence no restriction on k. All values of the energy are allowed. The allowed values of energy form a continuum and are given by

$$E = \frac{h^2}{8\pi^2 m} k^2 \qquad \text{[Using equation (3)]}$$

A freely moving electron therefore possesses a continuous energy spectrum.

$$k = \sqrt{\left(\frac{2mE}{\hbar^2}\right)} = \frac{p}{\hbar} = \frac{2\pi}{\lambda}$$

The k vector describes the wave properties of the electrons.

As $E \propto k^2$, the graph between E and k is a *parabola* as visualised in Figure 9.7.

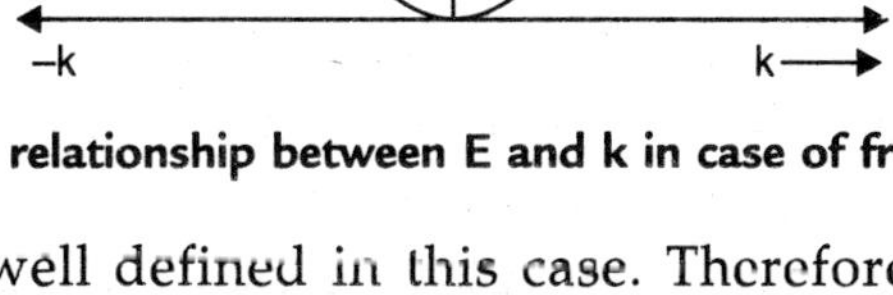

FIGURE 9.7 **Parabolic relationship between E and k in case of free electron**

The momentum is well defined in this case. Therefore according to uncertainty principle it is difficult to assign a position to the electron, *i.e.*, the electron position is indeterminate.

(*ii*) **Infinite Potential Well : A Confined Particle in One Dimensional Box:** A particle is trapped in one dimensional box according to the following potential distribution

$$V = 0 \quad \text{for} \quad 0 < x < L$$

$$V \to \infty \quad \text{for} \quad x \geq L \text{ and } x \leq 0$$

Schematically, this can be described as in Figure 9.8.

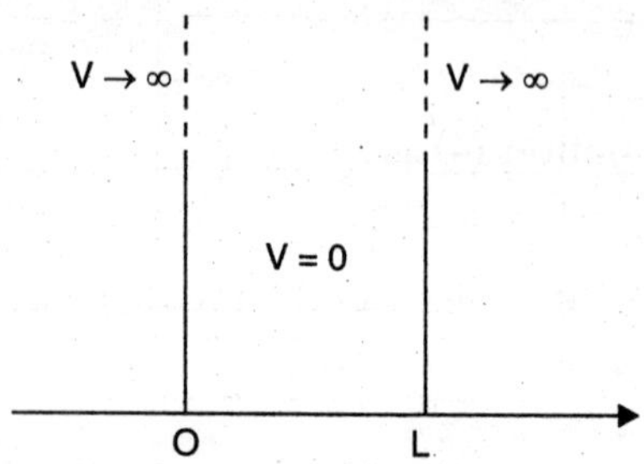

FIGURE 9.8 **Particle in an infinite quantum well**

This indicates that the probability of finding the particle outside the well is zero. The particle will need infinity energy to come out of the well. Therefore $\psi(x) = 0$ for $x \le 0$ and $x \ge L$ and finite inside the well. Hence the Schrödinger equation becomes

$$\frac{\partial^2 \psi(x)}{\partial x^2} + \frac{2mE}{\hbar^2}\psi(x) = 0 \qquad \text{...(1)}$$

The general solution of this second order linear differential equation will be

$$\psi(x) = A\,e^{ikx} + B\,e^{-ikx} \qquad \text{...(2)}$$

where A and B are constants and $k^2 = 2mE/\hbar^2$, where k is a real quantity. Now, considering the boundary conditions $\psi(0) = \psi(L) = 0$ and $\left.\frac{d\psi(x)}{dx}\right|_{x=0} = \left.\frac{d\psi(x)}{dx}\right|_{x=L} = 0,$

we get

$$\psi(x) = 2Ai \sin kx \qquad \text{...(3)}$$

and

$$kL = n\pi,\ n = 1, 2, 3, \ldots.. \qquad \text{...(4)}$$

These in turn gives

$$E_n = \frac{n^2h^2}{8mL^2}, \qquad n = 1, 2, 3 \qquad \text{...(5)}$$

Eq. (5) shows only those values of energy are possible which n is an integer.

So, $$E_1 = \frac{h^2}{8mL^2},\ E_2 = \frac{h^2}{2mL^2},\ E_3 = \frac{9h^2}{8mL^2},\ \ldots\ldots. \text{ etc.}$$

are allowed energy states. It is shown in Figure 9.9.

Energy of particle in one dimensional box can take only discrete values *i.e.*, it is quantised. The value of energy E_1 for $n = 1$ is called **zero point energy**, which signifies that there must be some movement of particles (atoms, molecules etc.) at absolute zero temperature.

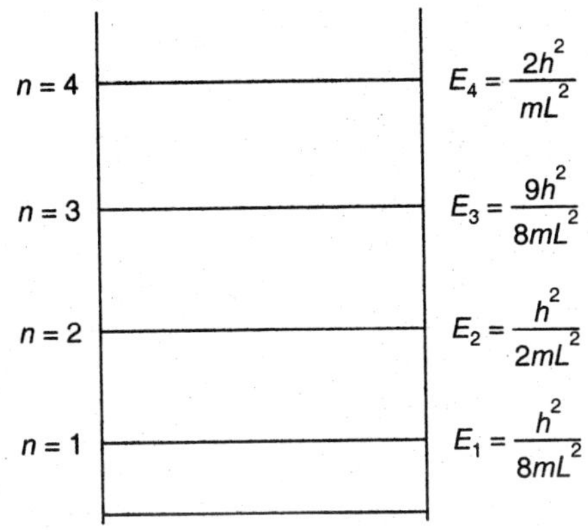

FIGURE 9.9 Energy level diagram for an electron confined to a one dimensional potential well. Note that the electron cannot have zero energy. The lowest allowed energy is E_1

The wave functions corresponding to the above allowed discrete energy levels can be obtained as follows

$$\psi_n(x) = A_1 \sin \frac{n\pi x}{L}, \; n = 1, 2, 3... \qquad ...(6)$$

where $A_1 = 2i\Lambda$, called the *normalised constant* and can be determined from the probability of finding the particle inside the *quantum well* as unity. Thus, we get

$$\int_0^L |\psi(x)|^2 = 1$$

$$\Rightarrow \qquad A_1 = \sqrt{\left(\frac{2}{L}\right)}$$

Thus, the wave function becomes

$$\psi_n(x) = \sqrt{\left(\frac{2}{L}\right)} \sin \frac{n\pi x}{L} \qquad ...(7)$$

and called the eigen function or characteristic function.

$$|\psi_n(x)|^2 = \frac{2}{L} \sin^2 \frac{n\pi}{L} x \qquad ...(8)$$

Analysis of the above results shows

1. While y_n may be negative, $|\psi_n|^2$ is always positive $|\psi_n|^2$ gives the probability of finding the particle at certain place within the well. Wave function and $|\psi_i|^2$ can be plotted as shown in Figure 9.10. For $n = 1$, the probability of finding particle is largest in the middle of the box for most of the time.

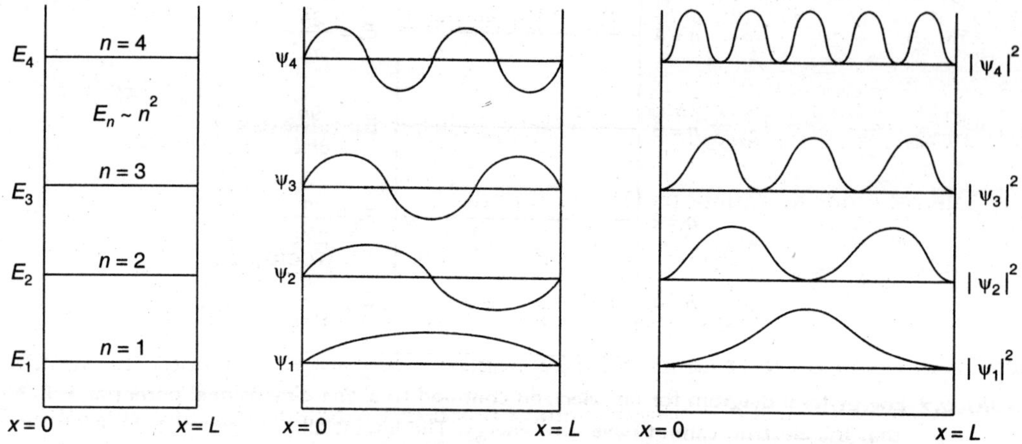

FIGURE 9.10 **Energy distribution, possible wave function, and probability distribution of electron in one dimensional box**

2. Equation (5) shows that energy is a function of quantum number and width of the well.

3. The particle (like electron) trapped in the potential well (box) can not take zero energy. Because if energy is zero, momentum also would be zero and the uncertainty principle requires that its wavelength λ be ∞. If its wavelength is infinity, then it cannot confired to a box. Therefore, the electron must possess a certain minimum amount of kinetic energy.

(*iii*) **Potential Step: Reflection and Tunnelling : Quantum Leak:** Let us consider a particle is moving along positive x-axis towards a finite potential step.

$$V = 0 \text{ for } x < 0 \text{ (region I)}$$

and

$$V = V_0 \text{ for } x \geq 0 \text{ (region II)}$$

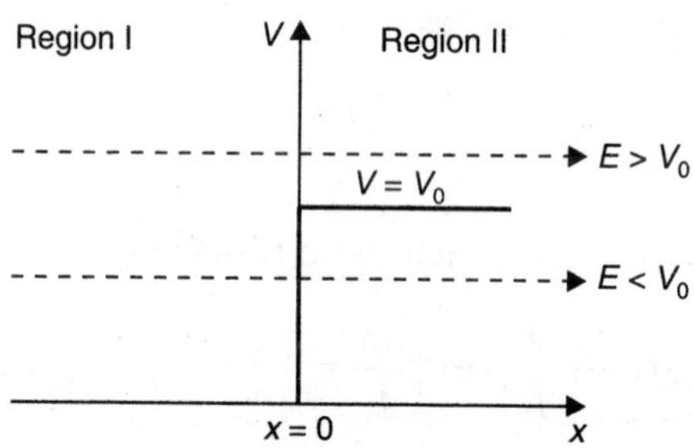

FIGURE 9.11 ***Step potential at x = 0***

Case I. $E > V_0$

If the incoming particle has energy greater than the potential step, then the Schrödinger equation will be

$$\frac{d^2\psi(x)}{dx^2} + k_1^2\,\psi(x) = 0 \text{ [in region } I\text{: } x < 0] \qquad \text{...(1)}$$

and $$\frac{d^2\psi(x)}{dx^2} + k_2^2\psi(x) = 0 \text{ [in region } II \text{ ; } x \geq 0] \qquad ...(2)$$

where $k_1^2 = \frac{2mE}{\hbar^2}$ and $k_2^2 = \frac{2m(E - V_0)}{\hbar^2}$; k_1, k_2 being real quantities.

The solutions of equations (1) and (2) are

$$\psi(x) = A\,e^{ik_1x} + B\,e^{-ik_1x} \text{ [for } x < 0 \text{ (region } I)] \qquad ...(3)$$

and $$\psi(x) = C\,e^{ik_2x} + D\,e^{-ik_2x} \text{ [for } x \geq 0 \text{ (region } II)] \qquad ...(4)$$

where A, B, C and D are arbitrary complex numbers. Physically, A denotes the amplitude of incident wave $\psi_{in}(x) = A\,e^{ik_1x}$, moving along positive x-direction. B is the amplitude of the reflected wave $\psi_{ref}(x) = B\,e^{-ik_1x}$ along negative x-direction in region I. C is the amplitude of the transmitted wave $\psi_{Tr}(x) = C\,e^{ik_2x}$ in region II moving along positive x-axis. Since no particles are incident from right $D = 0$. Also applying the boundary conditions of continuity for $\psi(x)$ and its derivative at $x = 0$, we get $(A + B) = C$ and $(A - B) = (k_1/k_2)\,C$, which gives

$$\frac{B}{A} = \frac{k_1 - k_2}{k_1 + k_2} \qquad ...(5)$$

and $$\frac{C}{A} = \frac{2k_1}{k_1 + k_2} \qquad ...(6)$$

The transmission coefficient (T) and reflection coefficient (R) are defined as the probability ratios of transmitted and reflected particles waves with respect to that of incident particle-wave respectively. Therefore, one can get the expressions of T and R as

$$T = \frac{k_2}{k_1}\left|\frac{C}{A}\right|^2 = \frac{4k_1k_2}{(k_1 + k_2)^2} \qquad ...(7)$$

and $$R = \left|\frac{B}{A}\right|^2 = \left(\frac{k_1 - k_2}{k_1 + k_2}\right)^2 \qquad ...(8)$$

It is to be noted that $\boxed{T + R = 1}$, which means the total probability always equal to unity.

These expressions are similar to the laws of reflection and transmission of light between the two media of varied refractive index. In this case of particle waves, the expressions physically mean that a quantum particle will be partially reflected even if it has enough energy to propagate past the step potential V_0. This condition is quite different from classical conditions.

Case II. $E < V_0$

When the incoming particle has energy less than the step potential. Schrödinger equation in region *I* will be same, but in region *II*, it will look like

$$\frac{d^2\psi(x)}{dx^2} - k_3^3\,\psi(x) = 0 \qquad \text{...(9)}$$

where $k_3^2 = \dfrac{2m(V_0 - E)}{\hbar^2} = -\,k_2^2$, k_3 being real decay constant. The solution will be similar to the previous one as

$$\psi(x) = A_1 e^{ik_1 x} + B_1 e^{-ik_1 x} \;\; [\text{for } x < 0 \text{ (region } I)] \qquad \text{...(10)}$$

and

$$\psi(x) = C_1 e^{-k_3 x} + D_1 e^{k_3 x} \;\; [\text{for } x \geq 0 \text{ (region } II)] \qquad \text{...(11)}$$

A_1, B_1, C_1 and D_1, are arbitrary complex numbers. Now, as $x \to \infty$, $e^{k_3 x} \to \infty$. Therefore, D_1 should be zero to prevent the particles from unphysically collecting at large positive x. Also applying the boundary conditions for $\psi(x)$ and its derivatives at $x = 0$, we get $(A_1 + B_1) = C_1$ and $(A_1 - B_1) = (ik_3/k_1)C_1$, which gives

$$\frac{B_1}{A_1} = \left(\frac{k_1 - ik_3}{k_1 + ik_3}\right)^2 \qquad \text{...(12)}$$

and

$$\frac{C_1}{A_1} = \frac{2k_1}{k_1 + ik_3} \qquad \text{...(13)}$$

Therefore, the reflection coefficient will be

$$R = \left|\frac{B_1}{A_1}\right|^2 = \left(\frac{k_1 - ik_3}{k_1 + ik_3}\right)^2 = 1 \qquad \text{...(14)}$$

The solution in region *II* ($x > 0$) is now an exponentially decaying function. This function is not automatically zero in the region of negative energy. If we find probability at $x = 0$ as

$$|C|^2 = |A|^2 \frac{4k_1}{\left(k_1^2 + k_3^2\right)^2} = |A|^2 \frac{4E}{E + (V_0 - E)} = |A|^2 \frac{4E}{V_0} \qquad \text{...(15)}$$

We see that this probability is finite (as $E < V_0$). Therefore the probability of finding the particle in the forbidden region of positive x can be done by integrating the wave function over the x values from 0 to ∞, which will be finite and proportional to $|C|^2$. This is an example of *tunnelling effect, a particle can be sometimes found where its energy is negative ($E < V_0$) and therefore, classically forbidden.* Actually, the wave penetrating a small distance from the boundary into region *II* is continually reflected till all the incident energy is turned back into region *I*. Due to such continual reflection, the amplitude of the wave penetrating into region *II* falls off exponentially. Classically, a particle of energy $E < V_0$ can never penetrate into region *II*, since its K.E. would be negative, but in quantum physical world there is a finite probability of the particle wave penetrating a short distance into the classically forbidden region *II* and decaying exponentially.

Now for the case of $E > V_0$, according to equation (6), $|C| > |A|$ since $k_1 > k_2$. So the amplitude of the transmitted wave is greater than that of incident wave. Also, as the kinetic energy of the particle is greater in region *I*, therefore, the de Broglie wavelength is shorter in region *I* with respect to region *II* as explained in Figure 9.12. On the other hand, for the case of $E < V_0$, the transmitted wave decays exponentially within a very short distance in region *II* as explained earlier and shown in Figure 9.12.

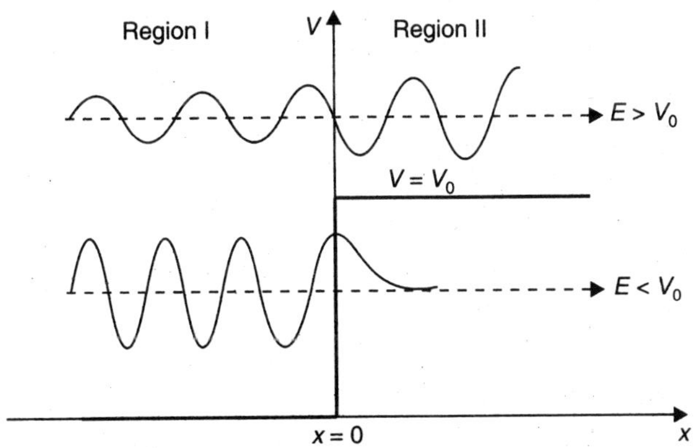

FIGURE 9.12 **Wave functions for a stop potential**

9.10 NUMERICAL PROBLEMS BASED ON WAVE MECHANICS

Problem 1. *Calculate de Broglie wavlength of an electron moving with velocity 2×10^4 m/s. [Given : mass of electron = 9.1×10^{-31} kg and Planck's constant = 6.6×10^{-34} Js].*

(*Amity Univer., May 2005*)

Hint:
$$\lambda = \frac{h}{mv} = \frac{6.6 \times 10^{-34}}{9.1 \times 10^{-31} \times 2 \times 10^{4}} = 3626 \text{ Å}.$$

Problem 2. *Show that the group velocity of de Broglie wave is given by*

$$v_g = \frac{d\nu}{d\left(\frac{1}{\lambda}\right)} = \frac{dE}{dp}.$$

Solution: The relation between the phase velocity and the group velocity of the de Broglie wave is given by

$$v_g = v_p - \lambda \frac{dv_p}{d\lambda}$$

Multiplying and dividing by λ^2 on R.H.S.

$$v_g = \lambda^2\left[\frac{v_p}{\lambda^2} - \frac{1}{\lambda}\frac{dv_p}{d\lambda}\right]$$

$$\Rightarrow \qquad v_g = -\lambda^2 \frac{d}{d\lambda}\left(\frac{v_p}{\lambda}\right)$$

$$\Rightarrow \qquad v_g = -\lambda^2 \frac{d\nu}{d\lambda} \quad \text{where} \quad \nu = \frac{v_p}{\lambda} = \text{frequency of the wave.}$$

Taking reciprocal of the equation

$$\frac{1}{v_g} = -\frac{1}{\lambda^2}\frac{d\lambda}{d\nu}$$

$$\Rightarrow \qquad \frac{1}{v_g} = \frac{d(1/\lambda)}{d\nu}$$

On reversing the equation $v_g = \dfrac{d\nu}{d(1/\lambda)}$

Multiplying and dividing by h on R.H.S.

$$v_g = \frac{d(h\nu)}{d(h/\lambda)} \quad \text{where } E = h\nu = \text{energy}$$

$$p = \frac{h}{\lambda} = \text{momentum.}$$

$$\boxed{v_g = \frac{dE}{dp}}$$

Problem 3. *Find group velocity in terms of energy and momentum.*

Hint: $v_p = \dfrac{E}{p}$ and $v_g = v_p - \lambda\dfrac{dv_p}{d\lambda}$

So $$v_g = \frac{E}{p} - \frac{d(E/p)}{d\lambda}$$

Problem 4. *If the phase velocity of a ripple is given by the relation $v_p = c_1 + c_2\lambda$, where c_1 and c_2 are constant. Find the group velocity of the ripple.*

Solution: We know that $v_g = v_p - \lambda\dfrac{dv_p}{d\lambda}$

$$v_g = c_1 + c_2\lambda - \lambda \frac{d}{d\lambda}(c_1 + c_2\lambda)$$

$$= c_1 + c_2\lambda - c_2\lambda \Rightarrow v_g = c_1.$$

Problem 5. *The phase velocity of a soap bubble is $c\lambda^{-1/2}$. Find the group velocity of soap bubble.*

Hint:
$$v_g = v_p - \lambda\frac{dv_p}{dt} = c\lambda^{-1/2} - \lambda\frac{d}{dt}\left(c\lambda^{-1/2}\right)$$

$$= c\lambda^{-1/2} + \frac{\lambda c}{2}\lambda^{-1/2}$$

$$= \frac{3c\lambda^{-1/2}}{2} = \frac{3}{2}v_p \Rightarrow \boxed{v_g = \frac{3}{2}v_p.}$$

Problem 6. *Use the uncertainty principle to deduce the lowest nature of energy of a particle of mass m enclosed in a box of size 'a'.*

Solution: The uncertainty relation is

$$\Delta x\, \Delta p \geq h$$

$$\Delta x\, \Delta p = h \qquad \text{[Using equality]}$$

$$\therefore \qquad \Delta p = \frac{h}{\Delta x}$$

The energy of the particle is given by

$$E = \frac{(\Delta p)^2}{2m}$$

$$E = \frac{h^2}{2m(\Delta x)^2}$$

According to the problem $\Delta x = a$

$\therefore$ Energy of the particle $E = \dfrac{h^2}{2ma^2}$.

Problem 7. *An electron is localised in a region of size (i) 1 cm (ii) 10^{-8} cm. Calculate order of kinetic energy of the electron in each case.*

Hint:
$$\Delta x\, \Delta p = h,\ \Delta p = \frac{h}{\Delta x},\ \text{K.E. } E = \frac{(\Delta p)^2}{2m} = \frac{h^2}{2m(\Delta x)^2}$$

$$(i)\ E = \frac{(6.6\times10^{-34})^2}{2\times9.1\times10^{-31}\times(10^{-2})^2}\ \text{J} \qquad (ii)\ E = \frac{(6.6\times10^{-34})^2}{2\times9.1\times10^{-31}\times(10^{-10})^2}.$$

Problem 8. *Show that if the measurement of the uncertainty in the location of the particle is equal to its de Broglie wavelength, the uncertainty in its velocity is equal to its velocity.*

$$(\Delta x = \lambda, \Delta v = v).$$

Solution: From uncertainty principle $\Delta x \, \Delta p = h$

Here $\Delta x = \lambda$

$\therefore$ Uncertainty is the measurement of the momentum

$$\Delta p = \frac{h}{\Delta x} \text{ or } m\Delta v = \frac{h}{\Delta x}$$

$$\Delta v = \frac{h}{m\Delta x} \quad \Rightarrow \text{ putting } \Delta x = \lambda$$

$$\therefore \quad \Delta v = \frac{h}{m\lambda}$$

$$\Delta v = \frac{h}{m} \cdot \frac{p}{h}$$

$$= \frac{h}{m} \frac{mv}{h}$$

$$\Rightarrow \quad \boxed{\Delta v = v.}$$

Problem 9. *Show that the uncertainty principle is not observed in daily life, while it is valid for microscopic world.*

Solution: Case I: Uncertainty principle for macroscopic world (daily life). Let us consider a sand particle of mass $m = 10^{-6}$ kg and its size is 10^{-4} m.

Therefore uncertainty in the measurement of the position of the sand particle is $\Delta x = 10^{-4}$ m.

Uncertainty in momentum of the sand particle

$$\Delta p = \frac{h}{\Delta x}$$

$$\Rightarrow \quad m \, \Delta v = \frac{h}{\Delta x}$$

$$\Delta v = \frac{h}{m\Delta x}$$

This gives the uncertainty in the velocity of the sand particle

$$\Delta v = \frac{6.626 \times 10^{-34}}{10^{-6} \times 10^{-4}} = 6.626 \times 10^{-24} \text{ m/s}$$

which is too small.

Therefore uncertainty principle is not effective in daily life.

Case II: Uncertainty principle in microscopic world (subatomic world):

Since the matter consists of atoms and molecules and atom comprises electrons. The mass of electron is $m_e = 9.1 \times 10^{-31}$ kg

The size of electron $= 1\text{Å} = 10^{-10}$ m

From uncertainty principle

∴ Uncertainty in the velocity of electron

$$\Delta v = \frac{h}{m\Delta x} = \frac{6.626 \times 10^{-34}}{9.1 \times 10^{-31} \times 10^{-10}}$$

$$\approx 0.74 \times 10^{7} \text{ m/s}$$

$$= 7.4 \times 10^{6} \text{ m/sec C.}$$

which is observable.

Thus this uncertainty in velocity of the microscopic produces unheard result. Hence uncertainty principle is effective in subatomic world.

Problem 10. *Find the de Broglie wavelength of a 0.01 kg pellet having a velocity of 10 m/s.*

Hint:
$$\lambda = \frac{h}{mv} = \frac{6.63 \times 10^{-34}\,\text{J.s}}{0.01\,\text{kg} \times 10\,\text{m/sec}} = 6.63 \times 10^{-33} \text{ m} = 6.63 \times 10^{-23} \text{ Å}.$$

Problem 11. *What is the uncertainty in the location of a photon of wavelength 3000 Å if this wavelength is known to an accuracy of one part in million?*

Hint:
$$p = \frac{hc}{\lambda c} = \frac{12.40 \times 10^{3}\,\text{eV Å}}{(3 \times 10^{3}\,\text{Å})c} = 4.13\,\frac{\text{eV}}{c}$$

The uncertainty in the photon momentum

$$\Delta p = \left| -\frac{h}{\lambda^2} \right| \Delta\lambda = p\frac{\Delta\lambda}{\lambda} = p \times 10^{-6} = 4.13 \times 10^{-6}\,\frac{\text{eV}}{c}$$

from which

$$\Delta x \geq \frac{h}{4\pi\Delta p} = \frac{hc}{4\pi c\Delta p} = \frac{12.4 \times 10^{3}\,\text{eV Å}}{4\pi c(4.13 \times 10^{-6}\,\text{eV}/c)} = 239 \times 10^{6} \text{ Å} = 23.9 \text{ mm}.$$

Problem 12. *What is de Broglie wavelength of an electron which has been accelerated from rest through a potential difference of 100 V.*

Hint:
$$\lambda = \frac{h}{\sqrt{2mqV}} = \frac{6.62 \times 10^{-34}}{(2 \times 9.1 \times 10^{-31} \times 100 \times 1.6 \times 10^{-19})^{1/2}} = 1.226 \text{ Å}.$$

Problem 13. *Calculate de Broglie wavelength associated with electron having 10 keV kinetic energy, where $m_e = 9.1 \times 10^{-31}$ kg and $h = 6.6 \times 10^{-34}$ Js.*

Hint:

$$\lambda = \frac{h}{mv} = \frac{h}{\sqrt{(2m_e E_k)}} = \frac{6.6 \times 10^{-34}}{(2 \times 9.1 \times 10^{-31} \times 10^4 \times 1.6 \times 10^{-19})^{1/2}} = 0.1223 \text{ Å}.$$

Problem 14. *An electron of energy 200 eV is passed through a circular hole of radius 10^{-6} m. What is the uncertainty introduced in the angle of emergence?*

Hint:

$$\frac{p^2}{2m} = E \Rightarrow p = \sqrt{2mE} = (2 \times 9.1 \times 10^{-31} \times 200 \times 1.6 \times 10^{-19})^{1/2}$$

$$\Delta p_x \approx \frac{h}{\Delta x} \quad \Delta x = 2r, \text{ where } r = 10^{-6} \text{ m}$$

$$\Delta p_x = \frac{6.63 \times 10^{-34}}{2 \times 10^{-6}} = 3.316 \times 10^{-28} \text{ kg m/sec}$$

$$\Delta\theta = \frac{\Delta p_x}{p} = 5.76 \times 10^{-6} \text{ radians.}$$

Problem 15. *The average period of that elapses between excitation of an atom and the time it emits radiation is 10^{-8} sec. Find the uncertainty in the frequency of light emitted.*

Hint:

$$\Delta E \Delta t = \hbar \Rightarrow \Delta E = \frac{6.63 \times 10^{-34}}{2\pi \times 10^{-8}} = 1.054 \times 10^{-26} \text{ J}$$

$$\Delta E = h\Delta\nu \Rightarrow \Delta\nu = \frac{\Delta E}{h} = 1.59 \times 10^{7} \text{ Hz.}$$

Problem 16. *Find the uncertainty in the kinetic energy of a proton confined to a nucleus of size of 10^{-14} m. (mass of proton = 1.66×10^{-27} kg).*

Hint:

$$\Delta p_x = \frac{h}{4\pi \Delta x} = 5.27 \times 10^{-19} \text{ kg m/s}$$

$$\Delta E_k = \frac{\Delta p_x^2}{2m} = 2.365 \times 10^{-11} \text{ J.}$$

Problem 17. *Which of the wave functions shown in Figure 9.13 have no physical significance for values of x as shown. Give reasons.*

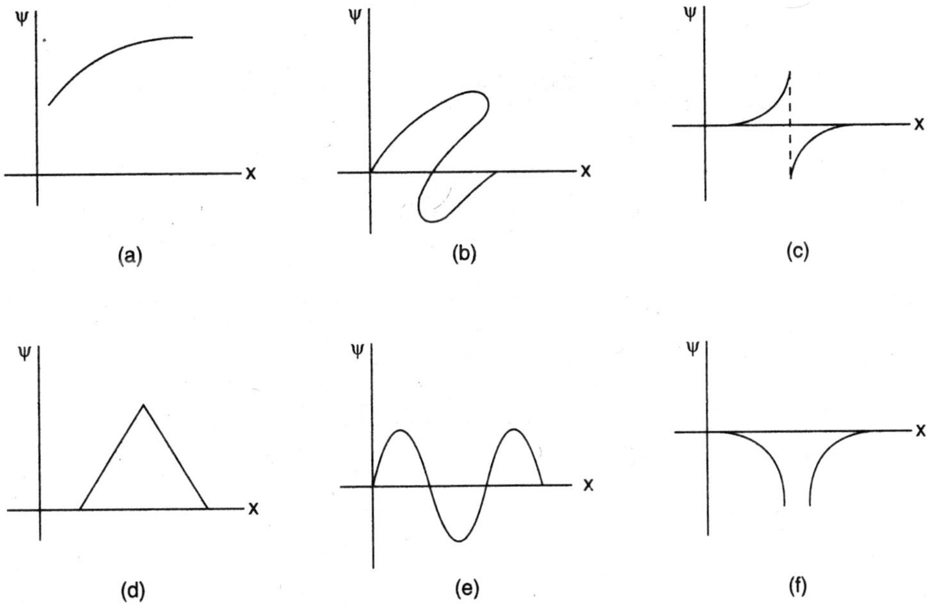

FIGURE 9.13

Ans. (*b*) Multivalued (*c*) Discontinuous and becomes infinite

(*d*) Discontinuous (*f*) becomes infinite.

Problem 18. *A wave function of a certain particle is given as* $\psi = A \sin^2 x$ *for* $\frac{-\pi}{2} < x < \frac{\pi}{2}$ *and* $\psi = 0$ *elsewhere. Find (i) the value of A, and (ii) the probability of finding the particle between* $x = -\frac{\pi}{4}$ *to* $x = +\frac{\pi}{4}$.

Solution: (*i*) Since the above function must obey the normalisation condition, we must have

$$\int_{-\pi/2}^{+\pi/2} A^2 \sin^4 x \, dx = 1$$

Now $$\sin^4 x = \left(\frac{1-\cos 2x}{2}\right)^2 = \frac{1}{4}[1 - 2\cos 2x + \cos^2 2x]$$

$$= \frac{1}{4}\left[1 - 2\cos 2x + \frac{1+\cos 4x}{2}\right]$$

$$= \frac{3}{8} - \frac{1}{2}\cos 2x + \frac{1}{8}\cos 4x$$

Now $$A^2 \int_{-\pi/2}^{+\pi/2} \left(\frac{3}{8} - \frac{1}{2}\cos 2x + \frac{1}{8}\cos 4x\right) dx$$

$$= A^2 \left(\frac{3x}{8}\right)_{-\pi/2}^{+\pi/2} = A^2 \frac{3}{8}\pi = 1 \Rightarrow A = \sqrt{\left(\frac{8}{3\pi}\right)}.$$

(ii) $$P_{12} = \frac{8}{3\pi}\int_{-\pi/4}^{\pi/4} \sin^4 x\, dx = \frac{8}{3\pi}\int_{-\pi/4}^{+\pi/4}\left(\frac{3}{8} - \frac{1}{2}\cos 2x + \frac{1}{8}\cos 4x\right) dx$$

$$= \frac{8}{3\pi}\left[\frac{3}{8}x - \frac{\sin 2x}{4} + \frac{\sin 4x}{32}\right]_{-\pi/4}^{\pi/4} = \frac{8}{3\pi} \times \left[\frac{3}{8}\times\frac{\pi}{2} - \frac{2}{4}\right]$$

$$= \frac{1}{2} - \frac{16}{12\pi} = 0.076.$$

Problem 19. *The minimum energy of a particles trapped in a one dimensional potential well is 4×10^{-18} J. What are the next three energies?*

Hint: $$E_n = \frac{n^2h^2}{8mL^2} = n^2E_1 \Rightarrow E_1 = 4 \times 10^{-18}\text{ J}$$

$$E_2 = 4E_1 = 16 \times 10^{-18}\text{ J},\ E_3 = 9E_1 = 36 \times 10^{-18}\text{ J},\ E_4 = 16E_1 = 64 \times 10^{-18}\text{ J}.$$

Problem 20. *A beam of electrons of energy 15 eV is incident at the boundary of a step potential of height 5 eV. Find the fraction of the beam reflected and transmitted.*

Solution: Fraction of the beam reflected $R = \left(\frac{k_1 - k_2}{k_1 + k_2}\right)^2$

$$\Rightarrow \quad R = \frac{\left(\sqrt{2mE} - \sqrt{2m(E-V)}\right)^2}{\left(\sqrt{2mE} + \sqrt{2m(E-V)}\right)^2} = \frac{\left(\sqrt{15} - \sqrt{10}\right)^2}{\left(\sqrt{15} + \sqrt{10}\right)^2} = 0.01 = 1\%$$

Fraction of the beam transmitted $T = \frac{4k_1k_2}{(k_1 + k_2)^2}$

$$\Rightarrow \quad T = \frac{4\sqrt{2mE}\sqrt{2m(E-V)}}{\left(\sqrt{2mE} + \sqrt{2m(E-V)}\right)^2} = \frac{4\sqrt{E(E-V)}}{\left(\sqrt{E} - \sqrt{E-V}\right)^2} = \frac{4\sqrt{15 \times 10}}{\left(\sqrt{15} + \sqrt{10}\right)^2} = 99\%$$

and $R + T = 1.$

EXERCISES

1. Differentiate between 'phase velocity' and 'group velocity'. *(Amity Univ., Noida, May 2005)*
2. What are de Broglie waves ? Show that the velocity of matter waves is found to be greater than the velocity of light. *(Amity Univ., May 2005)*
3. What was the aim of Davisson Germer's experiment ? Describe the experiment and show how far was it successful. *(Amity Univ., May 2006)*
4. Show that the de Broglie wavelength of a particle of rest mass m_0 and kinetic energy T is given by

$$\lambda = \frac{hc}{\sqrt{T(T + 2m_0c^2)}}$$ *(Amity Univ., May 2006)*

5. What do you understand by phase velocity and group velocity ? How is the group velocity of wavepacket associated with a particle related to the particle velocity ? *(Amity Univ., May 2006)*
6. State Heisenberg's uncertainty principle and describe an experiment to prove it. *(Amity Univ., May 2006, May 2005)*
7. What is Born's interpretation of wave function ? *(Amity Univ., May 2005)*
8. What are 'operators' in quantum mechanics ? Write down the formulae for (*i*) momentum operator, and (*ii*) kinetic energy operator with reference to a one-dimensional wave function. *(Amity Univ., May 2005)*
9. What do you understand by

   (*i*) eigen values and eigen functions of an operator

   (*ii*) expectation value of a dynamical variable. *(Amity Univ., May 2006)*
10. What is de Broglie principle?
11. Which considerations led de Broglie to propose the wave nature of particles?
12. Explain the wave packet.
13. Show that the phase velocity can exceed the velocity of light in vacuum.
14. Explain the terms, 'phase velocity' and 'group velocity'.
15. State and explain Heisenberg's uncertainty principle.
16. Use uncertainty relation to show that the electron cannot stay in the nucleus.
17. Explain the terms operators and expectation value.
18. Why should a wave function obey the normalisation condition?
19. Write down the one dimensional time dependent Schrödinger wave equation and explain the meaning of each terms.
20. Derive the time dependent Schrödinger equation.
21. Derive the time independent Schrödinger equation.
22. Give comparative account of the energy of a free particle and a particle confined to one dimensional infinitely deep potential well.

23. Set up the Schrödinger equation for a particle in an infinite deep one dimensional potential well and find expression for the wave function and energy of the particle.
24. A particle is approaching a finite step potential with an energy greater than the height of the step. Set up the Schrödinger equation for the problem and derive expression for reflection and transmission coefficients.
25. Describe Davisson and German experiment on diffraction of electron and explain how it explains the nature of an electron beam.
26. State and explain the Heisenberg's uncertainty principle with the suitable illustration.
27. Explain the physical signification of wave function.
28. Write Schrödinger equation for a particle in box. Solve it to obtain eigen values and eigen functions.
29. Derive time independent Schrödinger wave equation for a nonrelativistic free particle. Explain significance of the wave function.

SUGGESTION FOR FURTHER READING

1. J.J. Sakurai, Modern Quantum Mechanics, Pearson Education, New Delhi, 2009.
2. Ajoy Ghatak, Basic Quantum Mechanics, Macmillan India Ltd., New Delhi, 2002.
3. Devraj Singh, Fundamentals of Engineering Physics, Vol. 2, Dhanpat Rai and Co. (P) Ltd., New Delhi, 2009.
4. S.L. Kakani and H.M. Chandalia, Quantum Mechanics, Sultan Chand and Sons, New Delhi, 2004.
5. M.G. Arora, A Textbook of Quantum Mechanics, Anmol Publications Pvt. Ltd., New Delhi, 1997.
6. Eyvind H. Wichman, Berkeley Physics Course, Vol. 4, Quantum Mechanics, Tata McGraw-Hill Publishing Co. Ltd., New Delhi, 2008.
7. Devraj Singh, Engineering Physics, Vol. 2, Dhanpat Rai and Co., New Delhi, 2009.
8. G. Aruldhas, Quantum Mechanics, PHI Learning Pvt. Ltd., New Delhi, 2009.
9. B.K. Agarwal and Hari Prakash, Quantum Mechanics, Prentice Hall of India Pvt. Ltd., New Delhi, 2009.
10. Ajoy Ghatak and S. Lokhanathan, Quantum Mechanics, Macmillan India Ltd., New Delhi, 2005.

Chapter 10 ATOMIC PHYSICS

10.1 INTRODUCTION

Atomic physics (or **atom physics)** is the field of physics that studies atoms as an isolated system of **electrons** and an atomic nucleus. It is primarily concerned with the **arrangement of electrons around the nucleus** and the processes by which these arrangements change.

This includes ions as well as neutral atoms and unless otherwise stated, for the purposes of this discussion it should be assumed that the term *atom* includes ions.

The term *atomic physics* is often associated with nuclear power and nuclear bombs, due to the synonymous use of *atomic* and *nuclear* in standard English. However, physicists distinguish between atomic physics—which deals with the atom as a system of electron(s) and a nucleus—and nuclear physics—which considers **atomic nuclei** alone.

As with many scientific fields, strict delineation can be highly contrived and atomic physics is often considered in the wider context of *atomic, molecular and optical physics*. Physics research groups are usually so classified.

10.2 EARLIER ATOMIC MODELS AND THEIR SHORTCOMINGS

A number of models have been proposed by researchers, starting from time of Newton to Sommerfeld. However, for the current discussion, Bohr's model of the hydrogen atom and its modification by Sommerfeld will be worth mentioning.

Neils Bohr, proposed his atomic model in 1913 which could successfully explain the observed hydrogen and hydrogen like spectra. Followings are the postulates given by Bohr.

(*i*) The electron in an atom revolves in a fixed circular orbit about the nucleus, with the essential centripetal force supplied by the Coulomb interaction between the electron and the nucleus

i.e., $$\frac{Ze^2}{4\pi\epsilon_0 r^2} = \frac{mv^2}{r} = mr\omega^2 \quad ...(1)$$

Here, r is the radius of allowed orbit, v is the velocity of electron, Z is the atomic number, m is the mass of electron and ω is the corresponding angular velocity.

(*ii*) Out of a large number of possible orbits the electron will exist only in those for which the angular momentum of electron is integral multiple of $\frac{h}{2\pi}$ or $\hbar$, where h is Planck's constant.

$$mvr = m\omega r^2 = n\hbar \quad ...(2)$$

(*iii*) The electron, while moving in an allowed orbit does not radiate electromagnetic energy. The electromagnetic energy will be either emitted or absorbed only when an electron makes a transition from one allowed stationary state to another. The frequency of emitted/absorbed radiation is given by

$$\nu = \frac{E_i - E_f}{h} \quad ...(3)$$

where ν is the frequency of emitted or absorbed radiation, E_i and E_f are the energies of initial and final orbits.

One can obtain expression for radius of allowed orbits by solving equations (1) and (2),

$$\boxed{r_n = 4\pi\epsilon_0 \frac{\hbar^2 n^2}{me^2 Z}} \quad ...(4)$$

The equation (4) states that the radii of successive orbits are proportional to n^2 and have discrete values, which means that they are quantized.

The energy of electron in n^{th} Bohr orbit is given by

$$\boxed{E_n = -\frac{13.58\, Z^2}{n^2}\, eV} \quad ...(5)$$

As such the electron bound to nucleus can have any one of discrete set of allowed energy levels.

The Bohr theory could explain the experimental results, of hydrogen and hydrogen like atoms. However, this theory suffered from some difficulties :

(*i*) Bohr atomic model can predict only the transition frequencies, but tells nothing about the polarization and intensity of spectral lines.

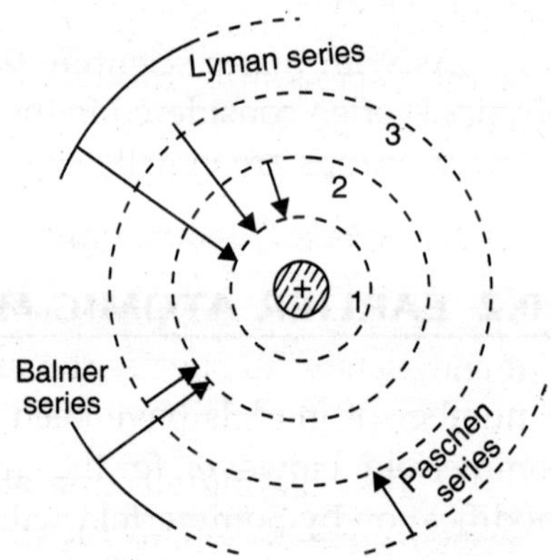

FIGURE 10.1 **Few spectral lines of Hydrogen (Z = 1) atom arising from transition between the levels**

(*ii*) According to the Bohr's theory, the frequency of the spectral lines in hydrogen are well defined. However, it has been observed that many of the lines in the Balmer and other series are not single lines at all. The two components in H_{α} lines with separation is ~ 0.13 Å have been reported. This is known as *fine structure of spectral lines*. The Bohr model could not explain the fine structure.

Extension of the Bohr's Theory: Sommerfeld Model: In 1916, Sommerfeld attempted to explain the fine structure. He made following assumptions for his model:

(*i*) In general, the path of an electron about the nucleus is ellipse. The nucleus is situated at one of the foci of the elliptical path.

(*ii*) In an elliptical orbit, the velocity of electron varies substantially at different points of the path, which causes relativistic variation in the mass of the moving electron. Hence, this model is sometimes, also referred as *relativistic atomic model*.

Shortcomings of Bohr Sommerfeld Atomic Models: The above discussed models were quite successful for Hydrogen and Hydrogen like atoms. However, these theories could not explain the following points:

(*i*) The sommerfeld theory could not predict the correct no. of lines in fine structure.

(*ii*) These theories could not explain the complex spectra of alkali metals.

(*iii*) The models were silent about the distribution of electrons in orbits around the nucleus.

(*iv*) Without giving the proper logic, the quantum numbers were introduced.

(*v*) The effect of magnetic and electric fields on spectral lines could not be explained.

In order to explain the complex spectra of atoms and their relation to the atomic structure, a new atomic model, commonly known as *Vector Atom Model* has been introduced.

10.3 VECTOR ATOM MODEL

The vector atom model basically deals with the *total angular momentum* of an atom, which is result of the combination of orbital and spin angular momenta. The two fundamental features of the vector atom model are (*i*) space quantization of orbits and (*ii*) spinning electron hypothesis.

10.3.1 Space Quantization

The angular momentum is a vector quantity, hence its direction must be specified to describe it completely. To specify the orientation or direction of an orbit, a reference is required. The direction of the magnetic field applied to the atom is chosen as the reference line.

The rotating electron about the nucleus forms a current loop which has a magnetic moment $\mu = I\mathbf{A}$, where I is the current in the loop and $\mathbf{A}$ is the area vector. The energy of loop-field system is given by $\mathbf{U} = -\mu.\,\mathbf{B} = -\,|\,\mu\,|\,|\,B\,|\cos\theta$, where θ is the angle between the magnetic moment μ and magnetic field **B**. As such classically any energy value between – μ**B** to + μ**B** is possible for the loop.

An electron orbiting around the nucleus in an atom possess angular momentum **L** which interacts with external applied magnetic field **B**. According to quantum theory, there are fixed directions of magnetic moment μ of magnetic dipole (formed by closed loop motion of electron about nucleus) with respect to the magnetic field **B**. The magnetic moment μ and angular momentum **L** can be related as

$$\boxed{\mu = \left(\frac{e}{2m_e}\right)\mathbf{L}} \qquad \text{...(1)}$$

This relationship is true from both classical and quantum mechanical point of view. Since discrete directions of μ are allowed, as such the direction of **L** will also be quantized in space. Here, the quantization refers the projection of **L** along the Z-direction ($\mathbf{L}_2$) which can have discrete values only. The orbital magnetic quantum number m_l gives the direction of **L** and tells the possible components of L in the Z-direction (the field direction). The phenomenon of quantization of **L** in the direction of magnetic field **B** is commonly known as **space quantization.**

The direction of **B** is along the Z axis. As such the component of **L** along Z-direction is given as $L_Z = m_l \hbar$, the m_l can table the values from $-l$ to $+l$ including zero.

The above discussed space quantization of orbits and possible components of **L** can be understood with the following example.

Let us calculate the allowed projections of **L** for $l = 2$. The **L** can be visualized as a vector lying on the surface of a cone (see Figure 10.2).

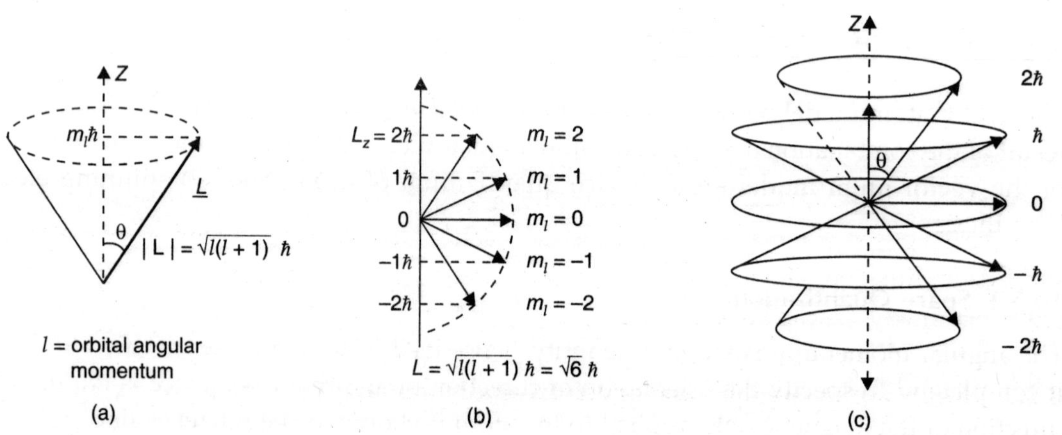

FIGURE 10.2 **Space Quantization**

The possible values of m_l for $l = 2$ can be $-2, -1, 0, 1, 2$ hence $L_Z = m_l\hbar = -2\hbar - \hbar, 0, \hbar, 2\hbar$. Figure 10.2(*b*) and 10.2(*c*) show space quantization of **L** for $l = 2$.

10.3.2 Spin Quantization

Goudsmit and Uhlenbeck in 1925 proposed that electron 'spins' about an axis through its centre of mass and further it has both angular momentum and a magnetic moment. The spin of electron is analogous to the planetary motion about the Sun in our solar system. Quantum mechanical treatment has demonstrated that the spin of electron should be quantized. As such a new quantum number *S* has been introduced. The spin angular momentum **S** is related with spin quantum number *s* as

$$s = \sqrt{s(s+1)}\,\hbar \qquad ...(1)$$

The spin can be either clockwise or anticlockwise as such *s* can have two values *i.e.*, $\pm \frac{1}{2}$

$$s = \sqrt{\frac{1}{2}\left(\frac{1}{2}+1\right)}\,\hbar = \frac{\sqrt{3}}{2}\,\hbar$$

Component of **S** along the direction of magnetic field is governed by quantum number m_s as

$$\mathbf{S}_z = m_s\hbar \qquad ....(2)$$

where m_s is the *spin magnetic quantum number*. The m_s can take $(2s + 1)$ values.

as such $\quad m_s$ can be $2 \times \frac{1}{2} + 1 = 2$ values

$$m_s = \pm \frac{1}{2}$$

Hence

$$\boxed{\mathbf{S}_z = \pm \frac{1}{2}\,\hbar} \qquad ...(3)$$

We have seen that the orbital and spin motions of an electron are quantized in magnitude as well as in direction. These motions are represented by quantized vectors, as such the atomic model is termed as *vector atom model*.

10.3.3 The Quantum Numbers and Spectroscopic Notations

The Bohr-Sommerfeld atomic theory uses four quantum numbers which are (*i*) Principal quantum number (*n*) (*ii*) Orbital quantum number (*l*) (*iii*) Spin quantum number (*s*) and (*iv*). Total angular momentum quantum number (*j*). In addition to these, three other quantum numbers have been introduced in vector atom model which are (*i*) Magnetic orbital quantum number (m_l), (*ii*) Magnetic spin quantum number (m_s) and (*iii*) Magnetic total angular momentum quantum number (m_j). The general description and possible values of these quantum numbers is given below.

(*i*) **Principal Quantum Number (*n*):** This quantum number represents the serial number of the atomic shells starting from the innermost. The '*n*' can have only positive integral values excluding zero.

$$n = 1, 2, 3, 4$$

(*ii*) **Orbital Quantum Number (*l*):** The Orbital quantum number '*l*' can have the values from 0 to $(n - 1)$, where '*n*' is the principal quantum number.

For example, if $n = 3$, then $l = 0, 1, 2$.

Conventionally we call

$l = 0$ *s* electron

$l = 1$ *p* electron

$l = 2$ *d* electron

$l = 3$ *f* electron

(*iii*) **Spin Quantum Number (*s*):** Spin quantum number *s* has only one value

i.e., $\frac{1}{2} \Rightarrow s = \frac{1}{2}$

(*iv*) **Total Angular Momentum Quantum Number (*j*):** The total angular momentum quantum number *j* is the sum of orbital angular momentum '*l*' and spin quantum number '*s*'. As such *j* can have $l \pm s$ or $l \pm \frac{1}{2}$ values.

(*v*) **Magnetic Orbital Quantum Number 'm_l':** As discussed previously in article 10.3.1, the magnetic orbital quantum number m_l can have $2l + 1$ values.

For example, $l = 2$, m_l can have $-2, -1, 0, 1, 2$ *i.e.,* total 5 values.

(*vi*) **Magnetic Spin Quantum Number 'm_s':** The magnetic spin quantum number m_s can take $(2s + 1)$ values. We know that $s = \frac{1}{2}$ and hence m_s can take $\left(2 \times \frac{1}{2} + 1\right) = 2$ values in total. The values of m_s, therefore are $-s$ and $+s$ or $-\frac{1}{2}$ and $+\frac{1}{2}$.

(*vii*) **Magnetic Total Angular Momentum Quantum Number m_j:** The total angular momentum **J** of an atom is a vector quantity and is the vector sum of orbital angular momentum **L** and spin angular momentum **S** *i.e.,* **J** = **L** + **S**. We define another quantum number m_j which is known as the magnetic total angular momentum quantum number and specify the orientation of **J** in space with respect to Z-axis.

According to quantum mechanics

$$\mathbf{J} = \sqrt{j(j+1)}\,\hbar \qquad \text{....(1)}$$

and

$$\mathbf{J}_z = m_j\hbar$$

The possible values of m_j are $2j + 1$ *i.e.*, m_j can take values from $-j$ to $+j$ including zero in integral steps.

Now

$$\boxed{\mathbf{J}_z = \mathbf{L}_z \pm \mathbf{S}_z} \qquad \text{...(2)}$$

which implies that $m_j = m_l \pm m_s$. Now m_l is an integer and m_s is $\pm\frac{1}{2}$ as such m_j will have the half integral values only. The m_j, m_l and m_s can have the maximum value j, l and s respectively.

10.4 SPIN-ORBIT COUPLING

In the previous sections, we have discussed that electron as a constituting element of an atom possesses orbital as well as spin angular momentum. Thus the total angular momentum of an atom will be the vector sum of the orbital and spin angular momenta of its electrons. The determination of total angular momentum of an atom depends on the process of coupling of the orbital and spin angular momenta. The two methods of coupling has been developed and described in following sections :

10.4.1 Russel-Saunder's or LS Coupling

In the LS coupling, the orbital angular momentum vectors and the spin angular momentum vectors of the optical electrons of an atom coupled strongly separately. The total angular momentum of the atom is thus the vector sum of the resultant orbital momentum vector **L** and spin angular momentum **S**. Mathematically this scheme can be represented as

$$\mathbf{L} = \Sigma \mathbf{l}_i, \quad \mathbf{S} = \Sigma \mathbf{S}_i \quad \text{and}$$

$$\boxed{\mathbf{J} = \mathbf{L} + \mathbf{S}} \qquad \text{...(1)}$$

The magnitude of total angular momentum is given as $\sqrt{j(j+1)}\,\hbar$, where quantum number **J** can have the values as

$$\mathbf{J} = |\,\mathbf{L}-\mathbf{S}\,|, |\,\mathbf{L}-\mathbf{S}\,| + 1 \,.....\, |\,\mathbf{L}+\mathbf{S}\,| \qquad \text{...(2)}$$

10.4.2 j-j Coupling

This coupling scheme is applicable for heavy atoms in which the interaction between the orbital and spin momenta of a single electron interaction among either is much stronger than the orbital or spin momenta of different electrons.

We know that the resultant angular momentum of an electron **J** = **l** + **S**. The vector sum of total angular momentum of all the individual electrons of atom will give total **J** values *i.e.*,

$$\mathbf{J} = \Sigma j_i \qquad ...(3)$$

10.5 ZEEMAN EFFECT

Zeeman in 1896, observed that the spectral lines are split up into components when the source emitting those spectral lines is placed in a strong magnetic field. He further observed that these components are polarized. The two types of Zeeman effect has been observed which are known as Normal Zeeman effect and Anomalous Zeeman effect. The schematic diagram of experimental arrangement for observing Normal Zeeman effect is shown in Figure 10.3.

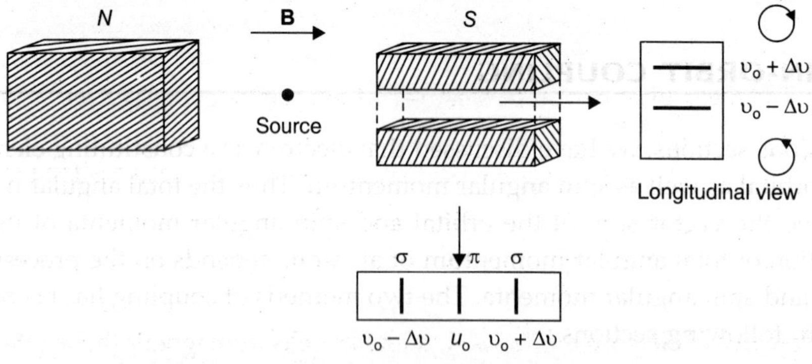

FIGURE 10.3 **Schematic diagram of Zeeman effect**

Figure 10.3 describes that a singlet spectral line split into two circularly polarized spectral lines, when viewed parallel to the magnetic field with frequencies $\upsilon_0 - \Delta\upsilon$ and $\upsilon_0 + \Delta\upsilon$. On the other hand, if the observation is made transverse to the magnetic field, three equally spared spectral lines of frequencies υ_0, $\upsilon_0 - \Delta\upsilon$ and $\upsilon_0 + \Delta\upsilon$ are observed, where $\Delta\upsilon$ is the change in frequency of the singlet-line, which is also known as *Zeeman shift*. Where υ_0 is the frequency of original line.

The interaction of the magnetic field **B** with the magnetic moment of the electron system results in splitting of the energy levels and hence in spectral lines. The magnetic moment of the electron may arise from its orbital as well as its spin motion. We know that the torque experienced by a magnet in a magnetic field is $\mu \times \mathbf{B}$, where μ is the magnetic moment. The potential energy of the magnetic dipole (the rotating electron around nucleus behaves like a tiny magnetic dipole) is given by $-\mu.\mathbf{B}$.

As such, if **B** is not too strong, the potential energy of orientation of an atom in a magnetic field **B** is given by $(-\mu_A \cdot \mathbf{B})$, where μ_A is the magnetic moment of atom. Let the energy of atomic level is E_0, in absence of **B**, then the energy of the system in presence of magnetic field will be given by

$$E = E_0 - [\mu_A \cdot \mathbf{B}] = E_0 - \mu_Z B \qquad ...(1)$$

where, μ_Z is the Z component of μ_A and **B** is parallel to the Z direction.

10.5.1 Normal Zeeman Effect

In the Normal Zeeman effect, a singlet line of frequency υ_0 splits into three plane polarized component lines having frequency $\upsilon_0 - \Delta\upsilon$, υ_0 and $\upsilon_0 + \Delta\upsilon$, when viewed at right angles to **B**. The component lines with frequency $\upsilon_0 \pm \Delta\upsilon$ are plane polarized with the electric vector perpendicular to **B**, while the central line has the same wavelength as the original is also plane polarized but the electric vector in this case is parallel to **B**.

The Normal Zeeman pattern observed due to transition between singlet ($s = 0$) states. The effect can be explained by *classical theory* as well as from the quantum theory of electron without considering the electron spin.

The current constituted by an electron of mass m and charge $-e$ in an atom will be given as

$$i = \frac{e}{T} \qquad ...(2)$$

Here the T is the orbital period of electron which can be expressed as

$$T = \frac{2\pi r}{v},$$ v is the velocity of electron and r is the radius of orbit.

Hence
$$i = \frac{ev}{2\pi r} \qquad ...(3)$$

Let the loop area of electron motion is A, then from electromagnetic theory, the magnitude of orbital magnetic moment

$$\mu_l = iA \qquad ...(4)$$

or
$$\mu_l = \frac{ev}{2\pi r} \times \pi r^2 = \frac{evr}{2\pi} \qquad ...(5)$$

The direction of μ_l will be opposite to that of its orbital angular momentum **L** as the charge of electron is negative.

$$L = mv\,r \qquad ...(6)$$

$$\frac{\mu_l}{L} = \frac{e}{2m} \qquad ...(7)$$

or in vector form
$$\mu_l = -\left(\frac{e}{2m}\right)\mathbf{L} \qquad ...(8)$$

The quantity $\frac{\mu_l}{L}$ is a constant and known as the *gyromagnetic ratio for electron.*

We know that $L = \sqrt{l(l+1)}\,\hbar$, hence

$$\mu_l = \left(\frac{e}{2m}\right)\sqrt{l(l+1)}\,\hbar$$

$$= \sqrt{l(l+1)}\,\frac{eh}{4\pi m}$$

$$\boxed{\mu_l = \sqrt{l(l+1)}\,\frac{e\hbar}{2m} = \sqrt{l(l+1)}\,\mu_B} \quad ...(9)$$

The quantity $\boxed{\mu_B = \frac{e\hbar}{2m}}$ is called as *'Bohr magneton'* which has the numerical value 9.27×10^{-24} A-m^2. The μ_B forms a natural unit for the measurement of atomic magnetic moments. Moreover the introduction of a magnetic field superimposes a precessional motion on the orbital motion of the entire orbit of the electron, about the direction of the magnetic field, with angular velocity $\omega = \frac{eB}{2m}$. The angular frequency is known as the **Larmor precessional** frequency and the above statement is referred as **'Larmor theorem'.**

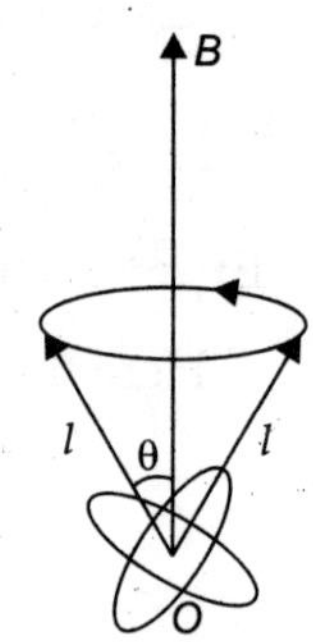

FIGURE 10.4

Quantum Mechanical Explanation of Zeeman Effect: In magnetic field the electron of atom with magnetic moment μ_l acquires additional energy due to its precessional motion which is given as

$$\Delta E = \mu_l \cdot \mathbf{B} = |\,|\mu_l|\,|\mathbf{B}|\,\cos\theta.$$

$$= \left|\frac{e}{2m}\mathbf{L}\right||\mathbf{B}|\cos\theta$$

$$= \frac{e}{2m}\,l\hbar\,B\cos\theta$$

$$= \frac{eB}{2m}\,\hbar\,l\cos\theta$$

$$= \omega\hbar\,l\cos\theta$$

$$= \hbar\,\omega\,m_l, \qquad m_l = l\cos\theta$$

$$\boxed{\Delta E = m_l\omega\hbar} \quad ...(1)$$

Now from quantum mechanics, the selection rule for the magnetic quantum number m_l is $\Delta m_l = 0, \pm 1$. This gives a normal triplet pattern. $\Delta m_l = 0$ gives π component whereas $\Delta m_l = \pm 1$ give two symmetrically spaced σ components.

Hence
$$\left.\begin{aligned} E_B &= E_0 && \text{for } \Delta m_l = 0 \\ E_B &= E_0 + \omega\hbar && \Delta m_l = +1 \\ E_B &= E_0 - \omega\hbar && \Delta m_l = -1 \end{aligned}\right\} \qquad ...(2)$$

Here E_B corresponds to the energy of spectral line in presence of **B** and $\mathbf{E}_0$ is the energy of the system without **B**.

In terms of frequency

$$\upsilon_B = \upsilon_0 \quad \text{for} \quad \Delta m_l = 0$$

$$h\upsilon = h\upsilon_0 + \frac{eB}{2m} \cdot \frac{h}{4\pi} \Delta m_l$$

$$\boxed{\upsilon = \upsilon_0 + \Delta m_l \frac{eB}{4\pi m}} \qquad ...(3)$$

Expression for Zeeman shift: The precessional frequency of electron in magnetic field is

$$\delta\omega = \frac{eB}{2m}$$

$$\omega = 2\pi\upsilon = 2\pi \frac{c}{\lambda}, \ \upsilon \text{ is the frequency of vibration of electron.}$$

Now on differentiation $\quad \delta\omega = -\frac{2\pi}{\lambda^2} c\, \delta\lambda, \qquad ...(4)$

$\Rightarrow \quad \delta\omega = 2\pi\, \delta\upsilon$

$\Rightarrow \quad \delta\upsilon = \pm \frac{eB}{4\pi m},$

from equation (4)
$$\delta\lambda = -\frac{\lambda^2}{2\pi} \delta\omega$$
$$= -\frac{\lambda^2}{2\pi c} 2\pi\delta\upsilon$$
$$= \pm \frac{\lambda^2 eB}{4\pi\, mc}$$

$$\boxed{\delta\lambda = \pm \frac{eB}{4\pi\, mc} \lambda^2} \qquad ...(5)$$

One can calculate the value of specific charge $\left(\frac{e}{m}\right)$ using equation (5)

$$\boxed{\frac{e}{m} = \left(\frac{4\pi c}{B\lambda^2}\right) \delta\lambda} \qquad ...(6)$$

$$= 1.757 \times 10^{11} \text{ c/kg.}$$

10.5.2 Anomalous Zeeman Effect

The anomalous Zeeman effect is observed due to transition between multiplet states. In this effect, the spectral lines are observed in more than three components after splitting. The anomalous Zeeman effect can be explained by considering the spin motion of electron.

Let **B** is small enough for Zeeman splitting, and hence the magnetic energy can be considered as a very small perturbation. Now due to the torque on magnetic moment, **J** will precess around **B** at a smaller rate compared to precession rate of orbital and spin angular momentum about **J**. Since **L** and **S** precess about **J**, μ_L and μ_S must precess about **J**.

According to the quantum theory, the spin and orbital magnetic moments are given as

$$\mu_S = -2\left(\frac{e}{2m}\right)\mathbf{S}, \mu_L = -\left(\frac{2}{2m}\right)\mathbf{L} \quad ...(1)$$

It is clear that each unit of spin angular momentum contributes twice is the unit of μ_L, hence the magnetic moment of atom μ_A is not antiparallel to **J**. The Figure 10.5 depicts the various momenta and their corresponding magnetic moments.

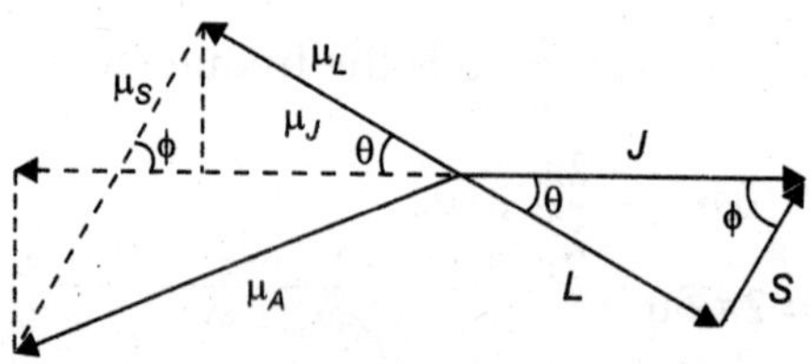

FIGURE 10.5

From the Figure 10.5

$$\mu_J = \mu_L \cos\theta + \mu_S \cos\phi$$
$$= \mu_L \cos(L, J) = \mu_S \cos(S, J)$$

$$\mu_L = -\frac{e}{2m} L\hbar$$

$$\mu_S = -2\left(\frac{e}{2m}\right)S\hbar$$

$$= -\left(\frac{e}{2m}\right)\hbar\,[L\cos(L, J) + 2S\cos(S, J)]$$

$$= -\left(\frac{e}{2m}\right)\hbar\left[L\left(\frac{J^2+L^2-S^2}{2LJ}\right) + 2S\left(\frac{J^2+S^2-L^2}{2SJ}\right)\right]$$

From quantum theory,

$$J^2 \to J(J+1)$$
$$L^2 \to L(L+1)$$
$$S^2 \to S(S+1)$$

$$= -\left(\frac{e\hbar}{2m}\right)\left[\frac{J(J+1)+L(L+1)-S(S+1)+2J(J+1)+2S(S+1)-2L(L+1)}{2\sqrt{J(J+1)}}\right]$$

$$= -\left(\frac{e\hbar}{2m}\right)\left[\frac{3J(J+1)+S(S+1)-L(L+1)}{2\sqrt{J(J+1)}\,\sqrt{J(J+1)}}\right]\sqrt{J(J+1)}$$

$$= -\left(\frac{e\hbar}{2m}\right)J\left[\frac{2J(J+1)+J(J+1)+S(S+1)-L(L+1)}{2J(J+1)}\right]$$

$$= -\left(\frac{e\hbar}{2m}\right)J\left[1+\frac{J(J+1)+S(S+1)-L(L+1)}{2J(J+1)}\right]$$

$$\boxed{\mu_J = -g\left(\frac{e}{2m}\right)\mathbf{J}} \quad \mathbf{J} = J\hbar$$

$$J = \sqrt{J(J+1)}$$

where,
$$g = 1 + \frac{J(J+1)+S(S+1)-L(L+1)}{2J(J+1)}$$

is know as *Lande 'g' factor* for *LS coupling*

In a weak magnetic field (parallel to Z direction) the allowed values of **J** are $M_J\hbar$ and

$$(\mu_J)_Z = -g\left(\frac{e}{2m}\right)M_J\hbar$$

$$= -g\left(\frac{e\hbar}{2g}\right)M_J$$

$$\boxed{(\mu_J)_z = -\mu_B M_J B} \qquad ...(2)$$

$\mu_B = \dfrac{e\hbar}{2m}$ is the **Bohr magneton.**

In a magnetic field, M_J can have $(2J+1)$ values *i.e.*, each energy level in presence of **B** can split into $(2J+1)$ magnetic sublevels of energy

$$\boxed{E_B = E_0 + \mu_B B M_J g} \qquad ...(3)$$

$$\boxed{\Delta E = \mu_B B M_J g}$$

This is additional energy due to presence of the magnetic field.

Examples of Normal and Anamalous Zeeman pattern.

1. **Zeeman effect in $^1F_3 - {}^1D_2$ transition:** For states $^1F_3$ and $^1D_2$, the L values will be 3 and 2. Hence under influence of **B** these states will split in $(2L + 1)$ substates *i.e.,* for $^1F_3$ there will be $2 \times 3 + 1 = 7$ and for $^1D_2$ there will be $2 \times 2 + 1 = 5$ substates

Multiplicity $2S + 1 = 1$ for $^1F_3$

$\Rightarrow$ $S = 0, \quad J = L$

and $2S + 1 = 1$ for $^1D_2$

$\Rightarrow$ $S = 0 \quad J = L.$

Therefore the value of Lande 'g' factor will be 1 in both the cases.

$$g = 1 + \frac{J(J+1) + S(S+1) - L(L+1)}{2J(J+1)}$$

$$= 1 + \frac{S(S+1)}{2J(J+1)} = 1 + 0 = 1 \qquad ...(1)$$

The energy difference between consecutive Zeeman lines

$$\Delta E = \frac{e\hbar}{2m} gB\Delta m_l$$

$$g = 1 \quad \Delta E = \frac{e\hbar}{2m} B\Delta m_l$$

$$\Delta m_l = 0, \pm 1$$

$$\Delta E = \frac{e\hbar}{2m} B.(\text{Lorentz unit})$$

The Lorentz unit is a unit of energy and used for expressing the splitting of energy levels.

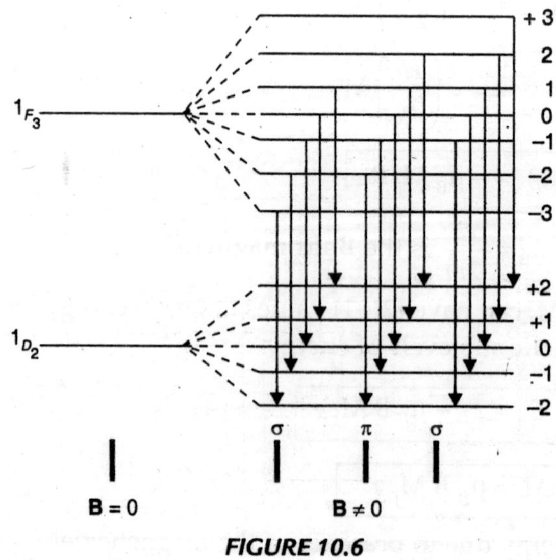

FIGURE 10.6

2. **Zeeman Effect in D_1 and D_2 lines of sodium:** The D_1 and D_2 lines in sodium are observed due to the transition between $^2P_{1/2} \rightarrow {}^2S_{1/2}$ states and $^2P_{3/2} \rightarrow {}^2S_{1/2}$ states. It is a doublet-doublet transition and hence anomalous Zeeman pattern will be observed. The term values, '*g*' factor and energy shift of spliting levels *i.e.*, gm_J are given in Table 1.

Table 10. 1

S.No.	State	L	S	J	Total m_j values	g	m_j	gm_j
1	$^2S_{1/2}$	0	$\frac{1}{2}$	$\frac{1}{2}$	2	2	$\pm\frac{1}{2}$	± 1
2	$^2P_{1/2}$	1	$\frac{1}{2}$	$\frac{1}{2}$	2	$\frac{2}{3}$	$\pm\frac{1}{2}$	$\pm\frac{1}{3}$
3	$^2P_{3/2}$	1	$\frac{1}{2}$	$\frac{3}{2}$	4	$\frac{4}{3}$	$\pm\frac{3}{2}, \pm\frac{1}{2}$	$\pm\frac{2}{3}, \pm 2$

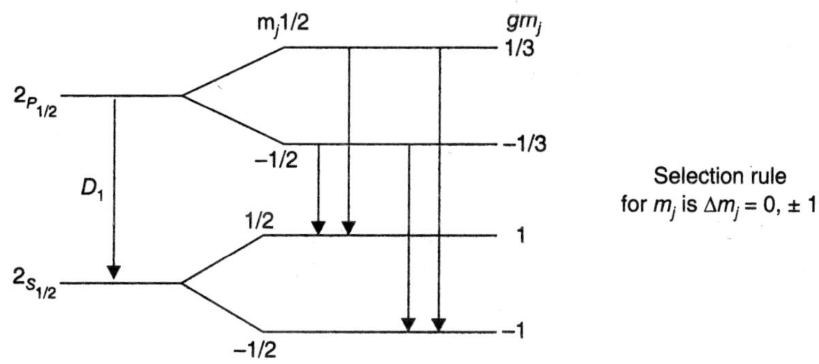

FIGURE 10.7

Hence we observe four component lines in D_1 line and six components for D_2 line of sodium under influence of weak magnetic field.

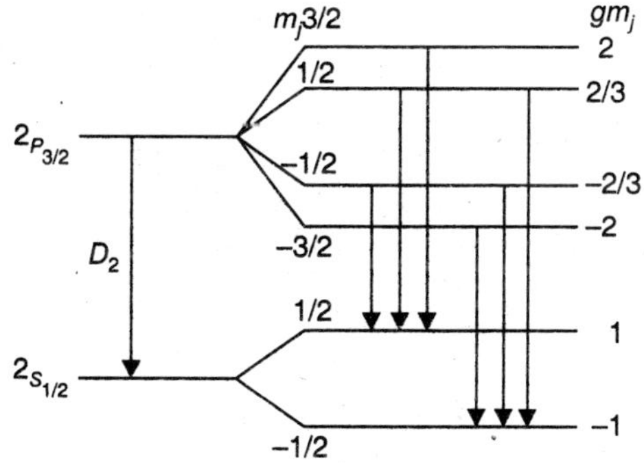

FIGURE 10.8

10.6 PASCHEN BACK EFFECT

A magnetic field will be said to be weak in case of Zeeman effect, if the total spread of the Zeeman pattern of each line is small relative to the spacing of the line themselves. It has been observed that if the **B** is made sufficiently strong, then the anamalous Zeeman splitting will revert to the normal Zeeman triplet through coalscence or disappearance of certain lines. This effect is known as *Paschen Back effect*. When magnetic field becomes very strong then the coupling between '*l*' and '*s*' breaks down and they quantized separately, discarding the significance of *J*. *L* and *S* precess separately about $\overline{B}$ independently. Therefore the energy change due to **B** will be result of precession of *L* about **B** and procession of *S* about **B** separately.

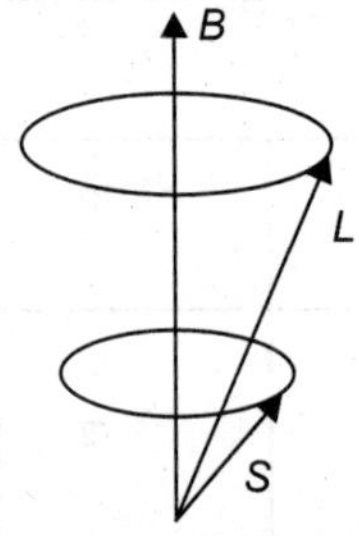

FIGURE 10.9

In huge magnetic field the spin-orbit interaction is swamped and hence the quantum numbers n, l, m_l, m_s with the axis parallel to **B**. Under these circumstances

$$(l)_Z = m_l\hbar \text{ and } (s)_z = m_s\hbar$$

and therefore

$$\mu_Z = -\frac{e\hbar}{2m}(m_l + 2m_s) = -\mu_B(m_j + m_s)$$

where $m_j = m_l + m_s$

Since μ_Z does not vary appreciably with further increase in **B**, the energy of the atom will be

$$\boxed{E = E_0 + \mu_B B(m_j + m_s)} \quad \text{...(1)}$$

and change in energy

$$\boxed{\Delta E = \mu_B B(m_j + m_s)} \quad \text{...(2)}$$

From equation (2) one can observe that **B** splits from single original level into several levels, each of which is characterized by $(m_j + m_s)$. Now $(m_j + m_s)$ can take any integral value from $(l + 1)$ to $-(l + 1)$, when $m_l = l$, and $m_s = 1/2$, *i.e.*, total $2l + 3$ levels in all. However if $l = 0$, there are only two levels with $m_j + m_s = \pm 1$. Here $(m_j + m_s)$ is known as strong field quantum number.

Consider transition from $l = 1$ to $l = 0$

The selection rules are $\Delta l = \pm 1, \Delta m_j = 0, \pm 1$

and $\Delta m_s = 0$

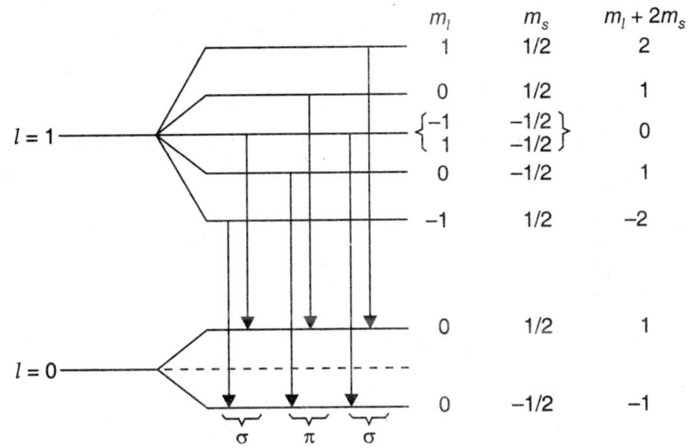

FIGURE 10.10

The allowed transitions lead to normal Zeeman triplet.

10.7 X-RAYS

In the year 1895, **Wilhelm Roentgen** found that a highly penetrating radiation of unknown nature is produced, when fast moving electrons impinge on matter. Since the nature of these radiations was unknown at the point of invention, they were named as X-rays. They were soon found to travel in straight lines, remain unaffected by electric and magnetic fields, pass radially through opaque materials, and cause phosphorescent substance to glow. The faster original electrons, the more penetrating were the resulting X-rays; and greater the number of electrons, the greater was the intensity of X-ray beam.

X-rays are electromagnetic waves. Electromagnetic theory predicts that an accelerated electric charge will radiate electromagnetic waves and rapidly moving electron suddenly brought to rest is certainly accelerated. Radiation produced under these circumstances is given the German name bremsstrahlung (or breaking radiation).

10.7.1 Production of X-Rays

X-rays are produced when fast moving electrons hit a suitable target after being accelerated under high potential difference.

The schematic arrangement for production of X-rays is shown in Figure 10.11. It consists of an evacuated bulb in which there is a filament F and target T. A high potential difference of 50-100 kV is maintained between the filament, which acts as cathode and target acting as anode. When the filament is heated by passing current, it emits electrons which are accelerated towards the target because of the potential difference. The target is shaped in such a way that its surface will be at an inclination of 45° to the impinging electron beam, so that X-rays are emitted in a direction perpendicular to that of the beam.

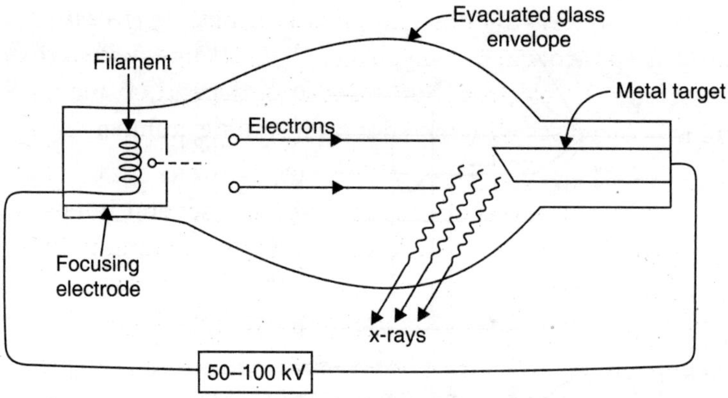

FIGURE 10.11 **X-rays are produced by bombarding a metal target**

When the accelerated electrons hit the target, they possess a kinetic energy, $E_k = eV$, where, e is the charge on the electron and V is the applied potential.

10.7.2 Properties of X-Rays

X-rays are having following properties:

1. X-rays are electromagnetic waves of very short wavelength ($\simeq$ 1 Å – 100 Å).
2. The speed of X-rays is vacuum is equal to velocity of light.
3. They ionise the gas through, which they pass.
4. They produce flourescence in several materials such as ZnS.
5. They can affect the photographic plate.
6. They remain undeflected by electric and magnetic fields.
7. They also show the phenomena like interference and diffraction just like light rays.
8. They are absorbed by the materials through which they traverse.
9. X-rays undergo reflection and refraction according to laws of visible light.

10.7.3 Continuous and Characteristic X-Rays

The X-ray emission consists of radiations of wavelength over a continuous range within a certain limit, but with varying intensity. A plot of the intensity of the various radiations so emitted as a function of wavelength is called the X-ray spectrum.

The X-ray spectrum of molybdenum is shown in Figure 10.12.

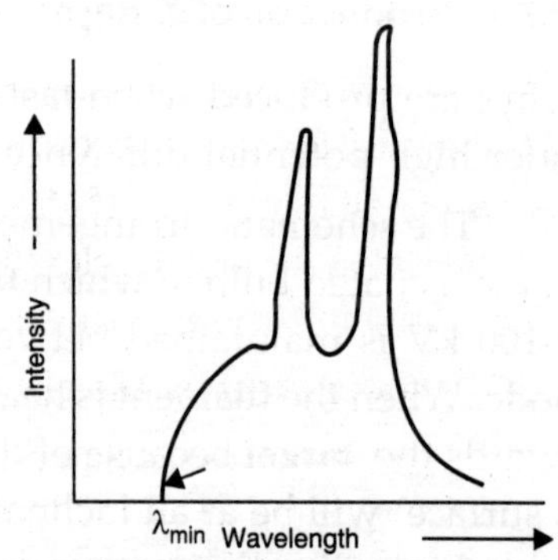

FIGURE 10.12 **X-rays spectrum of molybdenum**

The spectrum is **continuous** with a sharp, well defined cut off on the lower wavelength side at λ_{min}. No X-ray radiations with wavelength less than the cut off wavelength λ_{min} are found to be present in the X-ray emission. The value of cut off wavelength λ_{min}, changes with a change in the applied voltage V that accelerates the electron, but it does not depend on the material used for the target. But the intensity increases with increase in the atomic number of the element of which the target is made.

In addition to the continuous spectrum, there are several intensity peaks at certain wavelengths as shown in Figure 10.12. The radiation corresponding to these wavelengths are called the **characteristic X-rays**. The wavelength of the characteristic X-rays do not depend upon the applied voltage V, but depend on the material of the target.

During the collision with the target, most of the bombarding electrons lose their kinetic energy as heat. But, under certain circumstances, the kinetic energy of some of the electrons will be converted into electromagnetic radiations, which result in **continuous** and **characteristic X-ray spectra**, the explanation of which is given as follows :

EXPLANATION FOR THE CONTINUOUS X-RAY SPECTRUM

Among the electrons that collide with the target, some succeed in penetrating into the interior part of the target atoms and interact with the nuclei present there. This interaction gives rise to the **continuous spectrum,** which is also known as **bremsstrahlung** (which in German meaning braking radiation).

Let us consider one such electron with energy E_k approaching the nucleus. Since the nucleus is positively charged, the electron, because its negative charge, experiences an attractive force towards the nucleus. As a result, it is deflected from its straight line path (Figure 10.13). Such deviation is equivalent to a radial acceleration towards the nucleus. As per classical electromagnetic theory, an accelerated charge must emit electromagnetic radiations continuously, and of all frequencies, which explains the continuous emission spectrum. However, the fact that, there is a lower wavelength cut off (*i.e.*, λ_{min}), in the spectrum cannot be explained from the classical view. But the same can be explained with the help of photon concept of Einstein, according to which the accelerated charge radiates energy in terms of quanta's of energy $h\nu$. If $h\nu_1$, $h\nu_2$, $h\nu_3$, ... are the energies of the different photons emitted by the electron as it is accelerated, and also if E_k is the energy with which the electron emerges out of the influence of nucleus, then we can write,

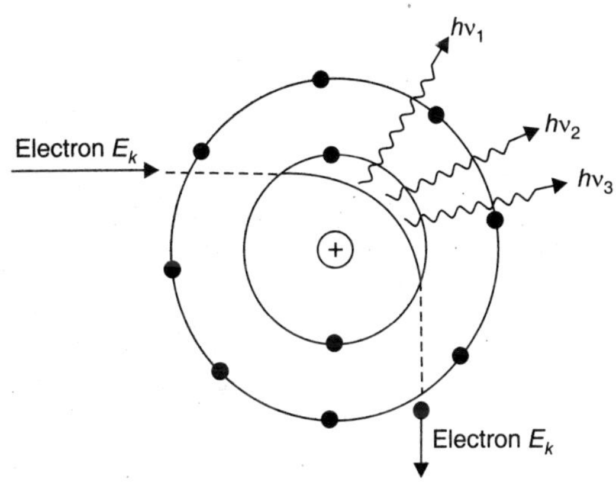

FIGURE 10.13

energy lost by the electron = $E_k - E'_k$.

Total energy radiated in the form of photons.

$$= h\nu_1 + h\nu_2 + h\nu_3 + ...$$

By the conservation of energy rule

$$E_k - E'_k = h\nu_1 + h\nu_2 + h\nu_3 + ...$$

(assuming no energy is acquired by the nucleus).

Under the limiting conditions, one can say that, a photon with highest possible energy is produced under the condition that

(*a*) The electron loses all its energy by the time, it emerges from the influence of nuclei, *i.e.*,

$$E'_k = 0 \text{ and}$$

(*b*) All the energy lost by the electron appears in the from of a single photon, emitted as an X-ray radiation.

$\therefore$ $$E_k = h\,\nu_{max}$$

But $$E_k = eV$$

where, *e* is the charge on the electron, and *V* is the applied potential

$\therefore$ $$h\nu_{max} = eV$$

or $$\nu_{max} = \frac{eV}{h}$$

Since $\nu_{max} = \dfrac{c}{\lambda_{min}}$, we can write

$$\boxed{\lambda_{min} = \frac{hc}{eV}} \qquad ...(1)$$

where, *c* is the velocity of light, λ_{min} is the shortest wavelength of X-rays radiated. Therefore, short wavelength limit is given by λ_{min}. From here, we can conclude that with increase in potential across the tube, the value of λ_{min} decreases. The relation represented by Equation (1) is known as **Duane and Hunt law.**

Substituting the values

$$h = 6.64 \times 10^{-34} \text{ J.s}, \; c = 3 \times 10^8 \text{ m/s}$$
$$e = 1.6 \times 10^{-19} \text{ C}$$

$$\lambda_{min} = \frac{6.64 \times 10^{-34} \times 3 \times 10^8}{1.6 \times 10^{-19}\, V} = \frac{12.4 \times 10^{-6}}{V} \text{ m}$$

$$= \frac{12400}{V} \text{ Å}$$

The experimental results are in good agreement with equation (1).

EXPLANATION FOR THE CHARACTERISTIC X-RAYS SPECTRUM

The energetic electrons, which succeed in penetrating into the target are subjected to the interaction with electrons and nuclei of the atoms in the target material. Those which interact with the nuclei give rise to continuous spectrum as explained earlier. The other electrons, in general, undergo multiple collisions with the atomic electrons and there by slowdown while losing their energy in the form of heat of the target. But occasionally, the collision may occur in such a way that, an incident electron transfers a large portion of its energy at once to an atomic electron in an interior orbit. The energy will be enough to knock the atomic electron out of the atom, which creates a vacancy in the corresponding energy state. Under such a condition, one of the electrons in the higher energy state (*i.e.*, the one in an outer shell) falls down into the vacant energy state, which results in the emission of a photon.

The energy of the photon-emitted will be equal to the energy difference between the two states between which the atomic electron undergoes transition. These are the photons which give rise to the characteristic *X*-rays peaks.

Now the atomic energy level pattern is characteristic of the particular type of element to which the given atoms belong to. Hence the energy of the photons that formulate the *X*-ray peaks, depend on the type of the target material used, and is independent of the energy of the incident electrons, which in turn means that it is independent of the variation in applied high voltage *V*.

10.7.4. Difference between Continuous and Characteristic Spectra

Continuous Spectra	*Characteristic Spectra*
1. This type of spectrum consists of radiation of all possible wavelengths within a range starting with a minimum wavelength limit.	The characteristic curve or spectrum has a few sharp peaks at certain wavelengths. These peaks appearing in continuous spectrum form line spectrum.
2. The fast moving electrons during deacceleration radiates energy in the form of *X*-rays.	The fast moving electrons remove an inner electron and energy radiates electrons makes a transition from higher state to lower state.
3. This type of spectrum consists of radiation with λ_{min} depending upon the potential applied.	The characteristic spectrum depends on the atom through which *X*-rays are emitted.

10.7.5 Energy Level Diagram *or* Understanding Characteristic Spectra in Terms of Bohr's Theory

The origin of characteristic *X*-rays can be easily understood in terms of Bohr's theory. The electrons producing *X*-rays are very energetic. These energetic electrons are capable of removing electrons even from the inner most *K*- or *L*-shel. When such an electron is removed, electron transition from the higher orbit occur. In accordance with Bohr's theory, this results in the emission of (characteristic *X*-ray) photons.

The detailed structure of characteristic radiations can reasonably be well understood in terms of an energy level diagram Figure 10.14. In discussing the X-ray spectrum, we make the following assumptions:

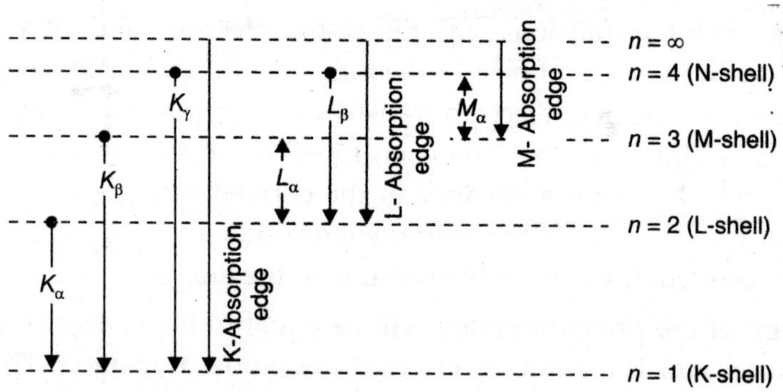

FIGURE 10.14

1. The electrons with a particular value of n have (almost) identical energy.
2. The energy differences between the energy of the N-shell ($n = 4$) and the energy of a free electron can (almost) be neglected.

Let us suppose the impinging electrons remove a K electron. This vacancy can be filled up by an electron from either of L, M or N shells or a free electron. This possible transition can result in the K_α, K_β line and the limiting line. The probability of K_α transition is maximum and hence the intensity of this line in K-series is more than the others. Similarly, radiations result when an election from the L shell ($n = 2$) is removed and the transitions results from higher orbits.

10.7.6 Moseley's Law

In 1913, Moseley carried out a systematic investigation of X-ray spectra of various elements. In his experimental observations, he found that, the frequencies of the K_α lines for various elements were proportional to the atomic numbers.

The actual relationship between the frequency ν of a particular category of spectral line, and the atomic number Z of the element which gives rise to the spectrum was given by Moseley as

$$\nu = a(Z - b)^2$$

where, a and b are constants.

The relation is known as **Moseley's law** which can be stated as,

"The frequency of any particular category of spectral lines in the X-ray characteristic spectra of various elements, are proportional to the square of the atomic number of the element giving rise to the spectrum."

10.8.1 Induced Absorption

In a material atoms are in ground state with energy (E_1) when a photon of energy $h\nu = E_2 - E_1$ is incident on the material, the atoms in the ground state absorb energy and jump over to excited state E_2. This process is called induced absorption as shown in Figure. 10.17. The frequency of photon absorbed is given as,

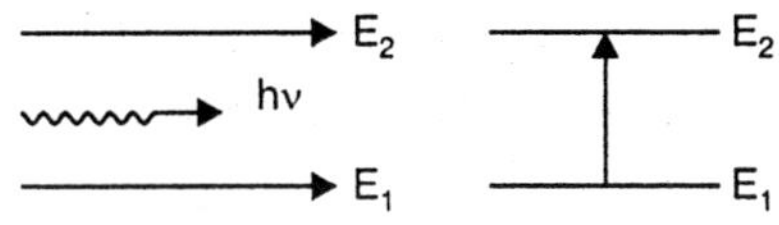

FIGURE 10.17 **Induced absorption**

$$\nu = \frac{E_2 - E_1}{h}$$

the atom in excited state can come to ground state by two process viz. spontaneous emission and stimulated emission.

10.8.2 Spontaneous Emission

The atoms in excited state E_2 are less stable and have very small life time of the order of 10^{-8} second so they come to ground state after 10^{-8} second along with the emission of a photon of energy $h\nu = E_2 - E_1$. This phenomenon is called spontaneous emission. The frequency is given by

$$\nu = \frac{E_2 - E_1}{h}$$

10.8.3 Stimulated Emission

In certain material the upper stage has longer life time of the order of 10^{-3} sec, known as metastable state and a material having metastable state is called as active material. If atom in excited state are compelled or forced to jump down to ground state by photon of energy $h\nu = E_2 - E_1$, then it jumps to ground state along with the emission of photon having energy $h\nu = E_2 - E_1$, this process is called stimulated emission both the photon in the same phase.

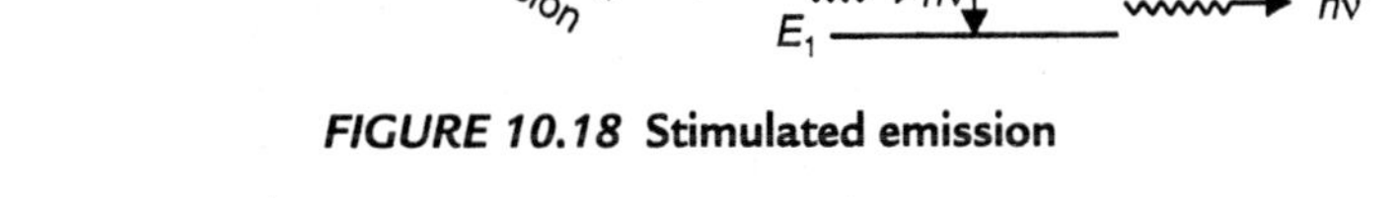

FIGURE 10.18 **Stimulated emission**

10.8.4 Einstein Coefficients

Let N_1 and N_2 represents the number of atoms in energy levels E_1 and E_2 respectively. Let density of radiation of particular frequency ν be represented by $u(\nu)$ then the absorption rate will be proportional to, (*i*) the density of radiation $u(\nu)$

The number of atoms N_1

Thus, absorption rate $= B_{12} N_1 u(\nu)$ ...(

where B_{12} is the coefficient of proportionality and is a characteristics of energy level.

Let us now consider the reverse process namely the emission of radiation at a particul frequency ν when the atom de-excites from level E_2 to E_1. Einstein pointed out that an atom i the excited level can make a transition to a lower energy level either through spontaneous through stimulated process. In spontaneous emission, the rate will be independent of energ density of radiation field $u(\nu)$ and depends only on the number of atoms present in the upp energy level.

Rate of spontaneous transition $= A_{21}N_2$ ...(

where A_{21} represents the probability of spontaneous transition.

But rate of stimulated emission will depend not only upon number of atoms in upp energy level but also the energy density.

∴ Rate of stimulated transition $= B_{21}N_2\, u(\nu)$ ...(

where $B_{21}\, u(\nu)$, given probability of stimulated transition, in steady state.

Rate of upward transition = Rate of downward transition using equation (1), (2) and (we get,

$$N_1 B_{12}\, u(\nu) = N_2 A_{21} + N_2 B_{21} u(\nu)$$

or $$(N_1 B_{12} - N_2 B_{21})\, u(\nu) = N_2 A_{21}$$

$$u(\nu) = \frac{N_2 A_{21}}{N_1 B_{12} - N_2 B_{21}}$$

$$u(\nu) = \frac{A_{21}}{\frac{N_1}{N_2} B_{12} - B_{21}}$$

Also from Maxwell Boltzmann's law.

$$\frac{N_1}{N_2} = e^{h\nu/k_B}$$

$$u(\nu) = \frac{A_{21}}{e^{h\nu/k_B} B_{12} - B_{21}}$$

$$u(\nu) = \frac{A_{21}/B_{21}}{e^{h\nu/k_B} - 1} \quad ...($$

Also from Planck radiation formula

$$u(\nu) = \frac{8\pi\nu^3}{c^3} \cdot \frac{h\nu}{e^{h\nu/k_B} - 1} \quad ...(5)$$

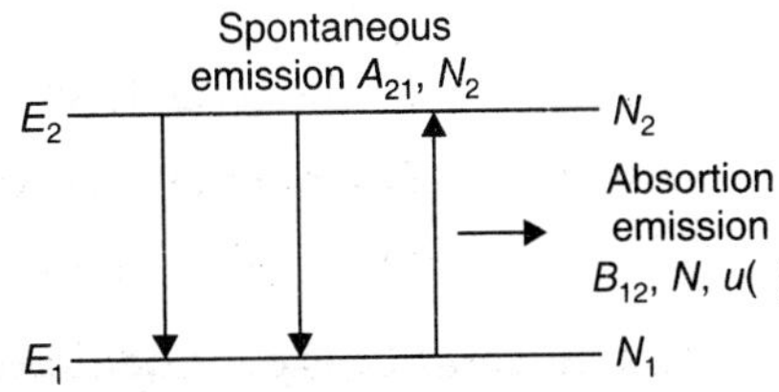

FIGURE 10.19 **Stimulated emission B_{21} N_2 $u(\nu)$**

Comparing equations (4) and (5) we get,

$$\frac{B_{12}}{B_{21}} = 1 \Rightarrow B_{12} = B_{21}$$

i.e., probability of stimulated absorption and stimulated emission is same.

Also $$\frac{A_{21}}{B_{21}} = \frac{8\pi h\nu^3}{c^3}$$

Putting this value in equation (4) we get,

$$u(\nu) = \frac{A_{21}}{B_{21}} = \frac{1}{e^{h\nu/k_B} - 1}$$

For normal optical source $T \sim 10^3$ K and $L = 6000$Å

$\frac{A_{21}}{B_{21}u(\nu)}$ >>> 1, thus for an optical frequency the emission is predominantly due to spontaneous transition and hence the emission from usual light source is incoherent.

10.8.5 Population Inversion

As we know earlier, the process of spontaneous emission is independent of external factor Einstein pointed out that two processes are equal probable under normal circumstances. Among two processes (spontaneous emission and stimulated emission) which predominate over other depend upon the number of atoms in two levels when $N_1 > N_2$, then absorption predominated over emission which is the case normally in normal materials. On the other hand when $N_1 < N_2$ then emission dominated over the absorption *i.e.*, stimulated emission is possible under this condition the absorption coefficient of material becomes negative. Thus if the light traverses through such a medium it gets amplified instead of attenuation, this type of material in which $N_2 > N_1$ is possible are called active material and this process is called *population inversion* thus for laser action population inversion is necessary.

We know that the population of a particular level is governed by Maxwell Boltzmann's equation. According to if there is a level corresponding to energy E_1 then number of atoms, N_1 in this level will be given by

$$N_1 = N_0 e^{-E_1/k_B}$$

where N_0 is the population is ground state, thus it is clear that the population of an energy level E_2 such that $E_2 > E_1$ will be given by

$$N_2 = N_0 e^{-E_2/k_B}$$

So, $N_2 < N_1$ under normal condition. But as we have pointed out that for laser action population inversion is necessary.

So we have to supply energy from outside in order to achieve this population inversion ($N_2 > N_1$). The device which is used for this porpose is called pump and this process is called pumping. It is notable point that population inversion is possible only in the materials which has metastable stage *i.e.*, Ruby, He-Ne mixture etc.

10.8.6 Laser Action : Light Amplification

There are three main things which are required for laser action which are as follows:

(*i*) Active material.

(*ii*) An specially prepared cylindrical tube in which light intensity can build up by multiple reflection. The ends of the tube are silvered one end is completly silvered while the other end is partially silvered so that an intense beam can emerge out of it.

(*iii*) Pumping system in order to achieve population inversion.

In order to understand laser action, let us consider that three energy levels E_1, E_2 and E_3 as shown in Figure 10.20 level, lifetime of which is ≈ 10^{-3} sec.

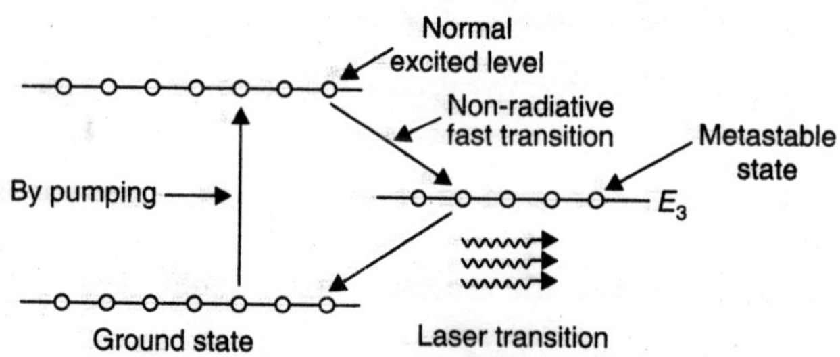

FIGURE 10.20 **Laser action or light amplification**

With the help of a pump some energy is given to the active material so that, they reach the level E_2, As E_2 is normal excited level, the lifetime of which is very small ≈ 10^{-8} sec, so they immediately relax down to the intermediate energy level E_3 which is metastable that is lifetime of E_3 is large ~ 10^{-3} sec thus the atoms stay for a long time in this level. But on the other hand the atoms are lifted to E_2. Only when these two condition are satisfied, the medium will act as amplifier this is the basic principle of laser when N_2 larger than N_1 *i.e.*, the number of atom at higher energy is more than the number of atoms of lower energy the process is known as *population inversion*, which is an artificial stimulation created in the medium. A non-equilibrium state is to be produced in which the population of the upper energy level exceeds to a large extent the population of the lower energy level. When the situation occurs, the population distribution between levels E_1 and E_2 is said to be inverted and the medium is said to have gone into the state of population inversion.

10.8.7 Pumping

The population inversion cannot be achieved thermally. It is achieved by exciting the medium with suitable form of energy, this process is called pumping. The commonly used methods for pumping are given as follows :

(*i*) **Optical Pumping:** In optical pumping light source is used to supply luminous energy most often this energy comes in the form of short flashes of light this method was first used by Maiman in the production of Ruby Laser and also widely used in solid state laser, the laser material is kept inside in a helical Xenon flash lamp of the type used in photography.

(*ii*) **Electric Discharge:** Another method of pumping is by direct electron excitation as occur in an electric discharge the method is preferred in gaseous ion laser for example Argon-ion. Laser the electric field of the order of several kV/m causes electrons emitted by the cathode, to be accelerated towards anode. Some of these electrons will collide with the atoms of the active medium, ionize the medium and raise it to the higher level. This produces the required population inversion.

(*iii*) **Inelastic atom-atom collision:** In an important class of lasers pumping by electrical discharge provides the initial excitation which raises one type of atoms. It is there laser atom which provide the population inversion required for the laser action. He-Ne laser is an example of this type.

(*iv*) **Direct Conversion:** A direct conversion of electrical energy into radiation occurs in light emitting diodes (LED's) and the semiconductor laser drives from them.

(*v*) **Chemical reaction:** In a chemical laser, the energy comes from a chemical reaction without any need for other energy source for example hydrogen can combine with flourine.

$$H_2 + F_2 \longrightarrow 2HF$$

This reaction generates enough heat to pump a CO_2 laser.

10.8.8 The Major Pumping Schemes

Atoms in general have a large number of energy levels among them, only three or four levels will be important for pumping process.

The transition between the two levels that generate stimulated emission is called the lasing transition. The top most level is called pumping level. Two important pumping schemes are widely used in the construction of lasers, these are known as three levels and four levels pumping schemes.

(*a*) **Three level pumping scheme:** Consider an atomic system which has three energy levels as shown in Figure 10.21 the state E_1 is the ground state and E_2 and E_3 are the excited states in the scheme, the energy states are such that atoms could readily be excited to upper most state E_3 (known as pumping level) this energy level is not a stable state. Hence atoms do not stay at this level and undergo downward transition either to E_1 or E_2 levels.

The probability of spontaneous transition $E_3 \to E_1$ is comparable to that of $E_3 \to E_2$. The level E_2 is a metastate, since the probability of spontaneous transition $E_2 \to E_1$ is extremely small.

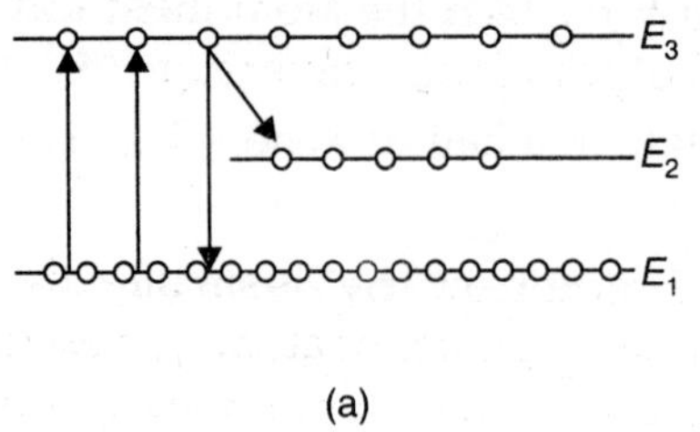

FIGURE 10.21 (*a*)

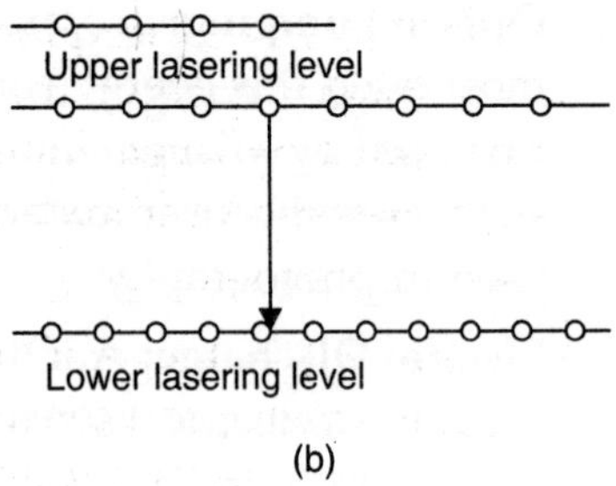

FIGURE 10.21 (*b*)

When the medium is exposed to radiation of frequency (ν_p) known as pumping frequency, $\nu_p = (E_3 - E_1)/h$, a large number of atoms will be excited to the higher energy level E_3 some of there atoms make spontaneous transition to the lowest level E_1 but many of them make spontaneous transition to the metastable level E_2 through a non-radiative transition.

As spontaneous transition from E_2 to E_1 occur rarely the atoms get trapped in energy state E_2. Because of pumping the process continue and after some time large number of atoms have energy E_2 when more than half of the ground state atoms accumulate at E_2, the population inversion is achieved between the states E_1 and E_2. Now a photon of energy $h\nu = E_2 - E_1$ can trigger stimulated emission of atoms at E_2. To achieve population inversion more than half of the ground state atoms must be pumped to the upper state. Therefore a very high pump power is required in this type of pumping scheme.

(*b*) **Four-level Pumping Scheme:** A typical four level pumping scheme is shown in Figure 10.22(*a*) and 10.22(*b*), pump frequency take the active centres from ground state energy E_1 to upper most energy E_4 from the energy level E_4, the atoms rapidly fall to the metastable state E_3. The population at this energy grows rapidly while the energy level E_2 is almost empty, therefore population inversion is achieved between the states E_2 and E_3 a photon of energy $h\nu = E_3 - E_2$ can start a chain stimulated emission, bringing the atoms in the energy state E_2 from there, the atoms undergo nonradiative transitions subsequently to the ground state E_1 and will again the available to participate in the process.

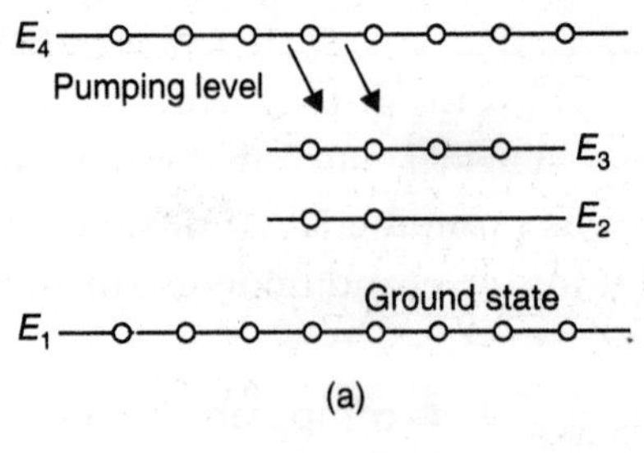

FIGURE 10.22 (*a*)

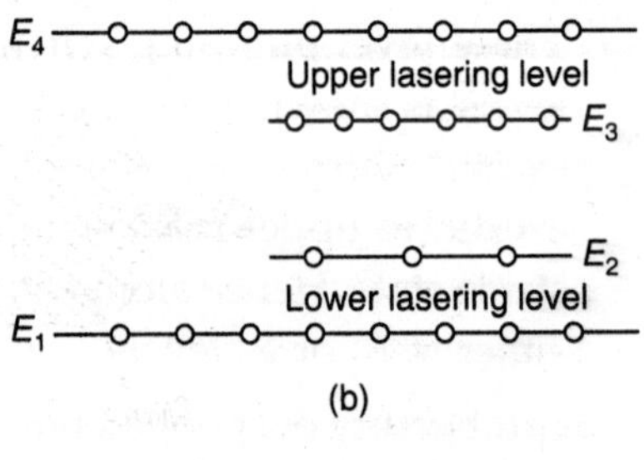

FIGURE 10.22 (*b*)

(*c*) **Two Level Pumping Scheme:** In case of two energy level E_1 (lower energy) and E_2 (higher energy), the time span Δt for which the atom must stay at upper level E_2, must be large enough to achieve population inversion. However according to Heisenberg uncertainty principle

$$\Delta E \, \Delta t \geq \hbar,$$

Δt will be longer if ΔE is small *i.e.*, E_2 is narrow if ΔE is smaller the pumping efficiency is smaller, as a consequence of this less number of atoms get excited through a sharp energy level supports the population inversion, enough population cannot be achieved at energy level E_2, in view of all small ΔE. The result will be that the upward transition would be accompanied by premature downward stimulated transition and the population at energy levels E_2 would not be possible to the required extent.

To obtain three or four level pumping scheme, the materials cannot be produced at one's will. One has to select the required materials from many available materials to suit a specific purpose.

10.8.9 The Ruby Laser

The first working laser was built in 1960 by Theodore Maiman, using a ruby crystal as the amplifying or active medium. Ruby belongs to the family of gems consisting of sapphire or alumina (Al_2O_3) with various types of impurities. For example pink ruby contains about 0.05% Cr atoms. Similarly, Al_2O_3 doped with Ti, Fe or Mn results in variously coloured sapphire. Most of these materials can be grown as single crystal.

Ruby crystals are available in rods several inches long, convenient for forming optical cavity (Figure 10.23). The crystal is cut and polished so that the ends are flat and parallel, with the end planes perpendicular to the axis of the rod. These ends are coated with a highly reflective material, such as Al or Ag, producing a resonant cavity in which light intensity can build up through multiple reflections. One of the end mirrors is construted to be partially transparent, so that a fraction of light will "leakout" of the resonant system. This transmitted light is the output of the laser. Of course, in designing such a laser one must choose the amount of transmission to be a small perturbation on the resonant system. The gain in photons per pass between the end plates must be larger than the transmission at the ends, as well as any other losses due to light scattering and absorption. The arrangement of parallel plates providing multiple interval reflections is similar to that used in Fabry-Perot interferometer; thus the silvered ends of the laser cavity are often referred as Fabry-Perot faces.

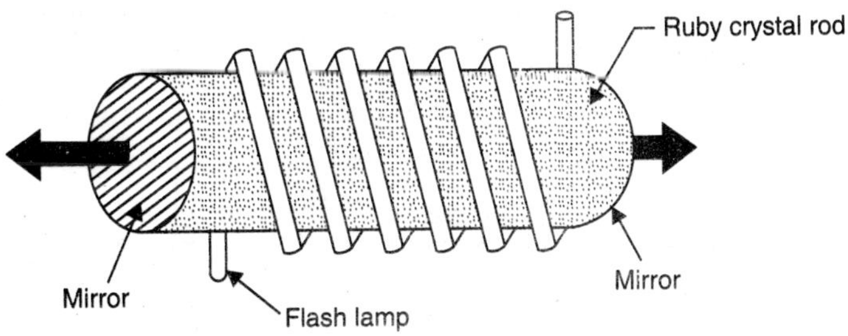

FIGURE 10.23 **Schematic diagram of a ruby laser. A ruby crystal rod is cut and polished so that its ends form mirrors to create the resonant cavity. A flash lamp supplies the necessary energy in the form of photons to pump the rod**

In case of ruby, chromium atoms in the crystal have their energy levels as shown in Figure 10.24, where only the energy levels which are important for stimulated emission are depicted.

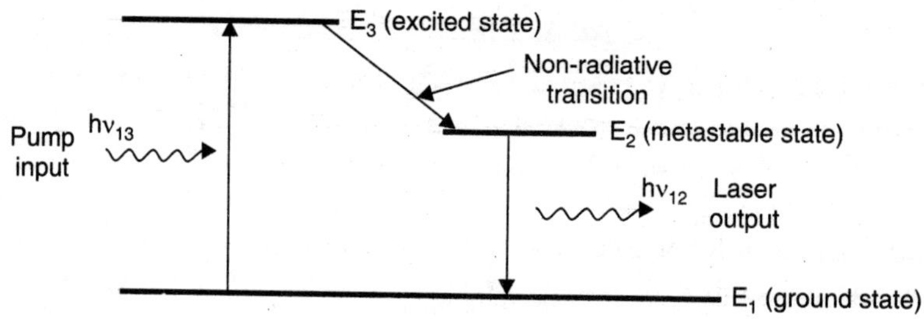

FIGURE 10.24 **Energy levels for chromium ions in ruby. The three level system include a ground level at E_1, an excited level E_3, and metastable state at E_2 where the excited electrons rapidly go. The mean lifetime of the metastable state is long enough to ensure that population inversion can be achieved between the levels E_1 and E_2**

This is basically a three-level system. Absorption occurs in the green part of spectrum exciting electrons from the ground state E_1 to the band of levels designated E_3 in Figure 10.24. Then electron decay rapidly to the level E_2. This transition is non-radiative. The level E_2 is very important for the stimulated emission process since electrons in this level have a mean lifetime of about 5 *ns* before they fall to the ground state. Because this lifetime is relatively long, E_2 is called a metastable state. If electrons are excited from E_1 to E_3 at a rate faster than the radiactive rate from E_2 back to E_1, the population of the metastable state E_2 becomes larger than that of the ground state E_1 (we assume that electrons fall from E_3 to E_2 in a negligibly short time).

In the experiment done by Maiman in 1960, population inversion is obtained by optical pumping of the ruby rod with a flash lamp such as the one shown in Figure 10.23. A common type of flash lamp is a glass tube wrapped around the ruby rod and filled with Xenon gas. A capacitor can be discharged through the Xenon-filled tube, creating a pulse of very intense light over a broad spectral range. If the light pulse from the flash tube is several milliseconds in duration, we might expect an output from the ruby laser over a large fraction of that time. However, the laser does not operate continuously during the light pulse but instead emits a series of very short spikes (Figure 10.25).

When the flash lamp intensity becomes large enough to create population inversion (the threshold pumping level), stimulated emission from the metastable level to ground level occurs with a resulting laser emission. Once the stimulated emission begins, the metastable level is depopulated very quickly. Thus the laser output consists of an intense spike lasting from a few nanoseconds to microseconds. After the stimulated emission spikes, population inversion build up again and a second spike results. This process continues as long as the flash lamp intensity is above the threshold pumping level.

In this situation, one can easily understand the metastable level never recieves a highly inverted population of electrons. Whenever the population of E_1 reaches the minimum required for stimulated emission, these electrons depleted quickly in one of the laser emision spikes.

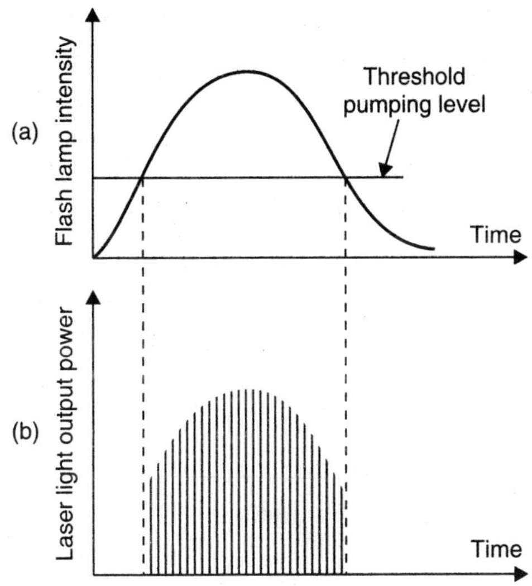

FIGURE 10.25 **Laser spikes in the output of ruby laser: (*a*) typical variation of intensity of the flash lamp with time the intensity is above the threshold pumping level only during a certain period of time, not during the entire duration when the lamp is powered. (*b*) Laser spikes occuring while the flash intensity is above the threshold pumping level. As soon as population inversion is achieved between the levels E_1 and E_2, the laser emits a pulse of light. This process results in the series of laser intensity spikes**

To prevent this, we must some how keep the coherent photon field in the ruby rod from building up (and thus prevent stimulated emission) until after a larger population inversion is obtained. This can be accompalished if we temporarily interupt the resonant character of the optical cavity. This process is called *Q*-switching, where *Q* is the quality factor of the resonant structure. A straight forward method for doing this is illustrated in Figure 10.26. The front face of the ruby rod is silvered to be partially reflecting but the back face is left unsilvered. The back reflector of the optical cavity is provided by an external mirror, which can be rotated at high speeds, when the mirror plane is aligned exactly

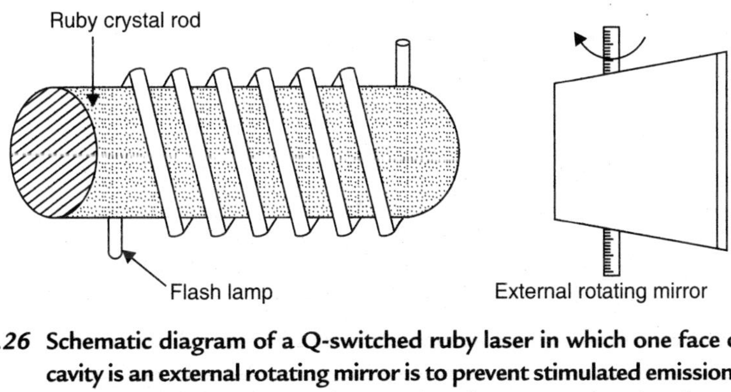

FIGURE 10.26 **Schematic diagram of a Q-switched ruby laser in which one face of the resonant cavity is an external rotating mirror is to prevent stimulated emission by interrupting the resonant nature of the laser cavity, thus preventing the photon field building up. This allows a larger population inversion to be achieved, and consequently a higher intensity laser light emission**

perpendicular to the laser axis, a resonant structure exists ; but as the mirror rotates away from this position there is no build up of photons through multiple reflections and no laser action can occur. Thus during a flash from the Xenon lamp, a very large inverted population build up while the mirror rotates off axis. When the mirror finally returns to the position at which light reflects back into the rod, stimulated emission can occur, and the large population of the metastable level is given up in one intense laser pulse. This structure is called a giant pulse laser or a Q-switched laser. By saving the electron population for a single pulse, a large amount of energy can be given up in a very short time. For example of the total energy of the pulse is 1 Joule and the pulse width as 100 *ns* (10^{-7} *s*), the peak pulse power is 10^7 J s^{-1} = 10 MW.

10.8.10 Helium-Neon Laser

A helium-neon laser, usually called a He-Ne laser, is a type of small gas laser. He-Ne lasers have many industrial and scientific applications, and are often used in laboratory demonstration of optics. Its usual operational wavelength is 633 nm, in the red portion of the visible spectrum.

The He-Ne laser was the first gas laser to be invented by Ali Javan, William Bennet Jr. and Donal Herriot at Bell Laboratory, USA, who in 1960 achieved continuous wave emissioin of the laser on the 1.15 μm wavelength line.

The active medium of the He-Ne laser is a mixture of helium and neon gas; in fact, He-Ne lasers perhaps could be labelled more properly as "neon lasers" since neon atoms are the active elements and helium only serves as a buffer gas (enhances lasing by causing the pumping efficiency to be increased).

The active/gain medium of the laser is a mixture of helium and neon gases, approximately in the ratio of 5 : 1, contained at low pressure (typically ~ 300 Pa) in a glass envelope. The energy of pump source of the laser is provided by an electrical discharge of around 1000 V through an anode and cathode at each end of the glass tube. The optical cavity of the laser typically consists of a plane, high reflecting mirror at one end of the laser tube, and a concave output coupler mirror of approximately 1% transmission at the other end.

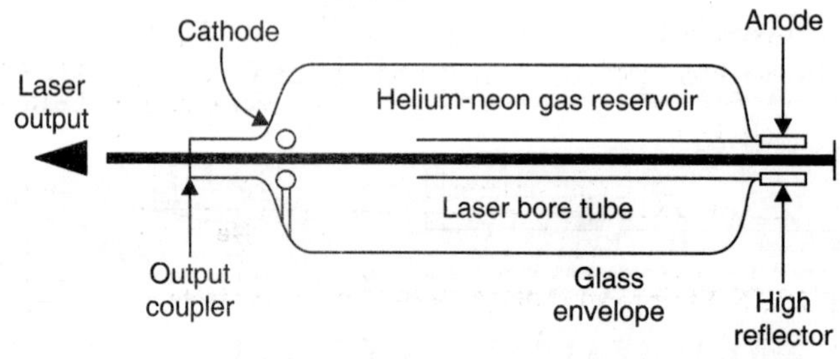

FIGURE 10.27 **Schematic diagram of a helium neon laser**

He-Ne lasers are typically small, with cavity length of around 15 cm upto 0.5 m and optical output powers ranging from 1 mW to 100 mW.

The red He-Ne laser wavelength is reported as 632.816 nm. This is in fact the wavelength in air, and corresponds to a vacuum wavelength of 632.991 nm. The precise operating wavelength lies within about 0.002 nm of this value, and fluctuates within this range due to thermal expansion of the cavity. Frequency stabilized versions enables the wavelength to be maintained within about 2 parts in 10^{12} for months and years of continuous operations.

The laser process in a He-Ne laser starts with collision of electrons from the electrical discharge with helium atoms in the gas. He + e^- $\longrightarrow$ He*, where (*) represents an excited state. This excited helium from ground state to the 2^3s_1 and 2^3s_0 long-lived, metastable excited states. Collision of the excited helium atoms with the ground state neon atoms results in transfer of energy to the neon atoms, exciting neon electrons into the 5*s* level. This is due to coincidence of energy levels between the helium and neon atoms.

The process is given by the reaction equation :

$$\text{He}^* + \text{Ne} \longrightarrow \text{He} + \text{Ne}^* + \Delta E$$

where ΔE is the smaller energy difference between the energy states of the two atoms, of the order 0.05 eV.

The number of neon atoms entering the excited states builds up as further collisions between helium and neon atoms occur causing a population inversion between the noen 5*s*, 3*p* and other electronic levels. Spontaneous and stimulated emission between the 5*s* ($^2P_{½}$) and 3*p* ($^2P_{½}$) states results in emission of 632.991 nm wavelength light, the typical operating wavelength of a He-Ne laser.

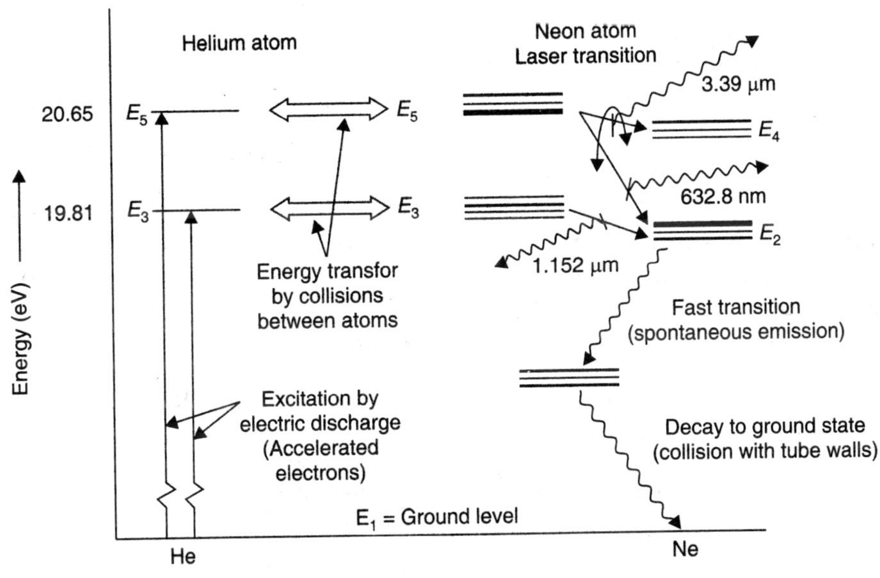

FIGURE 10.28 **Energy-level diagram for the He-Ne laser**

After this, fast radiative decay occurs from the 3*p* to the 2*p* ground state via collisions of the neon atoms with the container walls. Because of this last required step, the bore size of the laser cannot be made very large and the He-Ne laser is limited in both size and power.

With the correct selection of cavity mirrors, other wavelengths of laser emission of the He-Ne laser are possible. There are infrared transitions at 3.39 μm and 1.15 μm wavelengths and a variety of visible transitions, including a green (543.5 nm, the so called Green Ne Laser), a yellow (594 nm) and an electron and gaseous atoms. Since a helium atom weighs only about one-fifth as much as a neon atom, electrons transfer energy much more readily to helium atoms via collisions. This process results in helium excitation to detectable levels. There after, excitation of neon atoms to level 5*s* and 4*s* occurs by means of collisons with excited helium atoms. The slight differences in energy between the helium and neon levels are accounted for a change in the kinetic energy of the atoms during collision.

The process of pumping the active laser gas indirectly by transfer of energy via collisions with another excited gas is called "resonant excitation" and has been utilized in several types of gas lasers. For example, although lasing action has been observed in CO_2 alone, the pumping efficiency is greatly improved if a quantity of N_2 is added as a buffer gas.

If resonant transfer of energy from the He metastable atoms to Ne atoms occurs at a rate faster than the decay rate from 5*s* and 4*s* to lower levels, then population inversion may be achieved. Stimulated emission of radiation in neon occurs at six different wavelengths as visualised in Figure 10.28. Until 1985 lasing was practiced at only three of these lines: the familiar line at 632.8 nm in the red portion of the spectrum and two additional lines at 1.152 μm and 3.39 μm in the infrared.

The 3.39 μm line and 632.8 nm line share the same upper lasing level and are in competition for the same excited atoms that sustain the lasing process. When such a condition exists, the line having the greatest gain will loose, and the weaker line will not. In most cases, the lower energy transition predominate. With neon the gain is so great for the 3.39 μm that lasing can be obtained at this wavelength without a pair of mirrors for feedback. The phenomenon of achieving an output from a laser with high gain without the use of a pair of mirrors is termed as "superradiant lasing" which can occur in large He-Ne lasers.

The 1.15 μm line and 632.8 nm line share the same lower lasing line. If 1.15 μm line laser, atoms from 4*s* are transferred to the level 3*p*, raising its population and reducing the population inversion for the 632.8 nm line.

All three of these wavelengths may be present in the laser output simultaneously. If one is interested in a He-Ne laser whose output is at the 632.8 nm wavelength only, positive steps must be taken to suppress oscillations at the other two wavelengths. Both the 3.39 μm and 1.15 μm lines have greater gain than the 632.8 nm line and are capable of stealing power that otherwise would be available as output at the 632.8 nm wavelength.

Table 10.2 He-Ne Lasers

Wavelength	632.8 nm
Output power	0.5 – 50 mW
Beam diameter	0.5 – 2.0 nm
Beam divergence	0.5 – 3 m Radian
Coherence length	0.1 – 2 m
Power stability	5%/Hr
Lifetime	> 20000 Hrs.

10.8.11 Properties of Laser Beam

The ordinary light and laser light both are electromagnetic waves. But the laser beam is superior than ordinary light beam because of following characteristics: (*i*) Highly directional (*ii*) Highly intense (*iii*) Highly monochromatic (*iv*) Highly coherent.

(*i*) **Directionality:** Light from an ordinary source (sodium lamp, mercury lamp, sunlight etc.) spreads out in all possible directions. If light from an ordinary source is needed in a particular direction, an aperture is to be placed infront of the light source on the other hand, light from laser source travels only in one direction *i.e.*, laser beam is highly directional. The directionality of laser beam is expressed in terms of full angle beam divergence and this angle is twice the angle made by outer edge with the axis of beam. The outer edge is the point where intensity of beam falls to $1/e$ of its value at the centre. A beam of light of wavelength (λ) radiated from an aperture of diameter (d) propagates as a parallel beam up to distance d^2/λ called *Rayleigh range* beyond this it begins to diverge due to diffraction. Angular spread on one side is given as $\theta = \beta\lambda/d$ where β is constant of proportionality having value nearly unity.

For typical laser beam divergence is about 0.01 milliradian *i.e.*, 0.01 mm for every meter travelled by it from this one can imagine that spread of laser beam sent from earth to moon (384400 km away) will be just few kilometer.

(*ii*) **Intensity:** Intensity of light at any point is the amount of energy passing normally per unit area per second at that point, provided area being considered around that point as light from an ordinary source spreads out more or less uniformly in all direction so, if we look at 100 walt of ordinary lamp from a distance of 30 cm, the light entering the eye is hardly 1/1000 watt. On the other hand laser source gives out light as a narrow beam, so its whole of energy is concentrated in a small region and then accounts high intensity of laser. If we look at 1 watt laser directly it appears many thousand times more intence than 100 watt ordinary bulb.

The high intensity of laser beam can also be accounted by calculating and comparing the number of photon output from a laser source and from an ordinary source photon energy (in negligible range) $= \dfrac{hc}{\lambda}$

$$= \frac{6.6 \times 10^{-34} \times 3 \times 10^{8}}{5 \times 10^{-7}}$$

$$= 10^{-19}\ \text{J}$$

As power output from gas laser ~ 10^{-3} watt and from a pulsed solid state laser is ~ 10^{9} watt. So laser photon/sec $= \dfrac{10^{-3}\ \text{watt} - 10^{9}\ \text{watt}}{10^{-19}} = 10^{16}$ to 10^{28} where as the no. of photon emitted per second from a hot source of light at 1000 K for same area is about 10^{12} thus laser beam is highly intense.

(*iii*) **Monochromaticity:** The light from laser source is more or less strictly monochromatic, whereas light from ordinary monochromatic source has frequency spread of the order of MH_Z. Qualitatively the degree of monochromaticity is characterissed by spreads of frequency Δv of spectral line of frequency (v_0). The degree of nonmonochromaticity is measured by the quantity $\Delta v / v_0$. Thus lower is the value of $(\Delta v / v_0)$ higher will be the degree of monochromaticity and vice versa. For absolute monochromaticity Δv should be zero, which is not possible to attain.

For ordinary source of light,

$$\frac{\Delta v}{v_0} = 2 \times 10^{-5} \text{ and for laser beam } \frac{\Delta v}{v_0} = 10^{-12}$$

Hence laser beam is highly monochromatic.

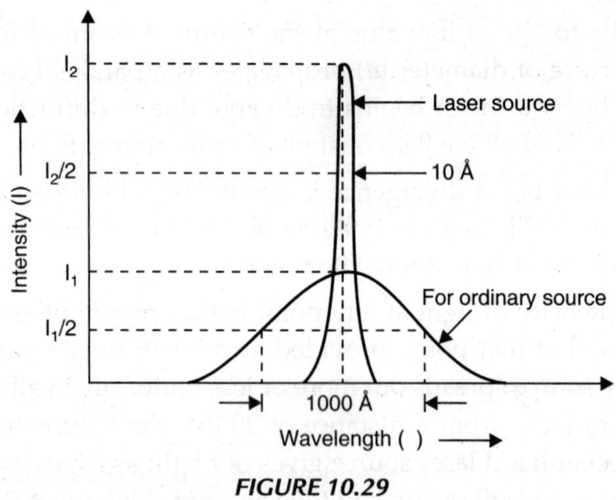

FIGURE 10.29

(*iv*) **Coherence:** Laser beam is highly coherent mass it has high degree of ordinary in the light field, which in turn is a measure of degree of phase correlation in the rediation field at different locations and at different times.

As we know that, the light emitted from source is the resultant of light radiations emitted due to transition of billions of atoms/molecules from higher energy state to lower energy state after initial exciation. As all the atoms/molecules operates independly and even does not radiates continuously, the wavefront so produced varies with space and time. Hence, there are two concept of coherence called spatial coherence and temporal coherence.

(*a*) **Spatial Coherence:** Spatial coherence is also known as lateral coherence it is measure of phase relationship between light field radiations at different points in space and describes how far apart two sources of two portion of the same source can be located along a transverse direction with respect to direction of observation. The concept of spatial coherence can be well explained by using Young's double slit experiment.

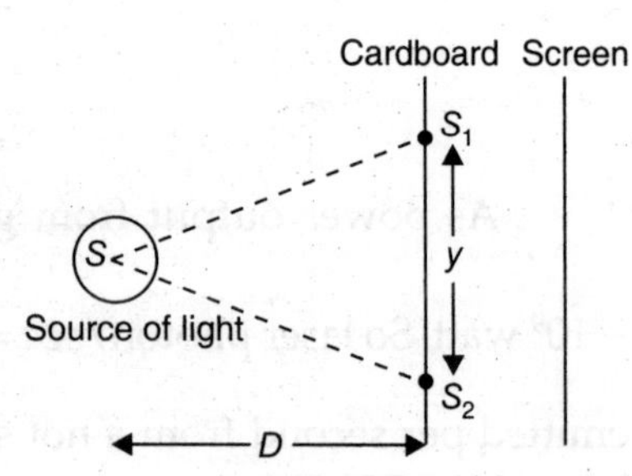

FIGURE 10.30

Consider a cardboard having two pin holes S_1 and S_2 at variable separation (y), as shown in Figure 10.30 the card board is placed at distance D from point source S.

The interference pattern produced is observed on screen placed to the right of slit S_1 and S_2. The distance between the slit is varied and maximum separation *i.e.*, $S_1S_2 = y_{max}$, for which sustained interference on the screen is observed, is calculated. This maximum distance is called coherence length of source.

If S is not source and has width 'a' as shown in Figure 10.30 then light from each point on the source will be spread out in angle $\approx \lambda/a$. Thus if angle $S_1SS_2 < \lambda/a$; than both the slits will receive light from each point on the source and this will be in coherence. From Figure 10.31 we get, $\angle S_1SS_2 = y/D$, therefore condition of coherence for S_1 and S_2 is

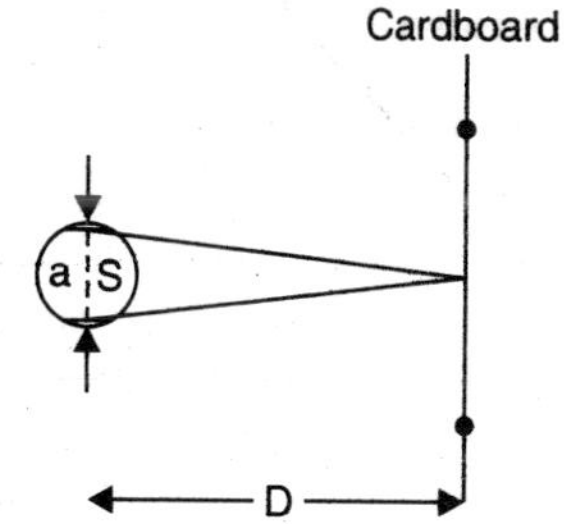

FIGURE 10.31

$$y/D < \lambda/a$$

or
$$y < \frac{\lambda D}{a}$$

$$y < \lambda/\theta, \text{ where } \theta = a/D$$

Therefore the limiting value of y is $y_{max} = \lambda/\theta$

where y_{max} is known as spatial coherence length.

(*b*) **Temporal Coherence:** Temporal coherence refers to phase relation between light field at a point and the light field at the same point at later time a perfectly coherent source emits light radiations of constant amplitude as shown in Figure 10.32.

However in active practice no coherent source emits ideal sinusoidal field for all values of time.

This is because atoms in excited state emits pulse of short duration of the order 10^{-10} sec and returns to initial state. After this, phase changes abruptly as shown in Figure 10.32.

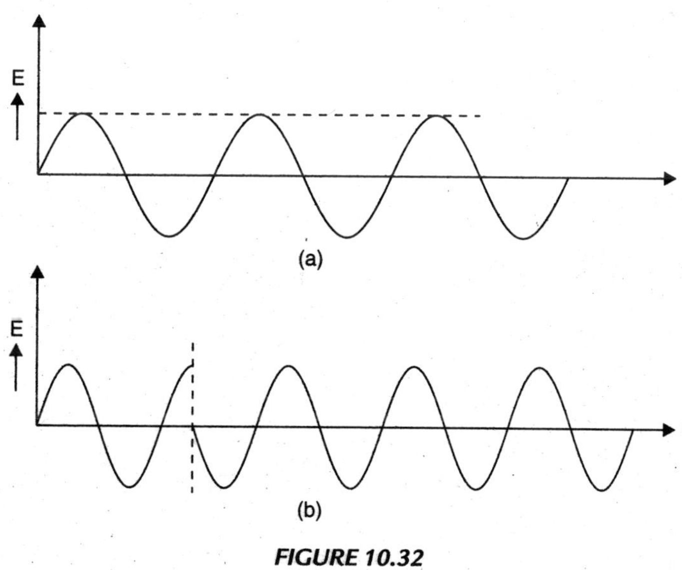

FIGURE 10.32

From the Figure 10.32 the interval for which field has definite phase relationship is known as coherence time and the length over which field remains sinusoidal is known as coherence length. Therefore,

Coherence length $l = c.\tau$

where c is velocity of light and τ is coherence time.

Michelson Interferometer can be used for measuring coherence length and coherence time adjust both the mirrors so as to obtain circular fringes. Let d be the difference in distance of mirrors from partially reflecting glass plate then path difference between interfering beam. So starting with equal distance of mirrors from partially reflecting glass plate *i.e.*, (position of zero path difference) increase the path difference continuously, contrast between dark and bright fringes falls off gradually. The path diffence where frignes vanishes or disappear is an estimated coherence light.

Coherence length from sodium light is estimated to be 3 cm, so coherence time,

$$\tau = \frac{l}{c} = \frac{3}{3 \times 10^{10}} \simeq 10^{-10} \text{ sec.}$$

[*For cadmium red light coherence length is 30 cm.]

10.8.12 Applications of Laser

Laser have found wide applications in following fields:

(*i*) **Material Processing:** For mechanical and Metallurgical engineers, a knowledge of laser technology is of great use, laser can cut, drill, weld, remove metal from surfaces and perform their operations even at surface in accessible by mechanical methods.

(*ii*) **Communication:** Laser play the essential role in using thin strands of glass fibres to transmit light signal that can be received and translated in to communication format. With the help of laser beam, the distance between the earth and moon has been measured within accuracy of an inch. Laser can be used in computers to transmit an entire memory bank from one computer to another.

(*iii*) **Medicine:** Surgeous use laser to burnup brain tumors and remove tattoos. Blood vessels are reconnected by laser welding. He-Ne laser is used to stimulate nerves in the wrist and ankles for treatment of paralysis. Eye surgeons use laser to perform operation of common eye diseases *e.g.*, glaucoma, cotaract and diabetic retinnopathy. Laser accupuncture is becoming popular for some disorders and relief pain.

(*iv*) **Applications in physics and chemistry:** Laser has initiated quite new fields of investigation in physics. An interesting example is of non-linear optics with special mension of harmonic generation and stimulated scattering. In the field of chemistry, laser used for both for diagnostic purposes and for producing inversible chemical change *i.e.*, laser photochemistry. In diagnostic techniques particualrly resonant Raman scattering and coherent antistokes, Raman scattering gave considerble information on the structure and property of poly atomic molecules. The most interesting chemical application for laser is in the field of photochemistry.

Biologists have been able to drill minute hole in the cell wall without damaging the cell itself.

(*v*) **Military Applications:** Due to great quality of energy which the laser can concentrate, it has been mentioned as potential 'death day' type of incendiary weapon for use against energy missiles much research has been done in the field of laser weaponry *e.g.*, development of higher energy output divices at great efficiency and actived destructive effects of laser beam.

10.9 NUMERICAL PROBLEMS BASED ON ATOMIC PHYSICS

Problem 1. *Calculate for He^+ (i) radius of the first Bohr orbit, (ii) velocity of electron moving in first, (iii) orbital frequency in the first orbit, (iv) kinetic energy and binding energy of electron in the grond state, (v) ionization potential and the first excitation potential, and (vi) wavelength of the resonance line emitted in the transition $n = 2 \longrightarrow n = 1$.*

Solution: (*i*) For helium atom Z = 2

$$r_n = a_0 \frac{n^2}{Z} \qquad (n = 1)$$

Radius of the first orbit

$$r_1 = \frac{a_0}{2} = \frac{0.529 \text{ Å}}{2} = 0.264 \text{ Å}$$

(*ii*) Velocity of electron

$$v_n = \frac{1}{4\pi\varepsilon_0} \frac{e^2}{\hbar} \frac{Z}{n} = (2.19 \times 10^6 \text{ m/s}) \frac{Z}{n}$$

$$v_1 = (2.19 \times 10^6 \text{ m/s}) \left(\frac{2}{1}\right) = 4.38 \times 10^6 \text{ m/s}$$

(*iii*) Orbital frequency

$$\omega_n = \left(\frac{1}{4\pi\varepsilon_0}\right)^2 \frac{me^4}{\hbar^3} \frac{Z^2}{n^3} = (4.14 \times 10^{15}) \frac{Z^2}{n^3}$$

$$\omega_1 = 16.56 \times 10^{15} \text{ rad/s}$$

(*iv*) Kinetic energy $\quad K_n = |E_n| = (13.6 \text{ eV}) \dfrac{Z^2}{n^2}$

$$K_1 - 54.32 \text{ eV}$$

Binding energy = 54.32 eV

(*v*) Ionization potential = 54.32 eV

First excitation energy is equal to the energy required to raise the electron from ground state ($n = 1$) to the first excited state ($n = 2$)

$$\Delta E = E_2 - E_1 = (13.58 \text{ eV})\, Z^2 \left(\frac{1}{1^2} - \frac{1}{2^2}\right) \qquad (Z = 2)$$

$$= 40.74 \text{ eV}$$

(*vi*) Wavelength of the resonance line

$$\frac{1}{\lambda} = R_{\infty} Z^2 \left(\frac{1}{n_f^2} - \frac{1}{n_1^2} \right) = R_{\infty} \left(\frac{1}{1^2} - \frac{1}{2^2} \right) = \frac{3}{4} R_{\infty} Z^2$$

$$\lambda = \frac{4}{3 R_{\infty} Z^2} = \frac{4}{3 \times (1.097 \times 10^7 \text{ m}^{-1}) \times 4} = 303.8 \text{ Å}$$

Problem 2. *Sodium doublets are produced by the transition* $3\,^2P_{3/2} \longrightarrow 3\,^2S_{1/2}$ *(D_1) and* $3\,^2P_{3/2} \longrightarrow 3\,^2S_{1/2}$ *(D_2). Calculate the Lande g factor for these levels.*

Hint: $$g = 1 + \frac{J(J+1) + S(S+1) - L(L+1)}{2J(J+1)}$$

(*i*) For $^2P_{3/2}$ state, $L = 1, S = 1/2, J = 3/2 \Rightarrow g = 4/3$

(*ii*) For $^2P_{1/2}$ state, $L = 1, S = 1/2, J = 1/2 \Rightarrow g = 2/3$

(*iii*) For $^2S_{1/2}$, state, $L = 0, S = 1/2, J = 1/2 \Rightarrow g = 2.$

Problem 3. *Show that the velocity of the electron in the first orbit of hydrogen atom is (1/137)c, where c is the velocity of light.*

Solution:
$$\left. \begin{aligned} m_e v r_n &= n\hbar \\ \text{and } r_n &= \frac{n^2\hbar^2}{k\, m_e\, e^2} \end{aligned} \right\} \Rightarrow v = \frac{n\hbar}{m_e\, r_n}$$

or
$$v_n = \frac{n\hbar}{m_e} \times \frac{k\, m_e\, e^2}{n^2\hbar^2} = \frac{k e^2}{n^2\hbar^2} = \frac{2\pi e^2 \times k}{nh}$$

$$= \frac{2 \times 3.142 \times (1.6 \times 10^{-19})^2 \times 9 \times 10^9}{6.6 \times 10^{-34}} \quad ; \quad n = 1$$

$$v_1 = 2.19 \times 10^6 \text{ m/sec}$$

$$\frac{v_1}{c} = \frac{2.19 \times 10^6}{3 \times 10^8} = \frac{0.73}{100} = \frac{1}{137} \Rightarrow v_1 = \frac{1}{137} c.$$

Problem 4. *How many different types of terms can two electron system consisting of d and f electrons possess ?*

Solution: For *d* electrons, we have $l_1 = 2$ and *f* electron $l_2 = 3$

$\therefore$ $$L = l_1 \pm l_2 = 2 \pm 3; \text{ Possible value} = 5, 4, 3, 2, 1$$

$$S = S_1 \pm S_2 = \frac{1}{2} \pm \frac{1}{2} = 0, 1$$

$\therefore$ $$r = 2S + 1 = 1, 3$$

Now $$J = L \pm S$$

$\therefore$ Corresponding to the values of L and S, J can have possible values

$$J = 5 \pm 0 = 5, \quad J = 4 \pm 0 = 4, \quad J = 3 \pm 0 = 3, \quad J = 2 \pm 0 = 2, \quad J = 1 \pm 0 = 1$$

we may write that these single terms for $S = 1$, we have

$$J = 5 \pm 1 = 6, 5, 4; \qquad J = 4 \pm 1 = 5, 4, 3; \quad 3 \pm 1 = 4, 3, 2;$$

$$J = 2 \pm 1 = 3, 2, 1; \quad \text{and} \quad J = 1 \pm 1 = 2, 1, 0$$

All these states are triplet state.

Problem 5. *Obtain L.S in terms of L, S, J. Calculate the possible values of L.S for L = 1 and S = 1/2.*

Solution:

$$J = L + S$$

$$|J|^2 = (L + S) . (L + S) = |\ L\ |^2 + 2\, L.S + |\, S\,|^2$$

$$L.S = \frac{1}{2}\left(|\,J\,|^2 - |\,L\,|^2 - |\,S\,|^2\right)$$

$$= \frac{1}{2}\left[J(J+1) - L(L+1) - S(S+1)\right]\hbar^2$$

For $L = 1, \quad S = 1/2, \quad J = L \pm S = 1 \pm 1/2 = 3/2, 1/2$

For $J = 3/2, \quad L.S. = \frac{1}{2}\hbar^2$

For $J = 1/2, L.S. = -\hbar^2$

Problem 6. *Calculate the normal Zeeman splitting of the line 6438 Å in a magnetic field of 0.5 T.*

Hint:

$$d\lambda = \frac{e\,B\lambda^2}{4\pi mc} = \frac{1.6 \times 10^{-19} \times 0.5 \times (6438 \times 10^{-10})^2}{4 \times 3.14 \times 9.1 \times 10^{-31} \times 3 \times 10^8} = 0.097 \text{ Å}$$

Problem 7. *What kind of Zeeman effect, normal or anomalous is observed in a weak magnetic field in the case spectral lines caused by the following transitions:*

(*i*) $^1P \longrightarrow {^1S}$, (*ii*) $^2D_{5/2} \longrightarrow 2P_{3/2}$ *and* (*iii*) $^3D_1 \longrightarrow {^3P_0}$ *and* (*iv*) $^5I_5 \longrightarrow {^5He_4}$.

Solution: (*i*) $^1P \longrightarrow {^1S}$. For the lower level: $L = 0$, $S = 0$, $J = 0$, $g = 1$. It does not split.

For the upper level: $L = 1$, $S = 0$, $J = 1$, $g = 1$, $M_J = -1, 0, 1$. It splits into three levels.

Normal Zeeman effect is observed.

(*ii*) $^2D_{5/2} \longrightarrow {^2P_{3/2}}$. For the lower level: $L = 1$, $S = 1/2$, $J = 3/2$, $g = 4/3$. It splits into four sub levels.

For the upper level: $L = 2$, $S = 1/2$, $J = 5/2$, $g = 6/5$. It splits into six sublevels.

Anomalous Zeeman Pattern is observed.

(*iii*) $^3D_1 \longrightarrow {^3P_0}$. Lower level: $L = 1$, $S = 1$, $J = 1$, $g = 1$. It does not split.

Upper level: $L = 2$, $S = 1$, $J = 1$, $g = 1/2$. It splits into three sublevels.

Normal Zeeman pattern is observed.

(*iv*) $^5I_5 \longrightarrow {}^5He_4$. Lower level: $L = 5, S = 2, J = 4, g = 19/20$.

Upper level: $L = 6, S = 2, J = 5, g = 19/20$. Since the two levels have identical Lande 'g' factor.

Normal Zeeman pattern is observed.

Problem 8. *For K line for Molybdenum (Z = 42) has a wavelength of 0.71 Å. Calculate the wavelength of K line of copper (Z = 29).*

Hint:

$$\nu = a(Z-1)^2 \quad \text{then} \quad \nu_1 = a\,(Z_1 - 1)^2 \text{ and}$$

$$\nu_2 = a\,(Z_2 - 1)^2 \;\Rightarrow\; \frac{\nu_1}{\nu_2} = \frac{(Z_1 - 1)^2}{(Z_2 - 1)^2}$$

and

$$\nu = \frac{c}{\lambda} \;\Rightarrow\; \frac{\nu_1}{\nu_2} = \frac{\lambda_2}{\lambda_1}$$

$\Rightarrow$

$$\lambda_2 = \lambda_1 \left(\frac{Z_1 - 1}{Z_2 - 1}\right)^2$$

$$= 0.71 \times 10^{-10} \left(\frac{42-1}{29-1}\right)^2 = 1.52 \times 10^{-10} \text{ m}$$

$$= 1.52 \text{ Å}.$$

Problem 9. *Calculate the energy of neutrons that produces first order Bragg diffraction at 20°, when incident on plane separated by a distance 2Å. (Mass of neutron = 1.67 × 10^{−27} kg).*

Hint:

$$2d \sin\theta = n\lambda \;\Rightarrow\; \lambda = \frac{2d\sin\theta}{n} = 1.368 \text{ Å}$$

and

$$p = \frac{h}{\lambda} \;\Rightarrow\; \frac{6.63 \times 10^{-34}}{1.368 \times 10^{-10}} = 4.847 \times 10^{-24}$$

$$E = \frac{p^2}{2m} \left[\text{K.E.} = \frac{1}{2} mv^2 = \frac{p^2}{2m}\right]$$

$$= 7.03 \times 10^{-21} \text{ J}.$$

Problem 10. *The K_α line for molybdenum has wavelength 0.7100 Å. Calculate wavelength of K_α for copper, given atomic number of molybdenum and copper are 42 and 29 respectively. Assuming screening constant ≅ 1.5.*

Hint: From Moseley's law $\frac{1}{\lambda} = R\,(Z-1)^2 \left[\frac{1}{1^2} - \frac{1}{2^2}\right]$

$$\frac{\lambda_{Cu}}{\lambda_{Mo}} = \frac{(Z_{Mo} - 1)^2}{(Z_{Cu} - 1)^2} \qquad Z_{Mo} = 42, \quad Z_{Cu} = 29$$

$$\boxed{\lambda_{Cu} = 1.52}$$

Problem 11. *An X-ray tube operated at 40 kV emits continuous X-ray spectrum with a short wavelength limit of 0.31 Å. Calculate Planck's constant.*

Hint:

$$eV = \frac{hc}{\lambda_{min}}$$

$$h = \frac{eV\lambda_{min}}{c} = 6.6 \times 10^{-34} \text{ J.s.}$$

Problem 12. *Ruby laser gives a laser beam of wavelength 6943 Å. Calculate the energy difference in eV between the two energy levels involved in this transition.*

Hint:

$$\Delta E = E_2 - E_1 = \frac{hc}{\lambda} = 1.79 \text{ eV.}$$

Problem 13. *Find the ratio of population density between the two states in a He-Ne laser capable of producing light of wavelength 632.8 nm at 300 K.*

Hint:

$$\frac{N_2}{N_1} = e^{-\Delta t/k_B T} \simeq 10^{-33}$$

Problem 14. *What is the minimum wavelength of X-rays emitted by an X-ray tube operating at 50 kV.*

Solution:

$$eV = \frac{hc}{\lambda_{min}}$$

$$\lambda_{min} = \frac{hc}{eV}$$

$$= \frac{6.6 \times 10^{-34} \times 3 \times 10^{8}}{1.6 \times 10^{-19} \times 50 \times 10^{3}} = \frac{6.6 \times 3}{8} \times 10^{-11}$$

$$= 0.2475 \times 10^{-10}$$

$$\boxed{= 0.25 \text{ Å}}$$

EXERCISES

1. What is the origin of magnetic moment in a neutral *H*-atom? Derive an expression for the orbital magnetic moment of *H*-atom. Name the unit for measurement of magnetic moment of atoms. What is the value of one unit. *(Amity University, May 2005)*
2. What is the difference between Paschen Back effect and Stark effect? *(Amity University, May 2005)*
3. Explain the terms 'optical pumping' and 'population inversion'. *(Amity Univeristy, May 2005)*
4. State and explain Bragg's law for X-ray diffraction. *(Amity University, May 2005)*

5. Write short notes on
   (*i*) Normal Zeeman Effect (*Amity University, May 2005*)
   (*ii*) Zeeman Effect
   (*iii*) Anomalous Zeeman Effect
   (*iv*) Bohr Magneton. (*Amity University, May 2006*)
6. What are X-rays and how are they emitted ? Differentiate between characteristic and continuous X-rays. Draw energy level diagram to illustrate the emission of X-ray spectral time. (*Amity University, May 2006*)
7. Differentiate between normal nad anomalous Zeeman effect. Obtain expression for Zeeman splitting of lines in a normal Zeeman pattern and draw the energy level diagram to show corresponding transitions. Write selection rules for such transitions. (*Amity University, May 2006*)
8. State and give the physical significance of various quantum numbers associated with electron in an atom.
9. What is vector model of atom ? State various quantum numbers and their signifcance associated with this model.
10. Explain Russel-Sander (L-S) and *j-j* couplings of orbital and spin angular momenta. How are the term symbols (atomic states) written making use of L-S coupling?
11. What is Zeeman effect? Use vector model (quantum theory) to obtain expression for normal Zeeman shift.
12. What do you mean by space quantization.
13. What is anomalous Zeeman effect ? Explain this effect taking example of splitting of sodium *D* lines.
14. What is Paschen Back effect. Obtain the splitting of ^2P and ^2S levels and indicate the allowed transitions in a strong magnetic field.
15. State and explain Moseley's law and explain one of its important applications.
16. Explain the emission of continuous and characteristic X-rays. Meniton the applications of Moseley's law.
17. Derive Bragg's law for X-rays diffraction in crystals.
18. Explain
    (*i*) Stimulated absorption,
    (*ii*) Spontaneous emission and
    (*iii*) Stimulated emission of radiation.
19. What are Einstein's coefficients *A* and *B* ? Obtain an expression for energy density of radiation under equilibrium condition in terms of Einstein's coefficient.
20. Explain the significance of Einstein's relations and Einstein's coefficients *A* and *B*. How did Einstein's prediction of stimulated emission lead to invention of laser.
21. Explain population inversion. What is the role of a metastable state in LASER absorption?
22. Explain (*i*) Active system, (*ii*) Pumping mechanism, and (*iii*) resonant cavity for a working LASER.
23. Explain the principle of a laser. State important characteristics of laser.

24. What do you mean by coherence? Explain spatial and temporal coherence.
25. Explain the working of a solid state ruby laser with a neat energy level diagram.
26. Describe the construction and working of He-Ne laser with energy level diagram.
27. Differentiate between a three level laser and a four level laser.
28. Why population inversion is essential for laser action? Explain the production of continuous level.
29. What is metastable state? How it help in LASER production? With neat figures, describe the construction and working of He-Ne laser. Mention its uses. Why is it superior to RUBY LASER.
30. Mention the important applications of LASER in the medical and engineering field.
31. Describe/Explain
    (*i*) Population inversion
    (*ii*) Characteristics of laser beam
    (*iii*) Interaction of radiation with matter.

SUGGESTION FOR FURTHER READING

1. A. Beiser, S. Mahajan and S. Rai Choudhry, Concepts of Modern Physics, Tata McGraw-Hill Education Pvt. Ltd., New Delhi, 2009.
2. S.L. Kakani and Subhra Kakani, Modern Physics, Viva Books Pvt. Ltd., New Delhi, 2009.
3. R.B. Singh, Introduction to Modern Physics, New Age International (P) Ltd., New Delhi, 2002.
4. H.B. Lal, Introductory Modern Physics, United Book Depot, Allahabad, 2003.
5. Sanjeev Puri, Modern Physics-Concepts and Applications, Narosa Publishing House, New Delhi, 2006.
6. N.K. Sehgal, K.L. Chopra and D.L. Sehgal, Modern Physics, Sultan Chand and Sons, New Delhi, 2007.
7. R. Murgeshan and K. Sivaprasath, Modern Physics, S. Chand and Co., Delhi, 2008.
8. J. Bernstein, P.M. Fishbane and S. Gasiorowicz, Modern Physics, Pearson Education, Delhi, 2005.
9. Devraj Singh, Engineering Physics, Vol. 1 and Vol. 2, Dhanpat Rai & Co. (P). Ltd., New Delhi, 2009.
10. Devraj Singh, Fundamentals of Engineering Physics, Vol. 1 and Vol. 2, Dhanpat Rai & Co. (P). Ltd., New Delhi, 2009.
11. W.T. Silfvast, Laser Fundamentals, Cambridge University Press, New Delhi, 2009.
12. A.K. Saxena, Principle of Modern Physics, Narosa Publishing House, New Delhi, 2007.

Chapter 11 SOLID STATE PHYSICS

11.1 INTRODUCTION

In recent decade solid state physics has seen many dramatic new developments and has become one of the largest independent branches of physics. It has simultaneously expanded into many new areas, playing a vital role in fields that were once the domain of the engineering and chemical sciences. A consequence of this explosive development is that no single university lecturer can today be expected to have a detailed knowledge of all aspects of this vast subject: likewise, it is impossible to conceive that could offer students a comprehensive understanding of the entire discipline and its many applications. In this chapter of solid state physics, we will study base topics of this branch of physics like free electron theory of metals, band theory of solids, semiconductors, photoconductivity and photovoltaics and superconductivity.

11.2 SOMMERFELD'S FREE ELECTRON THEORY OF METALS

In 1928, by applying quantum mechanical principles, *Arnold Sommerfeld* succeeded in overcoming many of drawbacks of the classical free electron theory, while retaining all the essential features of the classical free electron theory. His approach is based upon quantisation of electrical energy levels. He realised the role played by Pauli exclusion principle in restricting the energy values of electron. The theory proposed is known as **Quantum Free Electron Theory**.

The main assumptions of quantum free electron theory are,

1. The energy levels of the conduction electrons are quantised.
2. The distribution of electrons in various allowed energy levels occur as per Pauli exclusion principle.

   However, the following assumptions of classical electron theory continue to be applicable in quantum free electron theory also.
3. The electrons travel in a constant potential inside the metal but stay confined within its boundaries.
4. Both the attraction between the electrons and the lattice ions, and the repulsion between the electrons themselves are ignored.

11.2.1 Fermi Energy

Due to the quantization rules, a material in solid state possess a set of allowed energy levels. For a metal containing N free electrons, there will be N such allowed energy levels, which are separated by energy differences that are characteristic of the material. The allowed energy values for the conduction electrons in the metal are also quantised, are related to the energy levels of the metal. As per the Pauli exclusion principle, each energy level can accommodate a maximum of two electrons, one with spin up and the other one with spin down (Figure 11.1).

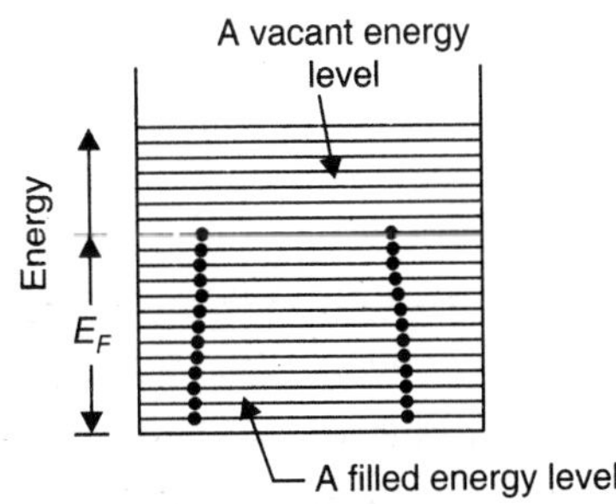

FIGURE 11.1 **Fermi Energy**

The allowed energy values of the electron are realised in terms of occupation of the allowed energy levels, under certain rules. When the filling up the energy levels is undertaken the universal rules that—any system which is free tends to go state of lowest energy–comes into picture.

Thus a pair of electrons, one with spin up, and other with spin down occupy the lowest level. The next pair of electrons occupy the next higher level, and so on, till all the electrons in metal are accommodated.

However, there will be many more allowed energy levels available for occupation by the electrons. **The energy of the highest occupied level at zero degree absolute is called Fermi energy, and the energy level is referred to as the Fermi level. The Fermi energy is denoted as E_F.**

When there is no external energy supply for the electrons, such as thermal energy or electrical energy, the electrons are free, and thus settle in the lowest allowed energy state available. Thus at a temperature of absolute zero, and when the metal is not in the influence of any external field, all the energy levels lying above the Fermi level are empty, and those lying below are completely filled. Since there are two electrons in each energy level, out of the N allowed energy levels, $\frac{N}{2}$ of them will be occupied.

11.2.2 Fermi Level

The Fermi level is an energy level mid way between top surface of valence band and bottom surface of conduction band as shown in Figure 11.2.

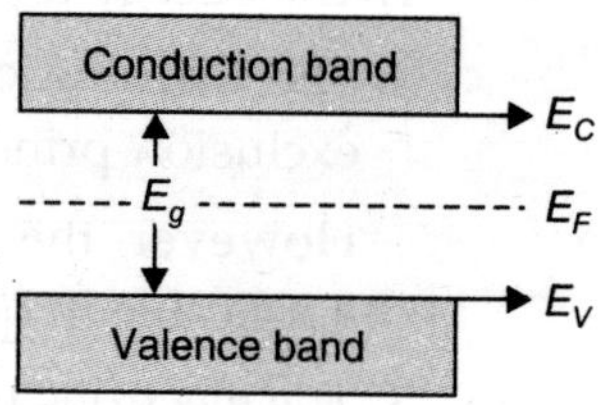

FIGURE 11.2 **Fermi Level**

If E_V is the energy corresponding to top surface of valence band, E_C is the energy corresponding to bottom surface of conduction band and E_F is the Fermi energy corresponding to Fermi level, then

$$E_F = \frac{E_C + E_V}{2} = E_C + \frac{E_V - E_C}{2}$$

$$= E_C + \frac{E_g}{2} \Rightarrow \therefore \quad E_F - E_C = \frac{E_g}{2} \qquad \text{[where } E_V - E_C = E_g \text{ (Energy gap)]}$$

11.2.3 Fermi Factor

When the temperature is not at absolute zero, the material will be receiving thermal energy from surroundings. At ordinary temperature, the amount of energy will be quite small*, because of which the electrons occupying energy levels below the Fermi level cannot absorb the thermal energy. It is because, there is no availability of unoccupied higher energy levels into which the electron can come into, due to the increase in energy by the *absorption of thermal energy*. However, there are unoccupied higher energy levels which are above the occupied energy levels at small energy differences. They are located near Fermi level. Those are the energy levels into which, the electrons in the energy levels near Fermi level, are capable of being excited**.

During thermal excitation (*i.e.*, $T > 0$), the electrons which absorb the thermal energy move into higher energy levels which were unoccupied at zero degree absolute (*i.e.*, $T = 0$). Though such excitations seem to be random, the occupation of various energy level obey a statistical distribution*** called *Fermi-Dirac distribution*, once the system is in thermal equilibrium (*i.e.*, at a steady temperature state).

The probability $f(E)$ that a given energy state with energy E is occupied at a steady temperature T, is given by

---

*The quantity of thermal energy is given by k_BT, where k_B is Boltzmann constant. At room temperature, $k_BT \simeq 0.025$ eV. Compare this with Fermi energy which is of the order of 5 eV.

**Since such electrons constitute only a small fraction of the total number of electrons in the metal as a whole, the specific heat of the material becomes low. Thus, reason for failure of classical free electron theory to account for low specific heat of metals is now understood.

***The statistical distribution deals with the question as to what is the probability occupation of a particular energy level.

$$f(E) = \frac{1}{e^{(E - E_F)/k_B T} - 1} = \frac{1}{\exp\left(\frac{E - E_F}{k_B T}\right) - 1}$$

$f(E)$ is called the Fermi factor and is defined as follows:

Fermi factor is the distribution function which gives the probability of occupation of a given energy state for a material in thermal equilibrium in terms of the Fermi energy, Boltzmann constant and the temperature.

The dependence of Fermi factor on temperature, and the effect on occupancy of energy level is shown in Figure 11.3.

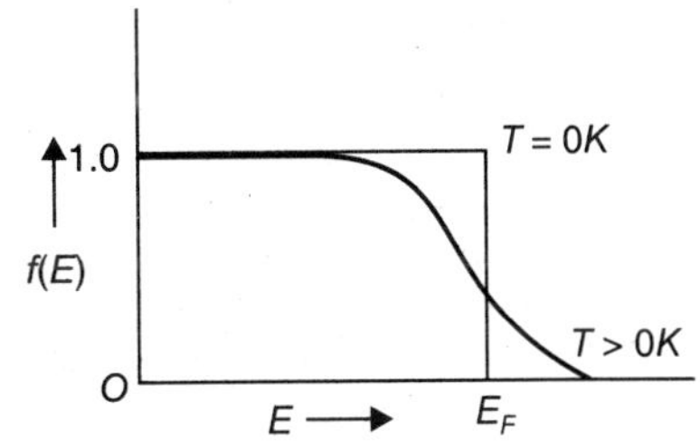

FIGURE 11.3 **Variation of f(E) with E**

Let us consider the different cases of distribution as follows :

(*i*) **Probability of occupation for $E < E_F$ at $T = 0$**

When $T = 0$, and $E < E_F$

$$f(E) = \frac{1}{e^{-\infty} + 1} = \frac{1}{0 + 1}$$

$\therefore\ f(E) = 1$ for $E < E_F$.

$\therefore$ At $T = 0$ K, all the energy levels below the Fermi level are occupied.

(*ii*) **Probability of occupation for $E > E_F$ at $T = 0$ K**

When $T = 0$, and $E > E_F$

$$f(E) = \frac{1}{e^{\infty} + 1} = \frac{1}{\infty}$$

$\therefore\ f(E) = 0$ for $E > E_F$.

$\therefore$ At $T = 0K$, all the energy levels above Fermi level are unoccupied.

In view of the above two cases, at $T = 0$K, the variation of $f(E)$ for different energy values, becomes a step function, as shown in Figure 11.3.

(*iii*) **Probability of occupation at ordinary temperatures**

At ordinary temperatures, the value of probability starts reducing from 1 for values of E close to but lesser than E_F.

The value of $f(E)$ becomes $\frac{1}{2}$ at $E = E_F$

Since for $E = E_F$, $e^{(E - E_F)/k_B T} = e^0 = 1$

$$f(E) = \frac{1}{e^{E - E_F/k_B T} + 1} = \frac{1}{1 + 1} = \frac{1}{2}$$

Further, for $E > E_F$, the probability value falls off to zero rapidly (Figure 11.3).

The above discussion on the variation $f(E)$ with E, brings forth the point that,

The Fermi energy is the most probable, or the average energy of the electron across which the energy transitions occurs at temperature above zero degree absolute, which result in the conductivity of the metal. This may be considered as the physical basis for the concept of Fermi energy.

11.2.4 Calculation of Fermi Energy at $T = 0$ K

Let the value of E_F of 0 K be denoted as E_{F_0}. Also, if we denote the number of electrons per unit volume, which possess energy only in the range E and $(E + dE)$, by $N(E)\, dE$, then, $N(E)\, dE$ is given by the product

density of available states in the energy range E and E + dE × probability of occupation of those energy levels by the electron.

But, probability of occupation of any given energy state by the electron is given by the Fermi factor $f(E)$.

$$\therefore \qquad N(E)\, dE = g(E)\, dE \times f(E)$$

The number of electrons per unit volume of the material n can be evaluated by integrating the above expression from $E = 0$ to $E = E_{max}$, where E_{max} is the maximum energy possessed by the electrons

$$\therefore \qquad n = \int N(E)\, dE$$

$$= \int_{E=0}^{E=E_{max}} g(E)\, f(E)\, dE$$

But at $T = 0$ K, the maximum energy that any electron of the material can have is E_{F_0}.

Hence $$E_{max} = E_{F_0}$$

Also $$f(E) = 1, \text{ at } T = 0 \text{ K}$$

$$\therefore \qquad n = \int_{E=0}^{E=E_{max}} g(E)\, dE \times 1$$

$g(E)\, dE$ is given by

$$g(E)\, dE = \frac{8\sqrt{2}\, \pi\, m^{3/2}}{h^3} E^{1/2}\, dE$$

where m is the mass of electron and h is the Planck's constant.

$$\therefore \qquad n = \frac{8\sqrt{2}\, \pi\, m^{3/2}}{h^3} \int_{E=0}^{E=E_{max}} E^{1/2}\, dE$$

$$= \frac{8\sqrt{2}\, \pi\, m^{3/2}}{h^3} \times \frac{2}{3} (E_{F_0})^{3/2}$$

$$= \left[\frac{16\sqrt{2}\, m^{3/2}}{h^3}\right]\left[\frac{\pi}{3}\right](E_{F_0})^{3/2}$$

$$\therefore \qquad (E_{F_0})^{3/2} = \frac{h^3}{(8m)^{3/2}}\left(\frac{3n}{\pi}\right)$$

or $$E_{F_0} = \left(\frac{h^2}{8m}\right)\left(\frac{3n}{\pi}\right)^{2/3} \qquad ...(1)$$

$$\therefore \qquad E_{F_0} = B\, n^{2/3} \qquad ...(2)$$

where $$B = \left(\frac{h^2}{8m}\right)\left(\frac{3}{\pi}\right)^{2/3} \text{ is a constant}$$

$$= 5.85 \times 10^{-38} \text{ J.}$$

11.2.5 Fermi energy at T > 0K

The Fermi energy E_F, at any temperature T in general can be expressed in terms of E_{F_0} through the relation

$$E_F \simeq E_{F_0}\left[1 - \frac{\pi^2}{12}\left(\frac{k_B T}{E_{F_0}}\right)^2\right]$$

Except at extremely high temperatures, the second term within the bracket is very small compared to unity.

$$\therefore \qquad \boxed{E_F \cong E_{F_0}}$$

Hence, the value of E_{F_0} can be taken to be essentially equal to E_F itself.

11.2.6 Fermi Temperature (T_F)

Fermi temperature (T_F) is the temperature at which the average thermal energy of the free electron in a solid becomes equal to the Fermi energy at 0K. But the thermal energy possessed by electrons is given by the product $k_B\, T$.

$\therefore$ When $T = T_F$, the equation, $E_{F_0} = k_B\, T_F$, is satisfied. But for practical purposes, $E_{F_0} \cong E_F$

$$\therefore \qquad k_B\, T_F = E_F \qquad ...(1)$$

or $$T_F = \frac{E_F}{k_B} \qquad ...(2)$$

The Fermi temperature is only a theoretical concept, since, at ordinary temperature, it is not possible for the electrons to receive thermal energy in a magnitude of E_F. For example, let us consider the case of metals. For metals, we know that E_F will be of the order of a few eV.

Let us consider the case in which $E_F = 3$ eV

$$= 3 \times 1.602 \times 10^{-19} \text{ J} = 4.8 \times 10^{-19} \text{ J}$$

$$\therefore \qquad T_F = \frac{4.8 \times 10^{-19}}{1.381 \times 10^{-23}} = 34800 \text{ K}$$

which is quite an exaggerated temperature to be realized in practice.

11.2.7 Fermi Velocity (V_F)

The energy of the electron, which are at the Fermi level is E_F. The velocity of the electrons which occupy the Fermi level is called the Fermi velocity v_F.

$$\therefore \qquad E_F = \frac{1}{2} m \, v_F^2$$

or

$$v_F = \sqrt{\left(\frac{2 E_F}{m}\right)}.$$

11.2.8 Comparison between Classical Free Electron Theory and Quantum Free Electron Theory

Similarities between the two theories

The following assumptions apply to both the theories:

1. The valence electrons are treated as though they constitute an ideal gas.
2. The valence electrons can move freely throughout the body of the solid.
3. The mutual repulsion between the electrons, and the force of attraction between the electrons and ions are considered insignificant.

Difference between the two theories

Classical Free Electron Theory	Quantum Free Electron Theory
1. The free electrons, which constitute the electron gas can have continuous energy values.	1. The energy values of the free electrons are discontinuous because of which the energy levels are discrete.
2. It is possible that many electron may possess same energy.	2. The free-electrons obey the Pauli exclusion principle. Hence no two electrons can possess same energy.
3. The pattern of distribution of energy among the free electron obey Maxwell-Boltzmann statistics.	3. The distribution of energy among the free electrons is according to Fermi Dirac statistics which imposes a severe restriction on the possible way in which the electrons absorb energy from an external source.

11.3 PERIODIC POTENTIAL IN CRYSTALLINE SOLID

In solids one deals with a large number of interacting particles and hence the problem of determining the electron wave functions and the energy levels is extremely complicated. A simple quantum mechanical picture of an electron in a crystal can, however, be obtained by assuming that the atomic nuclei are at rest in the crystalline state and that the electron is in a periodic potential which has periodicity of the lattice. The periodic potential may be considered to be caused by a fixed nuclei plus some average potential due to all other electrons. The solution of Schrödinger equation for this potential gives a set of state or levels, which may be occupied by the electrons according to Pauli's exclusive principle.

The potential energy of an electron at a distance r from an atomic nucleus of charge $Z\,e$ is given by

$$V = -\frac{Z\,e^2}{4\,\pi\,\varepsilon_0\,r}$$

To form a crystal, a number of such nuclei are brought close to each other. Therefore, the potential energy of an electron is the sum of potential energies due to the individual nuclei. Hence the variation of potential energy with distance for a one-dimensional crystal is shown in Figure 11.4.

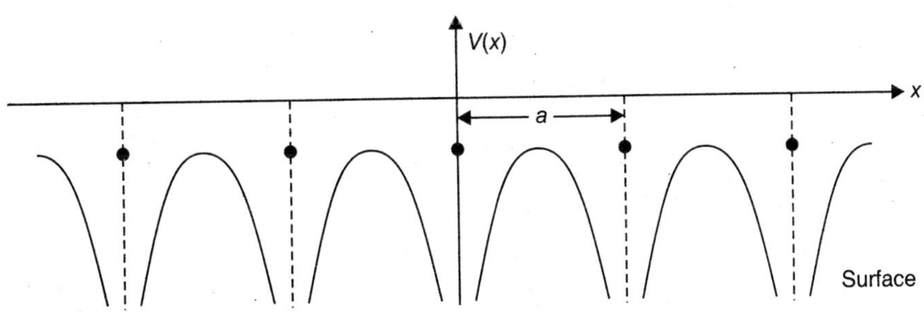

FIGURE 11.4 **Representation of potential experienced by an electron in a perfectly periodic crystal lattice with lattice parameter *a***

11.4 BLOCH THEOREM

In free electron theory, an electron moves in a constant potential V_0 and the Schrödinger equation for a one-dimensional case is

$$\frac{d^2\psi}{dx^2} + \frac{2m}{\hbar^2}(E - V_0)\,\psi = 0$$

The solution of this equation is plane waves of the type

$$\psi\,(x) = e^{\pm ikx}$$

where k represents the momentum of the electron divided by $\hbar$. From above equation, we get the kinetic energy of the electron

$$E_{kin} = E - V_0 = \frac{\hbar^2 k^2}{2m} = \frac{p^2}{2m}$$

The complete solution for the wave function containing the time is obtained by multiplying $\psi\,(x)$ by $e^{-i\omega t}$, where $\omega = \frac{E}{\hbar}$.

Now, consider the Schrödinger equation for an electron moving in one-dimension periodic potential. In this case, the potential energy of an electron satisfies the equation

$$V\,(x) = V\,(x + a)$$

where a is the period.

In this case, the Schrödinger equation is

$$\frac{d^2\psi}{dx^2} + \frac{2m}{\hbar^2}[E - V\,(x)]\,\psi = 0.$$

With reference to the solutions of this equation, there is an important theorem, which states that there exist solutions of the form.

$$\psi\,(x) = e^{\pm\, ikx}\, u_k\,(x)$$

where $$u_k\,(x) = u_k\,(x + a)$$

Thus the solution are plane waves modulated by the function $u_k\,(x)$, which has the same periodicity as the lattice. This theorem is known as the Bloch theorem or Floquet's theorem. The function defined by equation are known as *Bloch function*.

11.5 KRONIG-PENNEY MODEL (QUALITATIVE)

The free electron model implies that potential inside the solid is uniform in metals. A more accurate model allows for variation in potential energy due to the fixed positively charged lattice ions. In Figure 11.5, a one-dimensional lattice with ions separated by lattice parameter a, is presented.

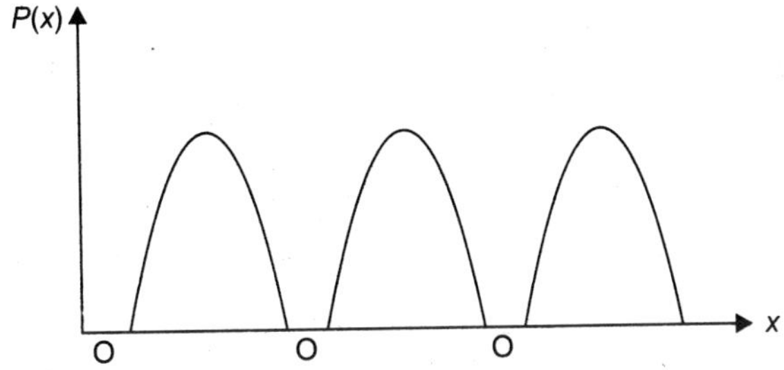

FIGURE 11.5 **One-Dimensional Lattice**

The highest potential is half way between the atom and goes to $-\infty$ as position of the ion is approached. This potential distribution is quite complicated and for mathematical solution of Schrödinger's equation a simpler model known as "*Kronig-Penney model*" is used. The Kronig-Penney model still displays the following essential features, where the function:

— has the same period as lattice.

— potential is higher between ions and lower near the lattice ion.

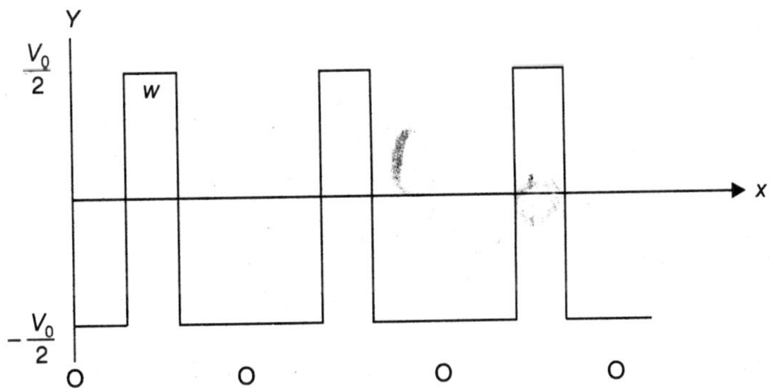

FIGURE 11.6 **Geometrical arrangement for Kronig-Penney Model**

The ions are located at $x = 0, a, 2a, 3a$ etc. The potential wells are separated from each other by potential barrier of height V_0 and width w. The Schrödinger's equation should be solved for potential distribution separately for $V = +V_0/2$ and $V = -V_0/2$ and compare the solutions at the boundary conditions.

The assumed wave function has the form

$$\psi(x) = U_k(x)\, e^{+ikx} \quad ...(1)$$

where $U_k(x)$ is a periodic function with the same period as the lattice.

The solution for the above equation exists if k is related to energy E by the following equation and assuming that as $w \to 0$, $V \to \infty$, $w\,V$ = constant

$$\cos ka = \frac{P \sin \alpha a}{\alpha a} + \cos \alpha a \quad ...(2)$$

where $P = \dfrac{4\pi^2 m a}{h^2} V_0 w$

$$\alpha = \frac{2\pi}{h}(2mE)^{1/2}$$

Here the relationship between E and k is not as easy as that of free electron. To find the exact E-k curve, we must first of all plot the right hand side of equation (2) as a function of αa and then by plotting the left hand side and using the fact that this always must be between + 1 to – 1, a solution will be found for the value of E for which right side is within these limits.

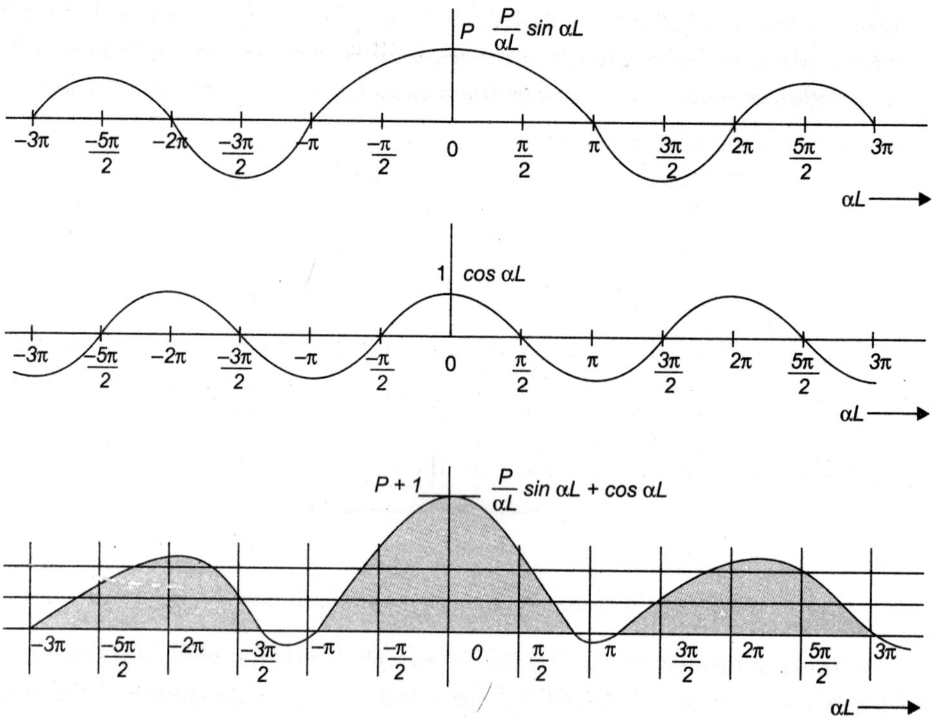

***FIGURE 11.7* Solution for Kronig-Penney Model**

The solution is defined only in the shaded region, *i.e.*, only certain values are allowed. This implies that there are only certain allowed energy bands in crystal solids.

For the solution of Kronig-Penney model mentioned above, if the value of P is large *i.e.*, $V_0 w$ is large, the function is given by right hand side of equation (2) crosses at a steep slopes. This means that the allowed bands are narrower and forbidden bands are wider. Let us observe some specific conditions.

1. As $P \to \infty$, the band reduces to a single energy level, *i.e.*, discrete energy spectrum for isolated atoms, which is the actual case. So P is proportional to the potential barrier of the given system.

$$\frac{\sin \alpha a}{\alpha a} = 0 \Rightarrow \sin \alpha a = 0$$

As $\quad P \to \infty \qquad \alpha a = n\pi$

$$\frac{2\pi a}{h}\sqrt{2mE} = n\pi$$

$$E = \frac{n^2 h^2}{8ma^2}$$

which is exactly same as the equation for an electron in an infinite potential well of width a. So, all the electrons are independent of each other and confined to infinite potential barrier.

2. When $P \to 0 \qquad \cos \alpha a = \cos ka$

$$\frac{2\pi a}{h}\sqrt{(2mE)} = ka$$

$$E = \frac{h^2k^2}{8\pi^2 m} = \frac{h^2k^2}{2m} \qquad \text{...(3)}$$

Equation (3) is exactly same as that of equation for a free electron. So the variation in the value of P from zero to infinity accounts for the whole range of conditions in the crystalline solid.

3. At the boundary conditions of the allowed band, *i.e.*,

$$\cos ka = \pm 1, \cos ka = \cos n\pi$$

$$ka = n\pi \text{ or } k = \frac{n\pi}{a} \text{ where } n = 1, 2, 3, \text{..... etc.}$$

Below E vs. k^2 curve is plotted for one-dimensional lattice structure from Kronig-Penney model. At the boundaries, we can see discontinuities in energy levels as the value of n changes. By transferring the zone curves by an appropriate multiple of $\frac{\pi}{a}$ to the axis E we can see the separation between energy levels very clearly. Each of these energy levels is termed as *Brillouin zone.*

From the Figure 11.8, we can say that in the middle of the Brillouin zones the E-k curve is identical to the free electron. But the boundaries of the zones, *i.e.*, at $k = \frac{n\pi}{a}$, their behaviour is totally different.

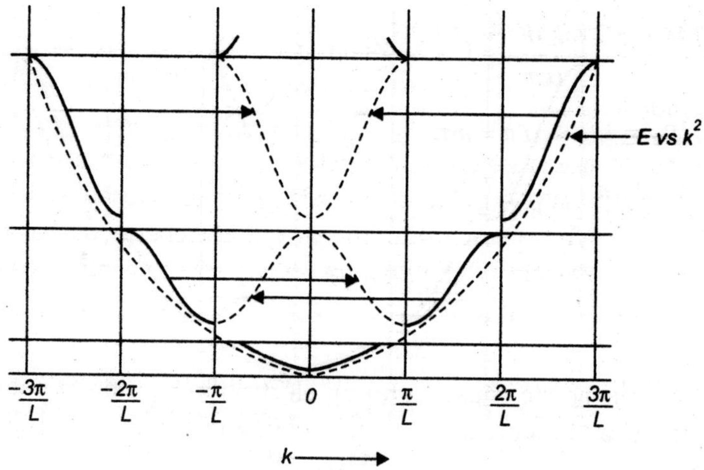

(a)

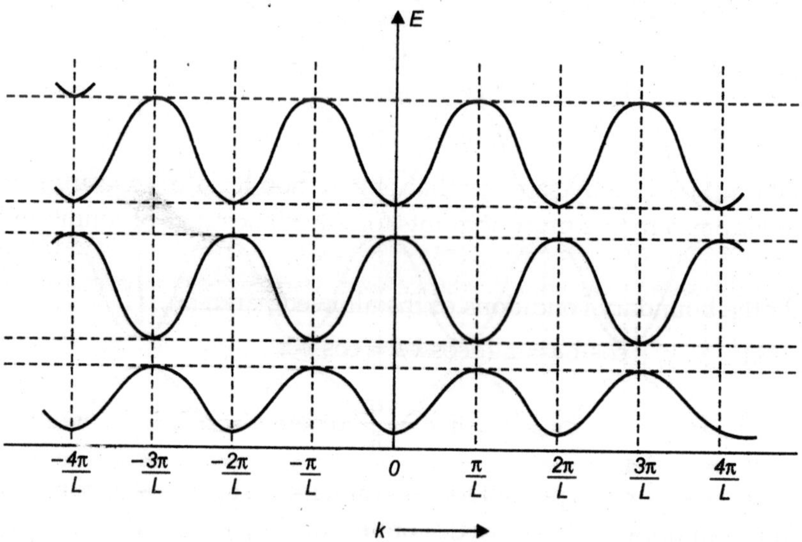

(b)

FIGURE 11.8 **(*a*) E-k curve for one-dimensioanl lattice structure from Kronig-Penney model for the normal and reduced zone cases. Also shown in the figure is the E - k curve for free electron. (*b*) plot of extended zone curves for the same lattice structure**

Electrons that occupy Brillouin zone can move freely inside a crystal. All the information for the Brillouin zones are contained in the first Brillouin zone if we use the reduced zone plot. So, the Brillouin zones are the allowed energy regions in *k*.

11.6 BAND THEORY OF SOLIDS

11.6.1 Energy Bands in Solids

When a large number of atoms are brought together to form a solid, a single energy level of an isolated atom is split into an energy band consisting of very closely spaced levels of slightly differing energy. It is known from quantum theory that the energy of the bound electron in an atom is quantised *i.e.*, the electrons can take on only discrete values of energy and we know that the electron is not localized at a given radius. Rather the electron location is obtained from a probability density function, which is the probability of finding the electron at a particular distance from the nucleus. Application of quantum mechanics and Schrödinger's wave equation is needed to find out the electron energy bands in a solid for assumed potential functions.

The probability density function for lowest energy state of a single isolated atom is shown in Figure 11.9 (*a*). When two atoms are brought close together as shown in Figure 11.9 (*b*), the wave functions of the electrons in the two atoms overlap and in the overlapping region the discrete single energy level splits into two levels [Figure 11.9 (*c*)] due to interacting potential so that each electron can occupy a distinct quantum state according to Pauli exclusion principle. In an assembly of N atoms, each of the energy level splits into N energy levels and forms an energy band of very closely spaced levels differing slightly in energy. For example, a solid containing 10^{19} atoms will have 10^{19} possible energy states. The maximum number of electrons that can be accommodated in these energy state is 2×10^{19} since only two electrons with opposite spin can occupy the same state in accordance with Pauli exclusion principle. If the width of the energy band be 2 eV, then the energy levels are separated by $\frac{2}{2} \times 10^{-19}$ *i.e.*, 10^{-19} eV. Since this spacing of energy level is extremely small, a quasi continuum level is formed.

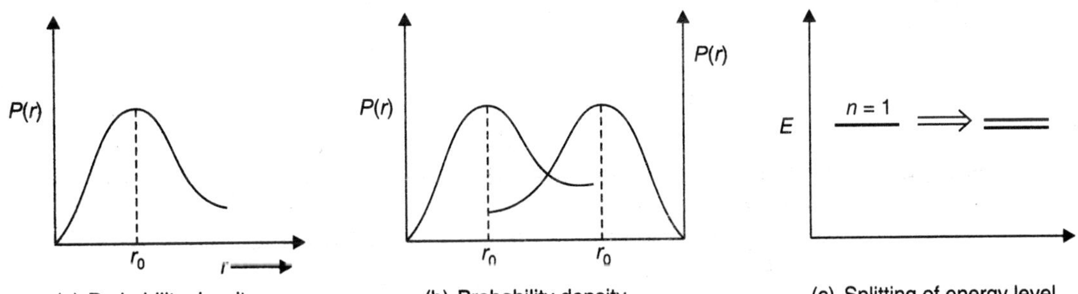

(a) Probability density function for lowest electron energy state of a single isolated atom.

(b) Probability density function of two atoms brought in close proximity to each other

(c) Splitting of energy level

FIGURE 11.9

Figure 11.10 shows the splitting of $n = 1$, $n = 2$, $n = 3$ i.e. K, L, M energy levels as the distance between different atoms is reduced. At first $n = 3$ (M-shell) energy level is affected and as the separation is reduced inner shells *i.e.*, $n = 2$ and $n = 1$ are gradually affected. Thus bands of allowed energies that the electrons may occupy are separated by bands of forbidden energy when the interatomic distance reaches the equilibrium value r_0. This is essence of energy-band theory of single crystal material.

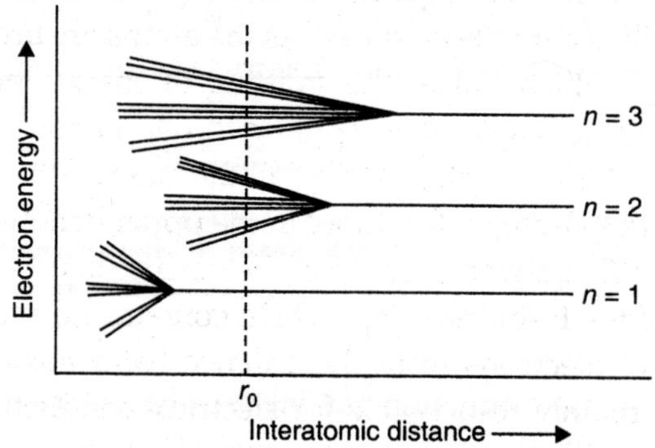

FIGURE 11.10 **Splitting of energy levels**

11.6.2 Valence and Conduction Bands

The energy band occupied by outermost electrons of an atom *i.e.*, valence electron is known as **valence band,** which is the highest occupied energy band. It may be completely filled or partially filled by electrons.

The next higher allowed band is called the **conduction band**, which is also may be completely empty or partially filled by electrons. In this band electrons are free to move.

The energy gap (E_g) between valence and conduction band is known as the **band gap**.

The splitting of energy band in a crystal like silicon is visualised in the Figure 11.11. Silicon has fourteen electrons in $1s^2$, $2s^2$, $2p^6$, $3s^2$, $3p^6$ configuration.

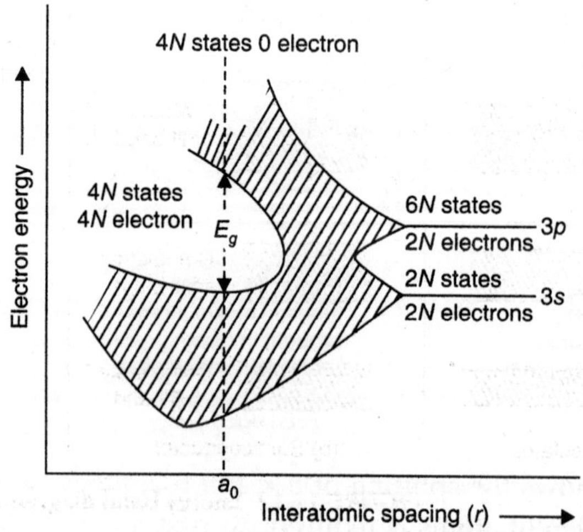

FIGURE 11.11 **Energy band of silicon**

Ten out of fourteen electrons occupy the innermost shells (K and L shells). The four remaining electrons in the outermost shell (M shell) which are called valence electrons. The 3 s states corresponds to $n = 3$ and $l = 0$, which contain 2 quantum states per atoms *i.e.*, 2 N states for N atoms. This state will contain 2 N electrons at $T = 0$ K. The 3 p state corresponds to $n = 3$ and $l = 1$, which contain 6 quantum states per atom *i.e.*, 6 N states and 2 N remaining valence electrons will occupy these states. As the atoms are brought closer, the 3s and 3p states interact and overlap, which causes the splitting of energy levels into bands. At equilibrium interatomic distance, the single band splits into two bands separated by the energy gap E_g. The lower band contains 4 N states. At 0 K, all the states in the lower band (valence band) will be occupied by valence electrons but those in the upper band will be empty (conduction band). The energy band gap E_g, between the top of the valence band and the bottom of the conduction band is the width of the forbidden gap, which contains no available states for the electrons to occupy. Excitation of electrons from the valence band through the forbidden gap into the conduction band is mainly responsible for electrical conduction.

11.6.3 Insulators, Semiconductors and Conductors

The distinction among insulators, semiconductors and conductors can be understood for their large variations in band structures and wide range of electrical conductivity.

A material having either completely full or empty energy band is an **insulator**. The conductivity of an insulator is very small as no free electron is available in the upper empty band for flow of current in this material. Between the full and empty band is a forbidden region whose width is so large that at physically realizable temperature electrons from the top of the highest filled band (valence band) can not be excited across the wide forbidden region to the bottom of the lowest empty band (conduction band). The width of the forbidden

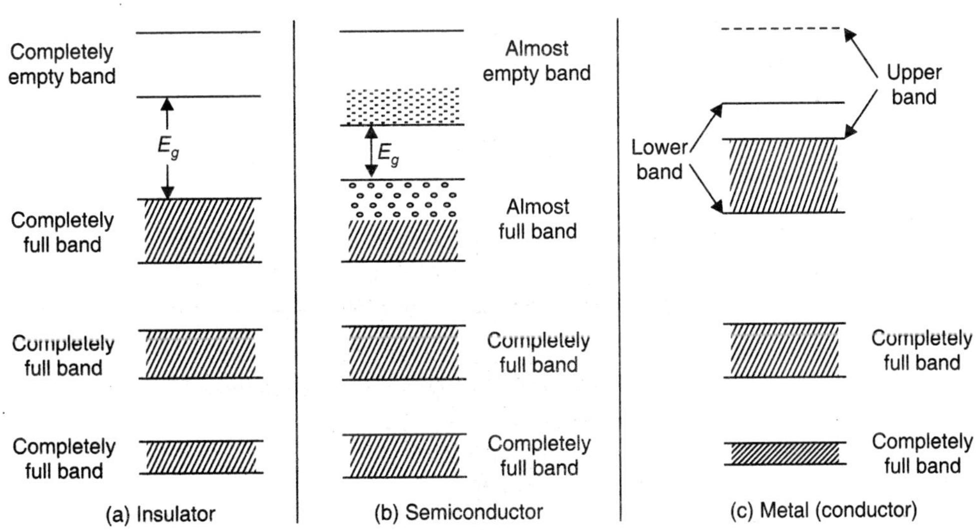

FIGURE 11.12 **Energy band diagrams**

region is called the *band gap energy* E_g whose value is of the order of 2.5 V to 6 V or larger for insulators. This situation is depicted in Figure 11.12 (*a*).

In *semiconductor*, the band gap energy (E_g) between full and empty band is small of the order of 1 eV. Electrons from states near the top of the lower full band will have an appreciable probability to be excited thermally across the small gap to states near the bottom of upper empty band. Thus a limited number of electrons will be available near the bottom of the upper empty band for conduction of electric current when an electric field is applied. The vacant states left behind by the thermally excited electrons will create positively charged empty states near the top of the lower full band (*valence band*).

At $T > 0$ K, valence electrons gains thermal energy to top into empty states alternatively filling one empty state and creating a new empty state. The movement of these positive charges are called *holes* in the valence band will also give rise to an electric current in a semiconductor. The electrical conductivity is much smaller than that of a metal owing to few number of free electrons and holes available in the energy bands. The conductivity is strongly temperature dependent and also a function of E_g. The semiconductor possesses a *negative temperature coefficient of resistance*, since the resistance decreases sharply with increase in temperature. Further the resistivity can be controlled over a wide range by means of doping *i.e.*, introducing proper impurity in the semiconductor.

The resistivity of semiconductor ranges for 10^{-2} to 10^6 Ω cm, that of metal is of the order of 10^{-6} Ω cm while insulators have resistivity greater than 10^{10} Ω cm.

The distinction between semiconductor and insulator is of degree only as all semiconductors are insulators at $T = 0$ K. Similarly all insulators are semiconductor at sufficiently high temperature, which is not realisable experimentally.

Figure 11.12 (*c*) shows the energy band diagram of metals in which a part of the upper region of valence band overlaps into a part of lower region of conduction band. Thus a large number of electrons and holes will be available for conduction. So these materials exhibits high thermal conductivity.

11.6.4 Intrinsic Semiconductor

An intrinsic semiconductor is a pure semiconductor. Silicon and germanium are the semiconductor material. To form a stable covalent bond, 8 electrons are required. The silicon atom at the centre has 4 valence electrons as shown in Figure 11.13 (*a*). It shares 4 electrons from the neighbour atoms to form the covalent bond. At absolute zero temperature, no energy is supplied to the crystal. All the electrons are engaged in forming the covalent bond and no free electrons are available. Hence there is no conduction.

Thus the semiconductor acts as insulator at 0 K. The conduction band is empty, as no conduction electrons are available. When thermal energy is supplied to the semiconductor, some of the covalent bonds are broken due to the energy supplied. These electrons jump from valence band (*VB*) to the conduction band (*CB*).

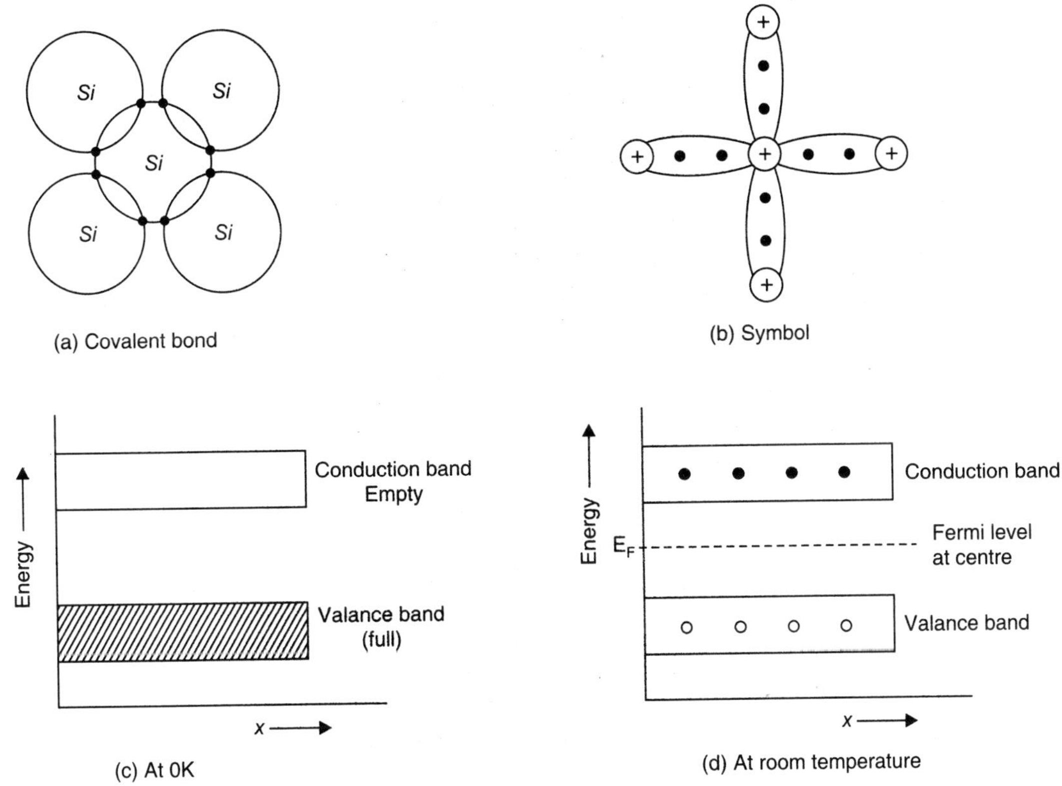

FIGURE 11.13 **Energy band diagram of intrinsic semiconductor**

Carrier Concentration and Fermi Level

We have assumed here that the width of the allowed energy band is comparable to the forbidden gap of width E_g. Conduction band electrons have energy lying between E_c and ∞, and, valence band electrons have energy lying between – ∞ to E_V. This is shown in Figure 11.14.

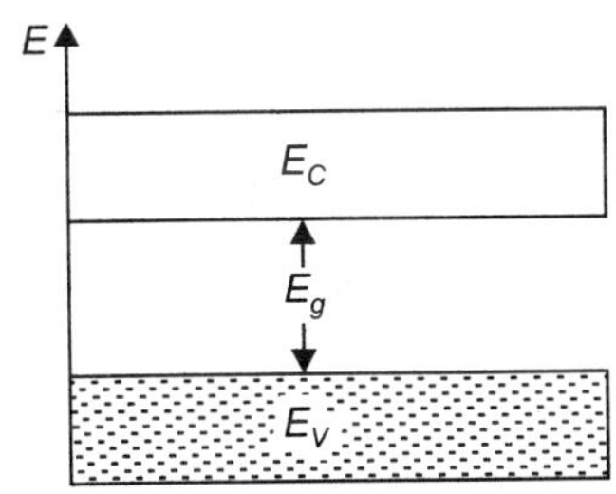

FIGURE 11.14 **Energy band diagram of an intrinsic semiconductor**

(*a*) **Electron Concentration in conduction band:** We know the total numbers of electrons per unit volume in conduction band is given by in terms of Fermi-Dirac distribution function $f(E)$ and density of state function $g(E)\,dE$ as

$$n_e = \int_{E_C}^{\text{top}} f(E)\, g(E)\, dE \qquad ...(1)$$

Here $f(E)$ decreases as we rise through the *CB* because $f(E) = 0$ for $E >> E_g$.

So we have set the upper limit for energy as infinity for *CB*.

Hence, equation (1) becomes

$$n_e = \int_{E_C}^{\infty} f(E)\, g(E)\, dE$$

Hence putting the values of $f(E)$ and $g(E)\, dE$, we have

$$n_e = \int_{E_C}^{\infty}\left[\frac{1}{e^{\{(E-E_F)/k_BT\}}+1}\right] \times \left[\frac{4\pi}{h^3}(2m_e^*)^{3/2}.(E-E_C)^{1/2}dE\right]$$

As $(E - E_F) >> k_BT$, we can ignore the unit term in the $f(E)$ and so

$$n_e = \frac{4\pi}{h^3}\,(2m_e^*)^{3/2} \int_{E_C}^{\infty} [(E-E_C)^{1/2} \times e^{\{(E_F-E)/K_BT\}}\, dE]$$

$$\Rightarrow \qquad n_e = \frac{4\pi}{h^3}(2m_e^*)^{3/2} \int_{E_C}^{\infty} [(E-E_C)^{1/2} \times e^{\{(E_F-E_C)/k_BT\}} \times e^{\{(E_C-E)/k_BT\}}\, dE] \qquad ...(2)$$

Let $y = (E - E_C)/k_B T$

$\Rightarrow \qquad dE = k_B T\, dy$ and $[E - E_C]^{1/2} = [y]^{1/2}\,[k_B T]^{1/2}$

Putting these into equation (2), we have

$$n_e = \frac{4\pi}{h^3}(2m_e^*)^{3/2}\,(k_B T)^{1/2} \times e^{\{(E_F-E_C)/k_BT\}} \int_0^{\infty} [(y)^{1/2} \times e^{-y} \times (k_BT)\, dy] \qquad ...(3)$$

We know $\int_0^{\infty} y^{1/2} \times e^{-y}\, dy = \left(\frac{\pi}{4}\right)^{1/2}$

Hence, from equation (3)

$$n_e = \frac{4\pi}{h^3}\,(2m_e^*)^{3/2}\,(k_B T)^{1/2} \times e^{\{(E_F-E_C)/k_BT\}} \times \left(\frac{\pi}{4}\right)^{1/2}$$

$$= 2\left(\frac{2\pi m_e^* k_BT}{h^2}\right)^{3/2} \times e^{\{(E_F-E_C)/k_BT\}} \qquad ...(4)$$

(*b*) **Hole Concentration in Valence Band (VB)**

Here the probability for a state of energy E to be unoccupied is given by $[1 - f(E)]$. Thus the total numbers of holes per unit volume is given by

$$n_h = \int_{bottom}^{E_V} [1 - f(E)]\, g(E)\, dE \qquad ...(5)$$

As earlier, we can change the lower limit as $-\infty$

Now $\qquad 1 - f(E) = 1 - \frac{1}{e^{\{(E-E_F)/k_BT\}}+1} = \frac{e^{\{(E-E_F)/k_BT\}}}{e^{\{(E-E_F)/k_BT\}}+1}$

However, for valence band, $E < E_F$, we have

$$e^{\{(E - E_F)/k_B T\}} << 1$$

Hence $$1 - f(E) \cong e^{\{(E - E_F)/k_B T\}} \qquad ...(6)$$

Thus, in valence band, $[1 - f(E)]$ decreases exponentially as we move downwards from top of the valence band, Now using equation (6) into equation (5), we have

$$n_h = \int_{-\infty}^{E_V} e^{\{(E - E_F)/k_B T\}} \times \left[\frac{4\pi}{h^3}(2m_h^*)^{3/2}(E_V - E)^{3/2}\, dE\right]$$

On solving as in above equation (3), we have

$$n_h = 2\left(\frac{2\pi m_h^* k_B T}{h^2}\right)^{3/2} \times e^{\{(E_V - E_F)/k_B T\}} \qquad ...(7)$$

(c) **Value of Fermi Energy**

In intrinsic semiconductor $n_e = n_h$

So equating equations (4) and (7), we have

$$2\left(\frac{2\pi m_e^* k_B T}{h^2}\right)^{3/2} \times e^{\{(E_F - E_C)/k_B T\}} = 2\left(\frac{2\pi m_h^* k_B T}{h^2}\right)^{3/2} \times e^{\{(E_V - E_F)/k_B T\}}$$

$$\Rightarrow \qquad (m_e^*)^{3/2} \times e^{\{(E_F - E_C)/k_B T\}} = (m_h^*)^{3/2} \times e^{\{(E_V - E_F)/k_B T\}}$$

$$\Rightarrow \qquad e^{\{(2E_F - E_C - E_V)/k_B T\}} = \left(\frac{m_h^*}{m_e^*}\right)^{3/2} \qquad ...(8)$$

$$\Rightarrow \qquad \left\{\frac{2E_F - E_C - E_V}{k_B T}\right\} = \frac{3}{2}\log\left(\frac{m_h^*}{m_e^*}\right)$$

$$\Rightarrow \qquad E_F = \frac{E_C + E_V}{2} + \frac{3}{4}\log\left(\frac{m_h^*}{m_e^*}\right) \qquad ...(9)$$

$$\left[\text{for } m_e^* = m_h^*, \log\frac{m_h^*}{m_e^*} = \log 1 = 0\right]$$

$$\Rightarrow \qquad \boxed{E_F = \frac{E_C + E_V}{2}} \qquad ...(10)$$

So Fermi level lies exactly in the middle of the top of valence band and the bottom of the conduction band.

Now putting the values of E_F into equations (4) and (7) separately, we have

$$n_e = 2\left(\frac{2\pi m_e^* k_B T}{h^2}\right)^{3/2} \times e^{\{(E_F - E_C)/k_B T\}} = 2\left(\frac{2\pi k_B T}{h^2}\right)^{3/2} \times \left[(m_e^*)^2\right]^{3/4} e^{\{(E_C + E_F)/2 - E_C/k_B T\}}$$

From equation (8),

$$n_e = 2\left(\frac{2\pi k_B T}{h^2}\right)^{3/2} \times [(m_e^*)\,(m_h^*)]^{3/4}\, e^{\{(E_V - E_C)/2k_B T\}}$$

$$= 2\left(\frac{2\pi k_B T}{h^2}\right)^{3/2} [m_e^*\, m_h^*]^{3/4}\, e^{\{-E_g/2k_B T\}} \qquad ...(11)$$

and

$$n_h = 2\left(\frac{2\pi k_B T}{h^2}\right)^{3/2} [m_e^*\, m_h^*]^{3/4}\, e^{\{-E_g/2k_B T\}} \qquad ...(12)$$

Hence $n_h = n_e = n_i$, the intrinsic carrier concentration of the semiconductor.

11.6.5 Extrinsic Semiconductor

A doped semiconductor is called an extrinsic semiconductor. These semiconductors are classified as *N*-type and *P*-type.

(*a*) **Negative type or N-type:** Covalent bond and energy band diagrams are shown in Figure 11.15. When the intrinsic semiconductor is doped with pentavalent impurity, negative type semiconductor is formed. The pentavalent impurities are antimony, arsenic and bismuth. The pentavalent atom at the centre has 5 valence electrons. This atom shares four electrons from the neighbour atoms. For the formation of stable covalent bond, only 8 electron need to rotate in the valence orbit. Thus one excess electron is produced by each impurity atom.

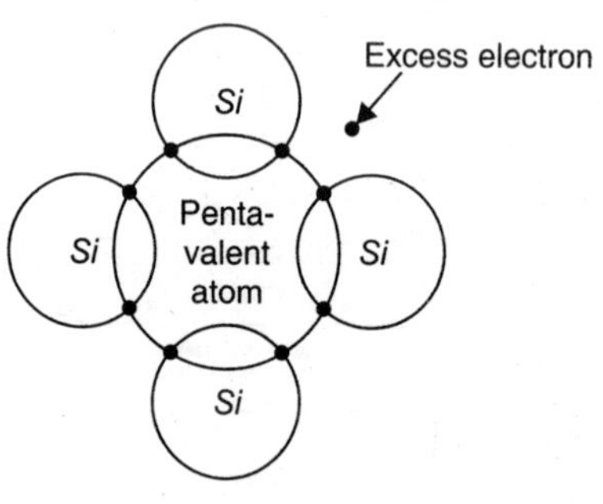

(a) Covalent bond

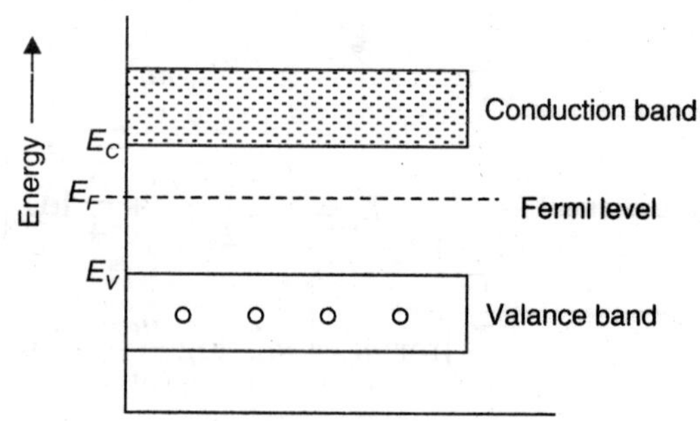

(b) Energy band diagram

FIGURE 11.15

Several impurity atoms donate several electrons. Since the impurity atoms donate electrons, they are known as donors. A few covalent bonds are broken at room temperature due to the thermal energy supplied. The vacancies are shown as holes in the valence band. The majority carriers are **electrons** and minority carriers are **holes**.

Fermi level corresponds to the centre of gravity of the electrons and holes. In the case of intrinsic semiconductor, the number of electrons are equal to number of holes. The Fermi level lies midway between the valance band and conduction band. In *N*-type semiconductor the Fermi level is lifted towards the conduction band as the conduction electrons are the majority carriers.

For *N*-type semiconductor a donor level E_d is formed below the Fermi level. At a temperature *T*, the density of conduction electron from equation (4) is given by

$$n_e = 2\left(\frac{2\pi m_e^* k_B T}{h^2}\right) e^{\{(E_F - E_C)/k_B T\}}$$

The density of vacancies in the donor levels of energy E_d is given by

$$N_d^+ = n_d[1 - f(E)] = n_d\left[1 - \frac{1}{e^{\{(E_d - E_F)/k_B T\}} + 1}\right]$$

$$= n_d\left[\frac{e^{\{(E_d - E_F)/k_B T\}}}{e^{\{(E_d - E_F)/k_B T\}} + 1}\right]$$

Here N_d^+ is concentration for ionised donor atoms *S*. Assuming the position a little above the donor level by a few '$k_B T$' the above expression can be approximated

$$N_d^+ = n_d\,[e^{\{(E_d - E_F)/k_B T\}}] \qquad \text{...(13)}$$

However, the electron concentration in the conduction band and the vacancy concentration in donor level should be equal. Hence equating equation (4) and (13), we have

$$n_e = 2\left(\frac{2\pi m_e^* k_B T}{h^2}\right)^{3/2} e^{\{(E_F - E_C)/k_B T\}} = n_d\,[e^{\{(E_d - E_F)/k_B T\}}]$$

or $$\log\left[2\left(\frac{2\pi m_e^* k_B T}{h^2}\right)^{3/2}\right] + \left\{\frac{E_F - E_C}{k_B T}\right\} = \log n_d + \left\{\frac{E_d - E_F}{k_B T}\right\}$$

$$\Rightarrow \quad \left\{\frac{2E_F - E_C - E_d}{k_B T}\right\} = \log\frac{n_d}{\left[2\left(\frac{2\pi m_e^* k_B T}{h^2}\right)^{3/2}\right]}$$

$$\Rightarrow \qquad E_F = \left\{\frac{E_C + E_d}{2}\right\} + \frac{k_B T}{2} \log \frac{n_d}{\left[2\left(\frac{2\pi m_e^* k_B T}{h^2}\right)^{3/2}\right]} \qquad ...(14)$$

At absolute zero temperature, $E_F = \left\{\frac{E_C + E_d}{2}\right\}$, this implies that the Fermi level lies at half way between donor level and the bottom of the conduction band.

With an increase in temperature, E_F decreases and at room temperature it comes below the donor level.

Substituting the value of E_F from equation (14) into equation (2) we have the free electron concentration in the conduction band as

$$n_e = 2\left(\frac{2\pi m_e^* k_B T}{h^2}\right)^{3/2} \exp.\left\{\frac{E_F - E_C}{k_B T}\right\}$$

$$= 2\left(\frac{2\pi m_e^* k_B T}{h^2}\right)^{3/2} \exp.\left[\frac{\left(\frac{E_C + E_d}{2}\right) + \frac{k_B T}{2}\log \frac{n_d}{\left[2\left(\frac{2\pi m_e^* k_B T}{h^2}\right)^{1/2}\right]} - E_C}{k_B T}\right]$$

$$= 2\left(\frac{2\pi m_e^* k_B T}{h^2}\right)^{3/2} \exp.\left[\left(\frac{E_d - E_C}{2k_B T}\right) + \frac{1}{2}\log \frac{n_d}{\left[2\left(\frac{2\pi m_e^* k_B T}{h^2}\right)^{3/2}\right]}\right]$$

$$= 2 \times \left(\frac{2\pi m_e^* k_B T}{h^2}\right)^{3/2} e^{\{(E_d - E_C)/2k_B T\}} \times \frac{n_d}{\left[2\left(\frac{2\pi m_e^* k_B T}{h^2}\right)^{3/2}\right]}$$

$$= (2\, n_d)^{1/2}\left(\frac{2\pi m_e^* k_B T}{h^2}\right)^{3/4} e^{\{(E_d - E_C)/2k_B T\}}$$

The ionization energy is given by $\Delta E = E_C - E_d$ of the donors, then, we have

$$n_e = (2\, n_d)^{1/2}\left(\frac{2\pi m_e^* k_B T}{h^2}\right) e^{\{-\Delta E/2k_B T\}} \qquad ...(15)$$

From equation (15), we say that *electron concentration* is proportional to the square root of the donor concentration of the semiconductor. With more increase E_F reaches up to the middle of the conduction band and valence band to make the material intrinsic.

(*b*) **Positive Type or P-Type:** When the intrinsic semiconductor is doped with trivalent atoms, positive type semiconductor is formed. The trivalent atoms are indium, gallium, boron and aluminium. The trivalent atom at the centre has 3 valence electrons. This atom shares 4 electrons from the neighbour atoms. 8 electrons are required to form the valence orbit. In other words the trivalent atom can accept one electron. This vacancy is known as a hole. The holes have positive charge. Millions of impurity atoms can accept million of electrons. Hence they are called as acceptors. The majority carriers are holes and the minority carriers are electrons. The covalent bond and energy band diagram are shown in Figure 11.16.

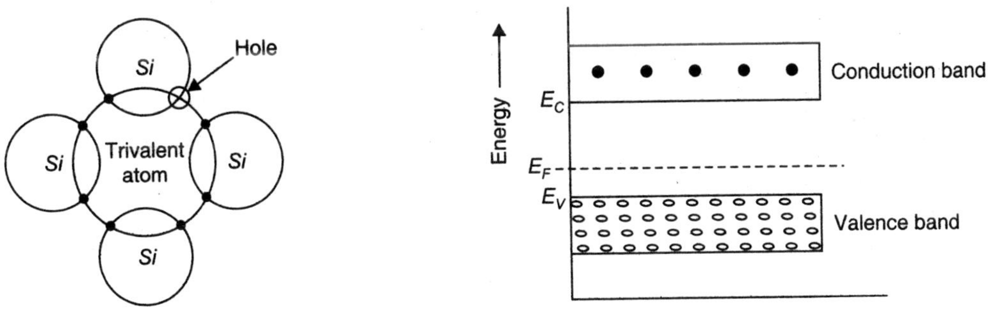

(a) Covalent bond (b) Energy band diagram

FIGURE 11.16

The electrons jumped from the valence band to the conduction band due to the thermal energy are represented in the conduction band. The valence orbit of each impurity atom has one hole. Thus the holes in the valence orbits of the impurity atoms are represented in the valence band. The Fermi level is shifted down as the majority carriers are the holes in the valence orbits.

For *P*-type semiconductors an acceptor level E_a is formed above the Fermi level and the valence band. At a temperature *T*, the density of holes is similar to equation (4) and is given by

$$n_h = 2\left(\frac{2\pi m_h^* k_B T}{h^2}\right)^{3/2} e^{\{(E_V - E_F)/k_B T\}} \quad ...(16)$$

The density of electrons in the acceptor levels of energy E_a is given by

$$N_a^- = n_a\,[f(E)] = n_a\left[\frac{1}{e^{\{(E_a - E_F)/k_B T\}} + 1}\right] \quad ...(17)$$

Here N_a^- is ionized acceptor atoms concentration. However, the hole concentration in the valence band and the electron concentration in the acceptor level should be equal.

Also, assuming $E_a - E_F >> a$ few $k_B T$, equating equations. (16) and (17), we have

$$2\left(\frac{2\pi m_h^* k_B T}{h^2}\right)^{3/2} e^{\{(E_V - E_F)/k_B T\}} = n_a\left[\frac{1}{e^{\{(E_a - E_F)/k_B T\}}}\right]$$

On simplification

$$E_F = \left\{\frac{E_a + E_V}{2}\right\} - \frac{k_B T}{2} \log \frac{n_a}{\left[2\left(\frac{2\pi m_h^* k_B T}{h^2}\right)^{3/2}\right]} \quad ...(18)$$

At absolute zero temperature

$$E_F = \left\{\frac{E_a + E_V}{2}\right\}$$

This implies that the Fermi level lies at half way between acceptor level and the top of the valence band.

With an increase in temperature, E_F increases and at room temperature it comes above the acceptor level slightly. Hence we have

$$\Rightarrow \qquad n_h = (2\, nd)^{1/2} \left(\frac{2\pi m_h^* k_B T}{h^2}\right) e^{(-\Delta E/k_B T)} \quad ...(19)$$

where $\Delta E = E_a - E_V$.

11.7 PHOTOCONDUCTIVITY

When light radiation falls on the crystal, electrical conductivity of the insulating crystal increases. This phenomenon is called photoconductivity.

When the energy of incident photon, is greater than the energy gap E_g, free electron-hole pairs are produced in the crystal by the absorption of the incident photons and these electrons and holes serves as the carriers of electrical current.

Impurities and imperfections in the crystal also contribute towards photoconductivity. If these are present in the crystal, then even the photons having energy below the threshold for the production of electron-hole pairs may be able to produce mobile electrons or holes. *Imperfections and impurities introduce discrete energy levels in the forbidden energy gap called traps.*

11.7.1 Variation of Photoconductivity with Illumination (Simple model of Photoconductor)

To study the variation of photoconductivity with illumination, consider a simple model of photoconductor.

Assumptions:

(*i*) When light radiations falls on the crystal, electron-hole pairs are produced uniformly throughout the crystal volume.

(*ii*) In the recombination process, electrons directly recombine with holes.

(*iii*) Mobility of holes is smaller than the mobility of electrons and therefore mobility of holes can be neglected.

Using these assumptions, rate of change of electron concentration can be written as

$$\frac{dn}{dt} = L - A\,n\,p$$

$$\Rightarrow \qquad \frac{dn}{dt} = L - A\,n^2 \quad [\text{Using } n = p] \qquad ...(1)$$

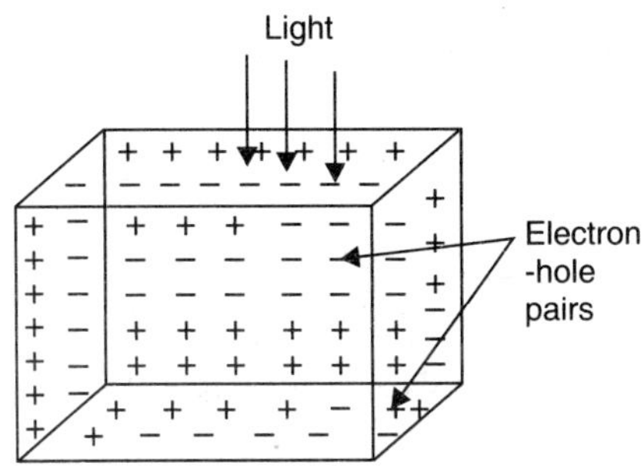

FIGURE 11.17 **A photo-conductor illuminated with light**

Here, L is the number of photons absorbed per unit volume per unit time, and $A\,n\,p$ is the recombination rate [we know that the recombination rate is proportional to the product "$n\,p$", A is the proportionality constant].

We know that drift velocity of the carriers, v, is proportional to applied electric field E, *i.e.*,

$$v \propto E \quad \Rightarrow \quad v = \mu E$$

Here μ is the constant of proportionality and is known as mobility of the medium. In steady state $\frac{dn}{dt} = 0$ and $n = n_0$.

Hence using equation (1)

$$L = A\,n_0^2$$

$$\Rightarrow \qquad n_0 = \sqrt{L/A} \quad \Rightarrow \quad n_0 \propto L^{0.5} \qquad ...(2)$$

and

$$\sigma = n_0\,e\,\mu = e\,\mu\,(L/A)^{0.5} \qquad ...(3)$$

Equation (3) predicts the variation of photoconductivity, σ, with light level L as $L^{0.5}$. However, practically the actual exponent ranges between 0.5 to 1.0 or more as shown in Figure 11.18.

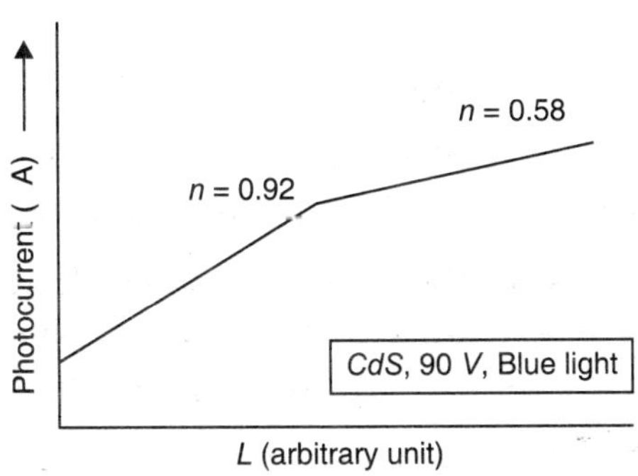

FIGURE 11.18 **Variation of photocurrent of cadmium sulphide with different intensity at 70 V bias voltage with blue light**

(*a*) **Decay :** The decay of photoelectrons after switching off the light suddenly is

$$\frac{dn}{dt} = 0 - A\,n^2 \qquad \text{...(4)}$$

$$\Rightarrow \qquad \int_{n_0}^{n} \frac{dn}{n^2} = -A \int_0^t dt$$

Solution of above equation is

$$\frac{1}{n} = At + \frac{1}{n_0} \qquad \text{...(5)}$$

where n_0 is the carrier concentration at $t = 0$ when light is just switched off.

Response time : t_0 : This is the time during which carrier concentration reduces to half of starting carrier concentration.

Hence at $t = t_0,\ n = n_0/2$

Using equation (5), $\dfrac{1}{n_0/2} = A\,t_0 + \dfrac{1}{n_0}$

$$\Rightarrow \qquad \frac{1}{n_0} = A\,t_0$$

$$\Rightarrow \qquad t_0 = \frac{1}{A\,n_0} = \frac{1}{A\sqrt{L/A}}$$

$$\Rightarrow \qquad t_0 = \sqrt{\frac{1}{LA}}$$

$$\Rightarrow \qquad t_0 = \frac{1}{L}\,\frac{L}{\sqrt{LA}} = \frac{1}{L}\sqrt{\frac{L}{A}} = \frac{n_0}{L}$$

or

$$t_0 = \frac{n_0}{L} \qquad \text{...(6)}$$

So

$$t_0 = \frac{1}{L}\left(\frac{\sigma}{e\mu}\right) = \frac{\sigma}{e\mu L} \qquad \text{...(7)}$$

$\therefore$ $t_0 \propto \sigma \Rightarrow$ Response time is directly proportional to σ at a given light intensity level.

Good photoconductors have large t_0. These are rarely found in practice.

(*b*) **Sensitivity or Gain Factor (G):** This can be calculated as the ratio of numbers of carriers crossing the unit area of specimen to the number of photons absorbed by that area in the specimen. It can be represented by '*G*'. So for a specimen of thickness, '*d*' and the cross-section area unity, we have

$$G = \frac{\text{Particle flux}}{L \times (d \times 1)} \qquad \text{...(8)}$$

We know flux is the rate of flow of carriers per unit area per unit time is given by 'F_n'. Also the current density

$$J_n = \sigma E = n_0 e \mu (V/d)$$

A potential V produces the particle flux F_n is given by

$$F_n = \frac{\mu V}{d} n_0$$

$$\Rightarrow \quad F_n = \frac{\mu V}{d} (L t_0) \quad \text{[from equation (6)]}$$

$$\Rightarrow \quad F_n = \frac{\mu V}{d} \times L \times \frac{1}{\sqrt{L A}}$$

Hence

$$F_n = \frac{\mu V}{d} \times L \times \frac{1}{\sqrt{L A}} \quad ...(9)$$

So from equation (8), $G = \frac{F_n}{L d}$

$$\Rightarrow \quad G = \frac{1}{L d} \times \frac{\mu V}{d} \times L \times \frac{1}{\sqrt{L A}}$$

$$\Rightarrow \quad G = \frac{\mu V}{d^2 (L A)^{1/2}} \quad ...(10)$$

Now, let T_e is the life time of the electron before illumination and it is nothing but the response time t_0, so

$$T_e = t_0 = \frac{1}{\sqrt{L A}}$$

Let T_d is the transit time of an electron between the electrodes and is given by

$$T_d = \frac{d}{V_d} = \frac{d}{\mu E}$$

$$\Rightarrow \quad T_d = \frac{d}{\mu V/d} = \frac{d^2}{\mu V}$$

So from equation (10)

$$\boxed{G = \frac{\mu V}{d^2 (L A)^{1/2}} = \frac{T_e}{T_d}} \quad ...(11)$$

The expression of G as given in equation (11) is quite general, however, this theoretical value of t_0 do not agree and in some instances the discrepancy is $\approx 10^8$ times. The result then indicates.

- Model is failure
- We should look out for some missing phenomenon not considered in this expression.

11.7.2 Effect of Traps on Photoconductivity

A trap is an impurity atom or imperfection in the crystal. Capable of capturing an electron or hole. In simple words, *it is an energy level, that can capture either electrons or holes.*

There are two types of traps.

(*i*) One type of traps helps electrons and holes to recombine and thereby assists in restoration of thermal equilibrium. These type of traps are called recombination centres. In the presence of traps, recombination proceeds at much higher rate.

(*ii*) Second type of traps does not contribute directly in recombination process, but restricts the freedom of motion of holes or electrons. This is shown in Figure 11.19.

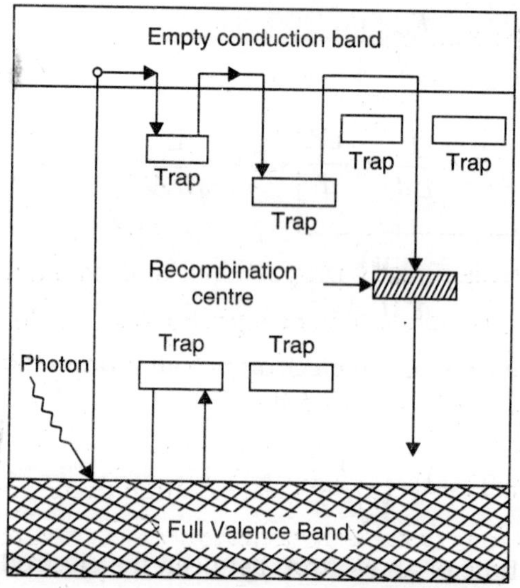

FIGURE 11.19 **Influence of traps in electron hole recombination**

To study the effect of second types of traps on photoconductivity, let us consider a crystal with N electron trap levels per unit volume. Rate of change of electron concentration 'n' is given by,

$$\frac{dn}{dt} = L - A\,n\,(n + N) + Bn_t \qquad ...(1)$$

where recombination coefficient A is same for both electron-hole recombination and electron trap capture. The term Bn_t represent the rate of thermal evaporation of trapped carrier back into conduction band and is neglected.

In steady state $$\frac{dn}{dt} = 0$$

Therefore from equation (1), we get

$$L - A\,n\,(n + N) = 0$$

or $$\frac{L}{A} = n_0\,(n_0\,N) \qquad ...(2)$$

Limiting Cases

(*i*) If illumination level L is low, n_0 will be small. So $n_0 << N$ and hence n_0 can be neglected in comparison to N. Therefore equation (2) becomes

$$\frac{L}{A} = n_0 N \quad \text{or} \quad n_0 = \frac{L}{AN} \qquad ...(3)$$

Therefore photocurrent density will be

$$J = \sigma\,E = n_0\,e\,\mu\,E$$

or $$\boxed{J = \frac{L}{AN}\,e\,\mu\,E} \qquad ...(4)$$

which shows that photocurrent is directly proportional to illumination level (L).

(*ii*) If illumination level L is high, n_0 will be high. So $n_0 >> N$. Hence equation (2) becomes

$$\frac{L}{A} = n_0^2 \text{ or } n_0 = \left(\frac{L}{A}\right)^{1/2} \qquad ...(5)$$

which is the same steady state current carrier concentration in the absence of traps.

When light incident on the crystal is switched off, $L = 0$. Therefore equation (1), we get

$$\frac{dn}{dt} = -\,A\,n\,(n + N)$$

or $$\frac{dn}{n(N + n)} = -\,A\,dt$$

Integrating both sides, we get

$$-\int \frac{dn}{n(n + N)} = A \int dt$$

$$\frac{1}{N}\log\frac{n+N}{n} = A\,t + C$$

At $t = 0, n = 0$

$$\therefore \quad \frac{1}{N}\log\frac{n_0+N}{n_0} = C$$

$$\therefore \quad \frac{1}{N}\log\frac{n+N}{n} = A\,t + \frac{1}{N}\log\frac{n_0+N}{n_0}$$

or
$$\frac{1}{N}\left[\log\frac{n+N}{n} - \log\frac{n_0+N}{n_0}\right] = A\,t$$

or
$$\log\frac{n+N}{n} - \log\frac{n_0+N}{n_0} = N\,A\,t$$

If number of traps per unit volume $N >> n$, then

$$\log\frac{N}{n} - \log\frac{N}{n_0} = N\,A\,t$$

or
$$\log\frac{N}{n}\times\frac{n_0}{N} = N\,A\,t \text{ or } \log\frac{n_0}{n} = N\,A\,t$$

or
$$\log\frac{n_0}{n} = -N\,A\,t \quad \text{or} \quad \boxed{n = n_0\,e^{-N\,A\,t}} \qquad ...(6)$$

At
$$t = t_0 = \frac{1}{NA} \qquad ...(7)$$

$$\therefore \quad \boxed{n = \frac{n_0}{e}}$$

Therefore response time in the absence of traps is $t_0 = \dfrac{1}{NA}$ in which carrier concentration reduces to $\dfrac{1}{e}$ of its steady state.

Equation (7) shows the response time is reduced by presence of traps.

Further
$$\sigma = n_0\,e\,\mu \qquad ...(8)$$

From equation (3) and (8), we get

$$\boxed{\sigma = \frac{L}{NA}\,e\,\mu}$$

which shows presence of traps reduced the photoconductivity.

11.7.3 Photoconduction in Semiconductors

The conductivity of a semiconductor increases when the radiation is made incident on it. This phenomenon is called *photoconductive effect*. This may be explained as follows:

The conductivity (σ) of a semiconductor increases with increase of concentration of charge carriers as is clear from the relation

$$\sigma = (n_e \mu_n + n_h \mu_h) e \qquad ...(1)$$

where n_e = free electron concentration per m^3

n_h = hole concentration per m^3

μ_n = free electron mobility

μ_h = hole mobility

e = elementary charge = 1.6×10^{-19} C.

When the radiant energy falls on a semiconductor, the covalent bonds get broken and electron-hole pair in excess of those generated thermally are created. These increased charge carriers decrease the resistance or increase the conductivity of the material and hence such a device is called a *photoresistor* or a *photoconductor*.

Figure 11.20 represents the energy level diagram of semiconductor having both acceptor and donor impurities (*i.e.*, of *PN* junction). If the photon of light is allowed to fall on the specimen, the following transitions are possible.

(*i*) An electron may jump from valence band to conduction band, thus creating an electron-hole pair. This process is called *intrinsic excitation.*

(*ii*) A donor electron may jump into conduction band or a valence electron may jump into an acceptor state. These two transitions are called the *impurity excitations.*

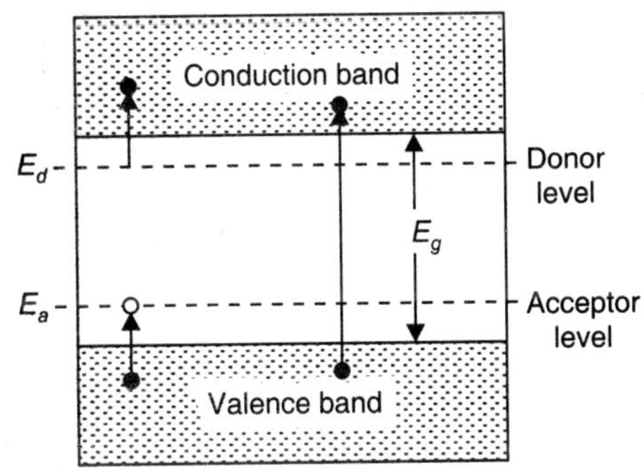

FIGURE 11.20 Energy level diagram of semiconductor

As the density of states in the conduction and valence band is much greater than the density of impurity states, therefore the *photoconductivity is mainly due to intrinsic excitation.*

Spectral Response: The minimum energy of a photon required for intrinsic excitation is the forbidden gap energy (E_g) of the semiconductor material. The long wavelength threshold of the material is defined as λ_c corresponding to the energy gap E_g and is given by

$$E_g = \frac{h_c}{\lambda_c} \; i.e., \lambda_c = \frac{h_c}{E_g}$$

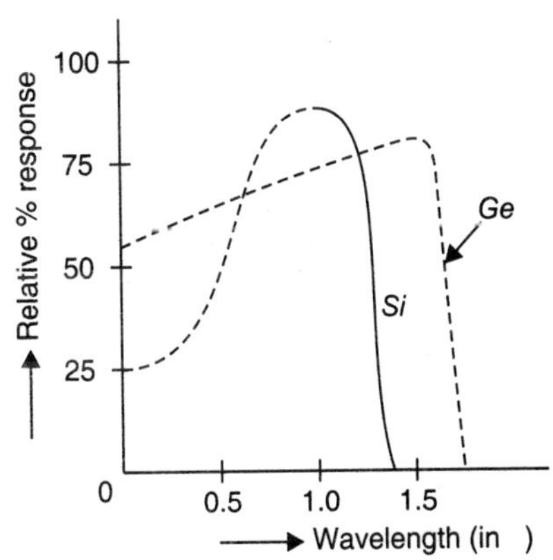

FIGURE 11.21 Spectral sensitivity curve

If E_g is expressed in eV and λ_c in μ, we get

$$\lambda_c = \frac{6.62 \times 10^{-34} \times 3 \times 10^{8}}{E_g \times 1.6 \times 10^{-19}}$$

$$= \frac{1.24}{E_g} \times 10^{-6}\,\text{m}$$

$$= \frac{1.24}{E_g}\,\mu\text{m}$$

For silicon E_g = 1.1 eV, therefore λ_c = 1.13 μm at room temperature. For germanium E_g = 0.72 eV, therefore λ_c = 1.73 μm at room temperature.

Figure 11.21 represents the spectral sensitivity curves for Si and Ge. It is noticed that the long wavelength limit is slightly greater than the calculated values of λ_c. This is due to impurity excitations. As the wavelength is decreased from value λ_c, the response increases, becomes maximum and again decreases.

The Photoelectric Current: If potential difference is applied across the semiconductor bar and light is allowed to fall on it, the charge carriers are generated and those move under the influence of applied potential difference. The charge carriers, which do not undergo recombination, reach the conducting contacts at the ends of the bar and hence constitute electric current. This current is given by

$$\boxed{i = \frac{e P_r \tau}{T_i}}$$

where P_r = rate of charge carriers produced by light

τ = average life time of newly created carriers

T_i = average transit time for carriers to reach the conducting contacts

e = charge on each carrier = 1.6×10^{-19} C.

11.7.4 Photoconductive Cells

The devices with the help of which light energy is converted to electrical energy are called *photoelectric cells*. Three types of photoelectric cells are: (*i*) photoemissive cells (*ii*) photo-voltaic cell and (*iii*) photoconductive cells.

Principle of Photoconductive Cells: It is based upon the principle that electrical resistance of semiconductors materials like selenium, lead sulphide etc. decreases when they are exposed to light radiations. The amount of this decrease in resistance (or increase in photoconductivity) depend on the light intensity and frequency of the incident light. So, if such a substance is inserted in the circuit and light is allowed to fall on it, its electric resistance (or photoconductivity) will change consequently, there will be a change of current in the circuit in steps accordingly to the variation of incident light intensity. The decrease in resistance (or increase in photoconductivity) of semiconductors on exposing to light may be explained as below:

If a photon striking the surface of such photosensitive material has energy E (= $h\,\nu$) greater than the energy gap between the valence and conduction bands of the material, sufficient energy will imparted to an electron to raise it to the conduction level. Consequently, a hole is left in the valence band. This electron-hole pair is free to serve as current carriers and hence, the conductivity of the material increases or the electrical resistance is reduced.

Construction and Working: The material generally used in photoconductive cells are cadmium sulphide (CdS), Cadmium selenide (CdSe) or lead sulphide (PbS). Commonly used material is CdS. The construction of one form of CdS cell and its circuit symbol is shown in Figure 11.22.

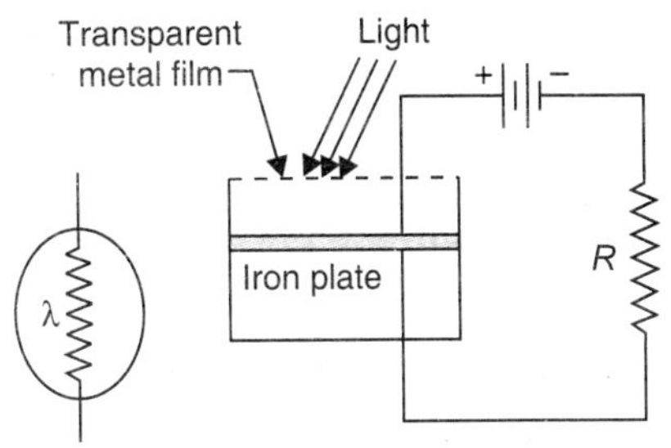

FIGURE 11.22 **CdS cell and its symbol**

In this cell, a thin film of CdS is deposited on one side of an iron plate and placed below a transparent film of metal. When light radiation of sufficient energy falls on transparent metal film, resistance of CdS layer get reduces and hence its conductance is increased. Consequently, a current starts flowing in the battery circuit connected between the iron plate and the transparent metal film. This external battery is included in the circuit to generate a direction and provide a path for the current flow. Voltages varying from a few volts to several hundred volts and applied depending upon the application of photoconducting cells.

Characteristics and Spectral Response: The following Figure 11.23 shows the illumination characteristics of a CdS cell. It depicts the relationship between illumination and resistance. It may be seen that when non-illuminated (*i.e.*, in absolute darkness), cells has a resistance in the ranges of 100 kΩ (very high) which is known as dark resistance. When illuminated with strong light, the cell resistance falls to only a few hundred ohms. So, the ratio of dark to light resistance of cell is about 1000 : 1.

The spectral response of CdS cell is shown in Figure 11.23 (*b*). It matches response of human eye. Like human eye, response is sensitive to visible.

Applications: As photocurrent increases linearity with light intensity and spectral response of CdS cell is similar to that the human eye, it is used for following purposes:

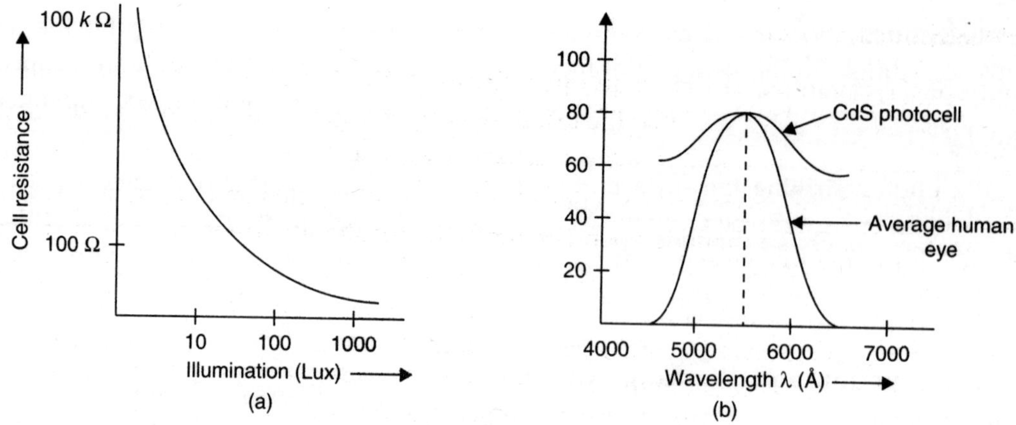

FIGURE 11.23

- To measure light intensity *i.e.*, to work as light meters
- To work as ON-OFF switch.
- In street light control
- In camera exposure setting
- In counting applications
- For relay control
- In burglar alarm
- As voltage regulator
- In aircraft and missile tracking system.

11.8 PHOTOVOLTAICS

Photovoltaics is the direct conversion of light into electricity at the atomic level. Some materials exhibit a property known as the photoelectric effect that causes them to absorb photons of light and release electrons. When these free electrons are captured, an electric current results that can be used as electricity.

The photoelectric effect was first noted by a French physicist, Edmund Becquerel, in 1839, who found that certain materials would produce small amounts of electric current when exposed to light. In 1905, Albert Einstein described the nature of light and the photoelectric effect on which photovoltaic technology is based, for which he later won a Nobel prize in physics. The first photovoltaic module was built by Bell Laboratories in 1954. It was billed as a solar battery and was mostly just a curiosity as it was too expensive to gain widespread use. In 1960s, the space industry began to make the first serious use of the technology to provide power abroad spacecraft. Through the space programs, the technology advanced, its reliability was established, and the cost began to decline. During the energy crisis in the 1970s, photovoltaic technology gained recognition as a source of power for non-space applications.

11.8.1 Photovoltaic Process

The height of the potential barrier in an open circuit, dark (non-illuminated) *P-N* junction always adjusts itself in such a way that the resultant current is zero and the electric field at the junction is in such a direction that it repels the majority carriers.

When light falls on diode surface, minority carriers get injected and hence the minority current increases. Since the diode is open circuited, the resultant current must remain zero. Therefore, majority current must increase by the same amount as the minority carrier current. This increase in majority current is possible only when the retarding electric field at the junction is reduced resulting in the lowering of the barrier height. Therefore, across the diode terminals, a voltage appears, which is equal to the decrease in the barrier potential. This constitutes the *photovoltaic e.m.f.* and is of the order of 0.1 V for Ge cell and 0.5 V for Si cell.

Expression for Photovoltaic e.m.f.: The photovoltaic e.m.f. V_p appears across the diode when the net current I in the diode is zero. Substituting $I = 0$ in the volt-ampere characteristics of a photodiode given by

$$I = I_s + I_0\,(1 - e^{Ve/\eta k_B T})$$

We get $$I_s + I_0 = (1 - e^{eV_p/\eta k_B T}) = 0$$ [replacing V by V_p]

or $$1 + \frac{I_s}{I_0} = e^{eV_p/\eta k_B T}$$

or $$\log\left(1 + \frac{I_s}{I_0}\right) = \frac{eV_p}{\eta k_B T}$$

Therefore, photovoltaic e.m.f. is given by

$$V_p = \frac{\eta k_B T}{e}\log\left(1 + \frac{I_s}{I_0}\right)$$

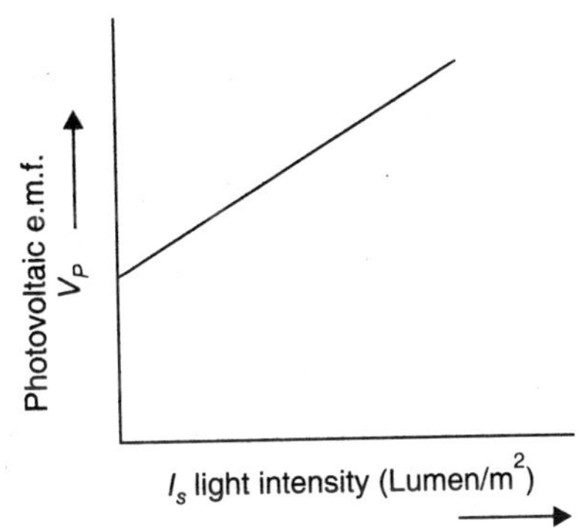

FIGURE 11.24 **V_p vs. I_s**

Since $\frac{I_s}{I_0} >> 1$, except for extremely small light intensities.

$$\therefore \qquad V_p = \frac{\eta k_B T}{e}\log\frac{I_s}{I_0}$$

The above equation shows that the photovoltaic e.m.f. V_p increases logarithmically with I_s and thus with illumination. Figure 11.24 shows graph for V_p versus light intensity.

11.8.2 Photovoltaic (PV) Cell/Solar Cells

Figure 11.25 illustrates the operation of a basic photovoltaic cell, also called a solar cell. Solar cells are made of the same kinds of the semiconductor materials such as silicon, used in the microelectronics industry. For solar cells, a thin semiconductor wafer is specially treated to form an electric field, positive on one side and negative on the other. When light energy strikes the solar cell, electrons are knocked loose from the atoms in the semiconductor material. If electrical conductors are attached to the positive and negative sides, forming an electrical circuit, the electrons can be captured in the form of an electric current *i.e.*, electricity. This electricity can then be used to power a load, such as a light or a tool.

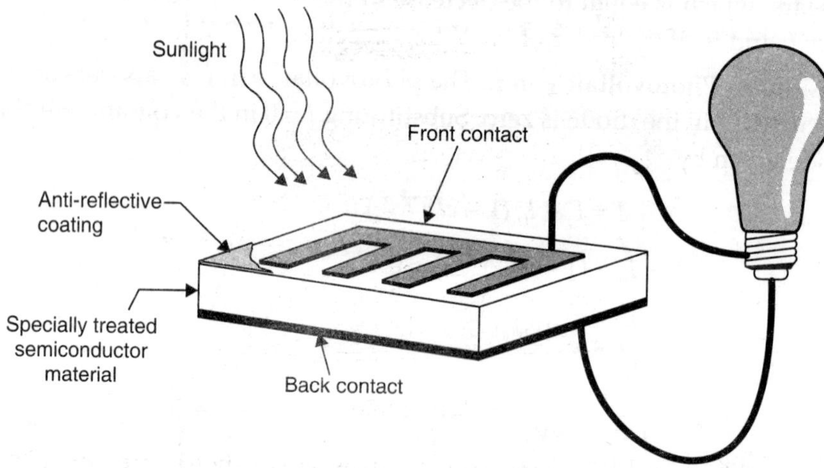

FIGURE 11.25 **Solar cell**

A number of solar cells electrically connected to each other and mounted in a support structure or frame is called *photovoltaic module*. Modules are designed to supply electricity at a certain voltage, such as a common 12V system. The current produced is directly dependent on how much light strikes the module.

Multiple modules can be wired together to form an array (Figure 11.26). In general, the larger the area of a module or array, the more electricity that will be produced. Photovoltaic modules and arrays produce direct current (dc) electricity. They can be connected in both series and parallel electrical arrangements to produce any required voltage and current combination. Today's most common *PV* devices use a single junction, or interface, to create an electric field within a semiconductor such as a *PV* cell. In a single junction PV cell, only photons, whose energy is equal to or greater than the band gap of the cell material can free an electron for an electric circuit. In other words, the photovoltaic response of single junction cells is limited to the portion of the Sun's spectrum, whose energy is above the band gap of the absorbing material, and lower-energy photons are not used.

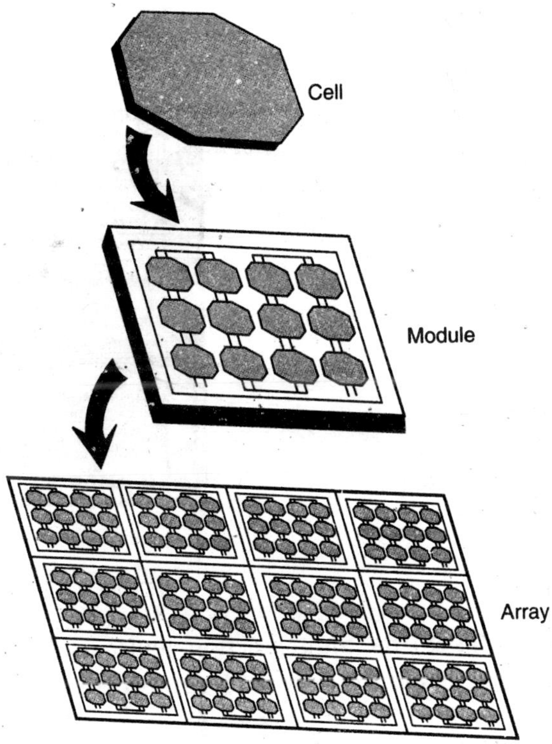

FIGURE 11.26

One way to get around this limitation is to use two or more different cells, with more than one band gap and more than one junction, to generate a voltage. These are referred to as "multi junctions" cells (also called "cascade" or "tandem" cells). Multi junction devices can achieve a higher total conversion efficiency because they can convert more of the energy spectrum of light to electricity.

As shown in Figure 11.27, a multi junction device is a stack of individual single-junction cells in descending order of band gap (E_g). The top cell captures the high energy photons and passes the rest of the photons on to be absorbed by lower band gap cells. Much of today's research in multi junction cells focuses on *gallium arsenide* as one (or all) of the component cells. Such cells have reached efficiencies of around 35% under concentrated sunlight. Other materials studied for multi junction devices have been *amorphous silicon* and copper indium diselenide.

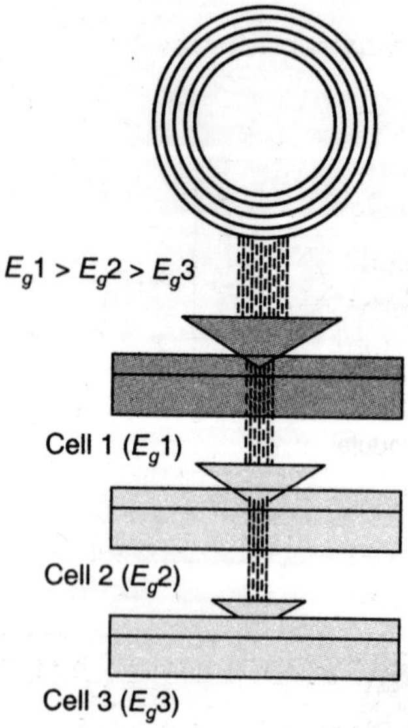

FIGURE 11.27

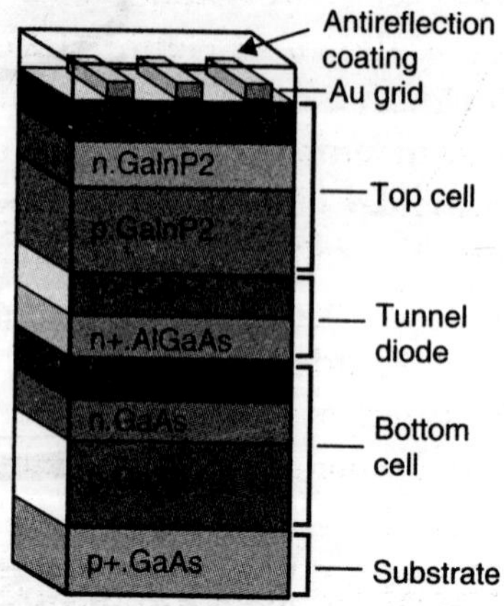

FIGURE 11.28

As an example (Figure 11.28), the multi junction device uses a top cell of gallium indium phosphide, "a tunnel junction" to aid the flow of electrons between the cells and a bottom cell of GaAs.

11.9 SUPERCONDUCTIVITY

Superconductivity is one of most exciting phenomena in Physics, both because of the vary nature of the phenomenon and also of its prospective applications of immense potential. It was the Dutch physicist Heike Kamerlingh Onnes, who discovered phenomenon of superconductivity. He was a professor of physics at Leiden in Western Netherlands. He made cryogenic laboratory of the Leiden University to become the cold research centre of the world of his time. His interest was in liquefying gases, the attempt which he succeeded in 1908. As an extension of his low temperature work, while studying the temperature dependance of resistance of metals, he made starting discovery that, the electrical resistance in certain metals such as lead and mercury is completely wiped out without any trace at low temperatures, very near to absolute zero in 1911. In 1913, he received the Nobel prize for his work.

Further work on the nature of superconductivity by W. Hans Meissner and Robert Ochsenfeld in 1933, revealed that, when a superconductor enters into superconducting phase, the magnetic flux lines are expelled from its interior. On the other hand, a sufficiently strong field brings the superconducting material to the normal phase and then flux lines can freely pass through the body.

A reasonably successful theory for explaining the phenomenon was provided by John Bardeen, Leon N. Cooper and Robert Schrieffer in 1957, and the *trio* was awarded the Nobel prize in 1972. (John Bardean picked the Nobel prize for the second time in Physics-first time with Brattain and Shockley in 1956 for inventing transistor).

11.9.1 The Temperature Dependence of Resistance of a Superconductor

The dependence of resistance of a superconductor on temperature is shown in Figure 11.29. The resistance of a superconductor in the non-superconducting state decreases with decrease in temperature as in the case of normal metal, and at a particular temperature T_c, the resistance abruptly drops to zero. T_c is called the *critical temperature* and signifies the transition from normal state to the superconducting state of the material under study. The critical temperature is different for different superconductors. Mercury (Hg) loses its resistance completely and turns into a superconductor at 4.2 K.

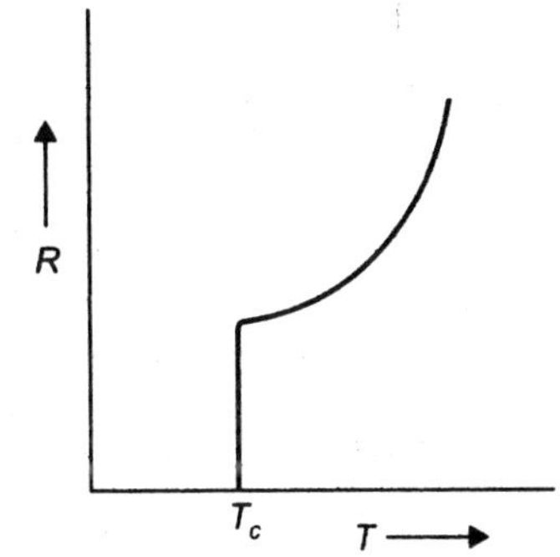

FIGURE 11.29 **R vs. T**

DEFINITION OF SUPERCONDUCTIVITY AND CRITICAL TEMPERATURE

The resistance offered by certain metals, alloys, and chemical compounds to the flow of electric current abruptly drops to zero below a threshold temperature. The phenomenon is called superconductivity and threshold temperature is called critical temperature.

11.9.2 Critical Magnetic Field: (Effect of Magnetic Field)

The existence of superconducting state of a metal in a magnetic field depends on the strength of the field in which it is placed and also on temperature. The condition for superconducting state to exist in metal is that some combination of temperature and field strength should be less than a critical value. *Superconductivity* is destroyed and normal conductivity is restored if the temperature of the specimen is raised above the critical value T_c or if a magnetic field greater than a critical value H_c is applied to the specimen. The minimum value of applied magnetic field when the material loses its superconductivity is called *critical field*. The variation of critical magnetic field with temperature can be represented by

$$\boxed{H_c = H_c(0)\left[1-\left(\frac{T}{T_c}\right)^2\right]}$$

where H_c = critical field strength at temperature T

$H_c(0)$ = Maximum field occurring at absolute zero

T_c = Critical temperature

The above relation defines a curve which divides the normal region in the magnetic field versus temperature diagram for a typical superconductor as shown in Figure 11.30.

$$\mathbf{B} = \mathbf{H} + 4\pi\mathbf{M} = 0 \quad ...(1)$$

$$\therefore \quad \chi = \frac{\mathbf{M}}{\mathbf{H}} = -\frac{1}{4\pi} \quad ...(2)$$

so that the magnetic susceptibility has a negative value. Normally, diamagnetic materials have susceptibilities less than one. Thus the superconductor is a perfect or ideal diamagnetic substance. This effect is known as Meissner effect.

Zero resistivity and perfect diamagnetism are the two independent essential properties of the superconducting state. This fact can be accounted for as follows:

From Ohm's law $\quad \mathbf{E} = \rho\,\mathbf{J}$ ...(3)

where **J** is held constant and ρ the resistivity goes to zero in superconductor, **E** must be zero, From Maxwell's equation

$$\text{Curl } \mathbf{E} = \nabla \times \mathbf{E} = -\frac{1}{C}\frac{\partial \mathbf{B}}{\partial t} = 0 \quad ...(4)$$

where **B** is the magnetic induction inside the material at time t. For a superconductor using equation (3) in equation (4) gives **B** as constant. However, the Meissner effect shows the other way. Thus the conditions defining the superconducting state are

$$\left.\begin{array}{lll} & \mathbf{E} = 0 & \text{from the absence of resistivity} \\ \text{and} & \mathbf{B} = 0 & \text{from Meissner effect} \end{array}\right\} \quad ...(5)$$

which are the two independent properties of superconducting state.

The Meissner effect can be understood using Figures 11.32 (*a*) and 11.32 (*b*). When a material is kept in magnetic field at $T > T_c$, it is magnetised [Fig. 11.32 (*a*)]. However, when, the temperature of the material is reduced suddenly to $T < T_c$, the flux is pushed out from the material [Figure 11.32 (*b*)].

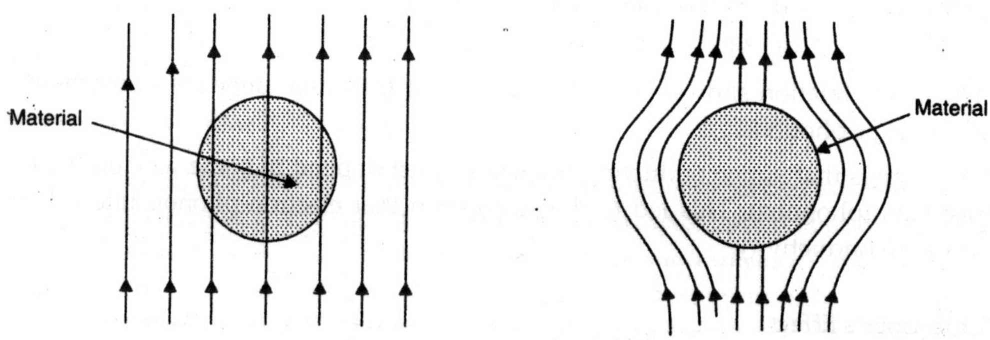

(a) Material cooled in field at $T > T_c$ (b) Magnetic field lines pushed out at $T < T_c$

FIGURE 11.32 **Meissner's effect**

11.9.1 The Temperature Dependence of Resistance of a Superconductor

The dependence of resistance of a superconductor on temperature is shown in Figure 11.29. The resistance of a superconductor in the non-superconducting state decreases with decrease in temperature as in the case of normal metal, and at a particular temperature T_c, the resistance abruptly drops to zero. T_c is called the *critical temperature* and signifies the transition from normal state to the superconducting state of the material under study. The critical temperature is different for different superconductors. Mercury (Hg) loses its resistance completely and turns into a superconductor at 4.2 K.

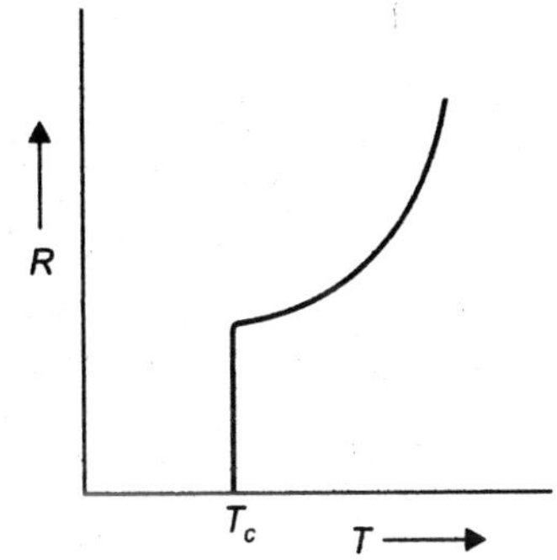

FIGURE 11.29 **R vs. T**

DEFINITION OF SUPERCONDUCTIVITY AND CRITICAL TEMPERATURE

The resistance offered by certain metals, alloys, and chemical compounds to the flow of electric current abruptly drops to zero below a threshold temperature. The phenomenon is called superconductivity and threshold temperature is called critical temperature.

11.9.2 Critical Magnetic Field: (Effect of Magnetic Field)

The existence of superconducting state of a metal in a magnetic field depends on the strength of the field in which it is placed and also on temperature. The condition for superconducting state to exist in metal is that some combination of temperature and field strength should be less than a critical value. *Superconductivity* is destroyed and normal conductivity is restored if the temperature of the specimen is raised above the critical value T_c or if a magnetic field greater than a critical value H_c is applied to the specimen. The minimum value of applied magnetic field when the material loses its superconductivity is called *critical field*. The variation of critical magnetic field with temperature can be represented by

$$H_c = H_c(0)\left[1-\left(\frac{T}{T_c}\right)^2\right]$$

where H_c = critical field strength at temperature T

$H_c(0)$ = Maximum field occurring at absolute zero

T_c = Critical temperature

The above relation defines a curve which divides the normal region in the magnetic field versus temperature diagram for a typical superconductor as shown in Figure 11.30.

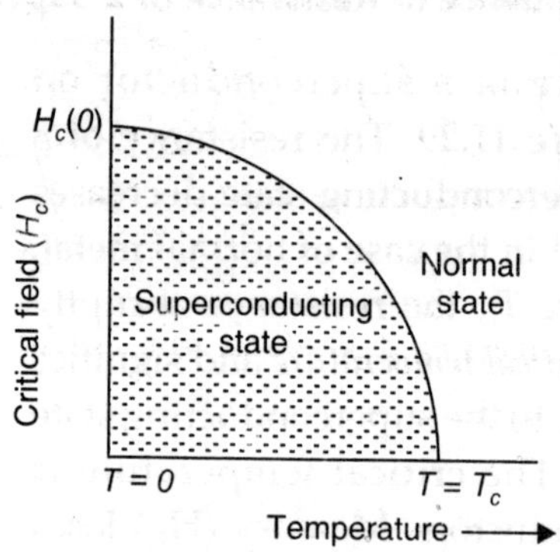

FIGURE 11.30 **Variation of critical field with temperature**

11.9.3 Critical Current Density

Magnetic field may arise due to current flowing in superconductor. The maximum current that can be passed through superconducting wire without destroying its superconducting state is known as *critical current*, I_c. This effect was observed in 1916 by Silsbee. If a superconducting wire of radius 'r' carries a current I, then induced magnetic field will be $H_1 = \frac{1}{2\pi r}$. Let H_c is the critical magnetic field associated with critical current I_c. Hence when H_I exceeds H_c, the specimen becomes normal. Hence critical current can be expressed as

$$\boxed{I_c = 2\,\pi\, r\, H_c}$$

So, I_c is maximum current that can pass through superconducting wire without disturbing superconducting state.

11.9.4 Persistent Current

If an electric current is set up in a perfect superconductor, it can persist for a very long time without any applied e.m.f. A current can be induced in ring of superconducting material by cooling it in magnetic field below a transition temperature and then switching OFF the field. When the field is switched OFF, the magnetic field outside the ring disappears but the flux inside the entire ring is trapped. The collapse of the magnetic field outside the ring induces a large persistent current inside the ring itself and maintains the trapped flux. Once the current starts even when the field is switched OFF, it persists for an indefinite period (Figure 11.31).

Once the current is set up, it persists for more than 10^5 years. Persistent current is one of most important properties of a superconductor.

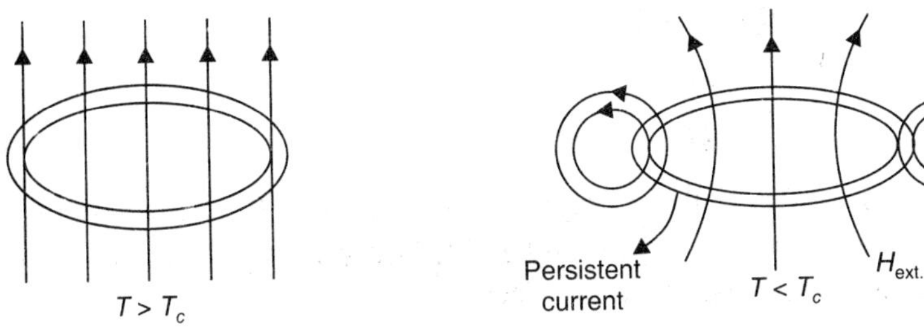

FIGURE 11.31 **Persistent current**

11.9.5 Isotope Effect

It has been observed that the critical temperature (T_c) of superconductors varies with the ionic or isotropic mass. The relation valid for some simple metals is given by

$$T_c \propto M^{-\alpha}$$

or

$$\boxed{M^{\alpha} T_c = \text{constant}}$$

where M is the atomic mass of isotope and α is roughly 0.5. Obviously, the existence of the isotope effect indicated that although superconductivity is an electronic phenomenon, it nevertheless depends in an important way on the vibrations of crystal lattice in which the electron move. Fortunately, not until after the development of the BCS theory was it discovered that the situation is more complicated than it had appeared. For some superconductors the exponent of M, *i.e.*, α is not $-1/2$, but near zero, *e.g.*, Ru and Zr and for at least one it is positive.

11.9.6 Debye Temperature

Debye temperature θ_D of phonon spectrum is given by

$$\theta_D M^{1/2} = \text{constant}, \; T_c \propto \theta_D \propto M^{1/2}$$

The above relation shows that superconducting transition depends upon the mass of the lattice ions or phonons.

This shows that superconducting transition must depend in some way on the mass of the lattice ions (phonons). This led to the suggestion that electron-phonon interaction be a cause of superconductivity.

11.9.7 Meissner's Effect

The second hallmark of superconductors, other than its zero resistance, was perfect diamagnetism, which was discovered by Meissner and Ochsenfeld in 1933. They found the magnetic field is pushed out of the superconducting material no matter how this field is created by an external magnet or by passing a current through the material itself. This means that the magnetic induction **B** inside the superconductor is always zero as long as it is in the superconducting state *i.e.*,

$$\mathbf{B} = \mathbf{H} + 4\,\pi\,\mathbf{M} = 0 \quad ...(1)$$

$$\therefore \quad \chi = \frac{\mathbf{M}}{\mathbf{H}} = -\frac{1}{4\pi} \quad ...(2)$$

so that the magnetic susceptibility has a negative value. Normally, diamagnetic materials have susceptibilities less than one. Thus the superconductor is a perfect or ideal diamagnetic substance. This effect is known as Meissner effect.

Zero resistivity and perfect diamagnetism are the two independent essential properties of the superconducting state. This fact can be accounted for as follows:

From Ohm's law $\quad \mathbf{E} = \rho\,\mathbf{J} \quad ...(3)$

where **J** is held constant and ρ the resistivity goes to zero in superconductor, **E** must be zero, From Maxwell's equation

$$\text{Curl } \mathbf{E} = \nabla \times \mathbf{E} = -\frac{1}{C}\frac{\partial \mathbf{B}}{\partial t} = 0 \quad ...(4)$$

where **B** is the magnetic induction inside the material at time t. For a superconductor using equation (3) in equation (4) gives **B** as constant. However, the Meissner effect shows the other way. Thus the conditions defining the superconducting state are

$$\left.\begin{array}{ll} \mathbf{E} = 0 & \text{from the absence of resistivity} \\ \text{and} \quad \mathbf{B} = 0 & \text{from Meissner effect} \end{array}\right\} \quad ...(5)$$

which are the two independent properties of superconducting state.

The Meissner effect can be understood using Figures 11.32 (*a*) and 11.32 (*b*). When a material is kept in magnetic field at $T > T_c$, it is magnetised [Fig. 11.32 (*a*)]. However, when, the temperature of the material is reduced suddenly to $T < T_c$, the flux is pushed out from the material [Figure 11.32 (*b*)].

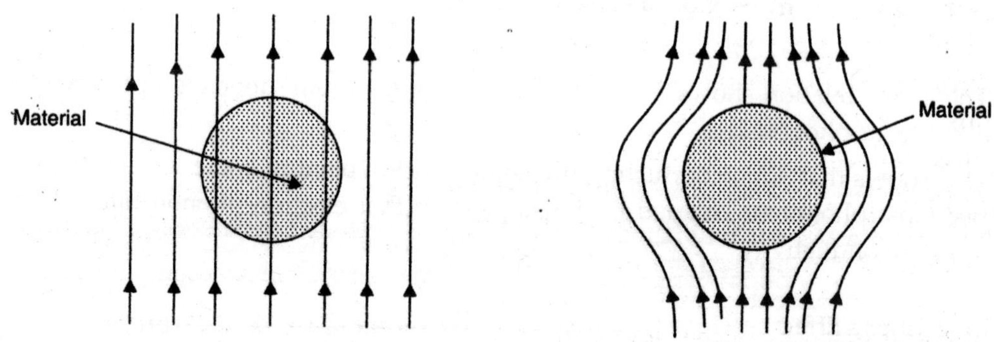

(a) Material cooled in field at $T > T_c$ (b) Magnetic field lines pushed out at $T < T_c$

FIGURE 11.32 **Meissner's effect**

Experimental Verification of Meissner's Effect. Consider a superconducting material above its critical temperature. A primary coil and secondary coil, are wound on the material (Figure 11.33). The primary coil is connected to a battery and a key *K*. The secondary coil is connected to a ballistic galvanometer (B.G.). When the key *K* is pressed, the primary circuit is closed and a current flows through the primary coil which sets up a magnetic field in it. The magnetic flux immediately links with the secondary coil, and hence a momentary current is driven through the B.G., which shows deflection. Since the primary current is steady, the magnetic flux will also become steady, and the flux linkage with the secondary become uncharging. When there is no further change in flux linkage in the secondary coil, the current will no more be driven in the secondary circuit.

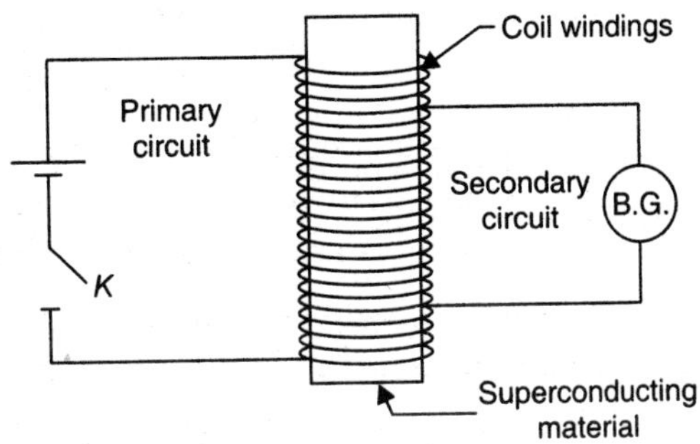

FIGURE 11.33 **Meissner's effect (Experiment)**

Now the temperature of the superconductor is decreased gradually. As soon as the temperature crosses below the critical temperature, the B.G. suddenly shows deflection, indicating that the flux linkage with the secondary coil has changed. The change in flux linkage is attributed to the expulsion of the magnetic flux from the body of the superconducting material as shown in Figure 11.32 (*b*).

11.9.8 Type I and Type II Superconductors

Superconducting materials are classified as either soft (Type I) or hard (Type II).

Type I Superconductor: A superconductor, which exhibits complete Meissner's effect is called *Type I superconductor*. Type I materials are often pure metals and have low values of critical magnetic field and critical current. Below T_c, the material is perfectly diamagnetic. Figure 11.34 shows the graph of Type I superconductor. If the magnetic field produced exceeds H_c, the material becomes normal and current cannot be sustained by itself. This was noted by Silsbee and is known as Silsbee effect.

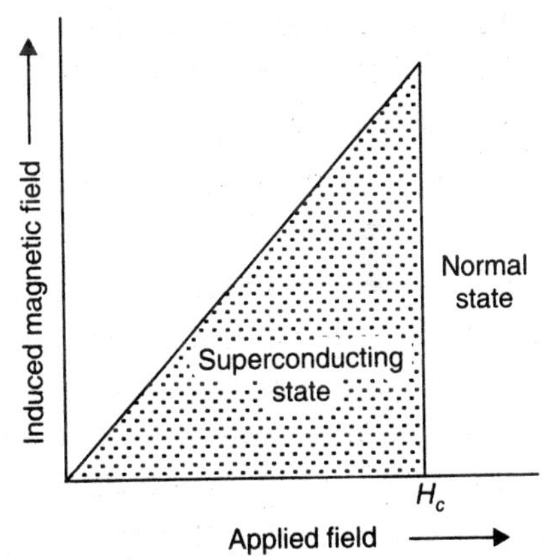

FIGURE 11.34 **Type I Superconductor**

Type II Superconductors: Type II superconductors are generally alloys or compounds, such as high temperature superconducting oxides like YBCO or BSSCO, which are able to carry very high current density in high transverse magnetic field without becoming normal. At low magnetic fields, Type II materials too will exhibit perfect diamagnetism. Figure 11.35 shows the graph for Type II superconductors.

However, another state exists in which flux penetrates the materials in clusters of flux lines. In this case, screening currents circulate around each flux bundle like small vortices. The centre of each vortex is normal, whereas the region of zero to low magnetic field is superconducting. Type II superconductors are a new kind of magnetic phase ; they are characterised by a lower critical field H_{c_1} at which the magnetic flux begins to enter ; and an upper critical field H_{c_2} at which superconductivity disappears. High temperature ceramic superconductor such as YBCO are examples of Type II superconductors.

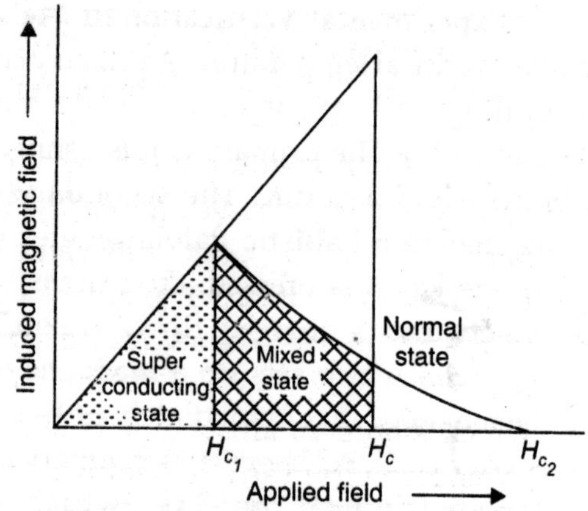

FIGURE 11.35 **Type II Superconductors**

11.9.9 The difference between Type I and Type II superconductors

Type I Superconductor	*Type II Superconductor*
1. Has only one critical field H_c.	Has two critical fields H_{c_1} and H_{c_2}.
2. Named as soft superconductors.	Named as hard superconductors.
3. Below H_c the material exhibits complete Meissner effect.	In the vertex region H_{c_1} and H_{c_2} Meissner effect is incomplete.

11.9.10 Explanation of Superconductivity (BCS Theory)

In 1957, Bardeen, Cooper and Schrieffer gave a theory to explain the phenomenon of superconductivity, which is known as *BCS* theory.

The *BCS* theory is based upon the formation of **Cooper pairs,** which is purely a quantum mechanical concept. During the flow of current in a superconductor, when an electron comes near a positive ion core of the lattice, it experiences an attractive force, because of the opposite charge polarity between electron and the ion core. The ion core will be displaced from the position due to this interaction, which is called *lattice distortion*. Now, an electron which comes near that place will also interact with the distorted lattice, which tends to reduce the energy of the electron. This process is looked upon as equivalent to interaction between the two electrons via the lattice.

The lattice vibrations are quantized in terms of what are called **phonons.** Thus the process is called *"electron-lattice electron interaction via the phonon field."* Because of the reduction of energy during the interaction, it is treated as equivalent to establish an attractive force between the two electrons, which is shown by Cooper to become maximum if the two electrons have equal and opposite spins and opposite momentum. The attractive force thus established may exceed the Coulomb repulsive force between the two electrons at temperatures below the critical temperature, leading to the formation of Cooper pairs.

Cooper pair is a bound pair of electrons formed by the interaction between the electrons with opposite spin or momentum in a phonon field.

As per quantum mechanical rules, a wave function could be associated with a Cooper pair by treating it as a single entity. Such a wave function has a property that it extends over a fairly large volume with finite value for its amplitude all over the region. As a result, the wave functions associated with similar Cooper pairs start overlapping. Typically for a given Cooper pair, such an overlapping may extend over 10^6 other pairs. Thus it extends virtually over the entire volume of the superconductor. This leads to a union of vast number of Cooper pairs in which motion of all such pairs are under strong correlation — one aiding the motion of the other. It results in an effect equivalent to the entire union moving as a single unit. The resistance encountered by any single Cooper pair is simply overcome by the cooperative action of other pairs in the union. Thus we can define superconductivity as follows:

When the electron flow in the form of Cooper pairs in materials, they do not encounter any scattering and the resistance factor vanishes or in other words, conductivity becomes infinity, which named as superconductivity.

In addition to the isotope effect, the idea that the electron-lattice interaction plays a crucial role in superconductivity is supported by the fact that, the best of the conductors such as gold, silver and copper do not exhibit superconductivity. The reason attributed is that the electrons in those metals, move so freely in the lattice that, the electron-lattice interaction is virtually absent. This rules out the possibility of formation of Cooper pairs and also that of occurrences of superconductivity in the material.

11.9.11 Applications of Superconductivity

The capability of retaining superconductivity at high fields, high current densities make superconductors useful in a variety of applications in electric power engineering, transportation, medical diagnostics and microelectrics.

Some futuristic applications of superconductors include transmission lines that carry power without resistance medical diagnostic equipment that eliminate the need for surgery, levitating trains and so on.

Some of the current uses of superconductors and some that hold the most promise for near future are:

(*i*) Magnet resonance imaging (*MRI*) machines enhance medical diagnostics by imaging internal organs often eliminating the need for invasive surgeries. *MRIs* which are currently made from low temperature superconductors, will be smaller and less expensive when made with *HTS*.

(*ii*) Maglev trains that seem to float on air because of superconducting magnets. These trains have been under development in Japan for two decade, the nearest prototype may exceed 547 km/h.

(*iii*) Energy storage in flywheel systems will ensure the quality and reliability of power transmitted to utility customers. In addition, energy storage provides utilities with cost saving by allowing them to store energy when the demand for the electricity is low and generating power is cheap. This energy is then dispended when demand is high and power production is more expensive.

(*iv*) Power transmission cables that carry current without energy losses, will increase the capacity of the transmission system saving money, space and energy.

(*v*) Current controllers help utilities deliver reliable power to their customers. *HTS* fault current limiters detect abnormally high current in utility grid. Then they reduce the fault current so that the system equipment can handle it.

(*vi*) Generators with superconducting wire in place of iron magnets will be smaller and lighter. New generators may also generate more power from less fuel.

(*vii*) Motors made with superconducting wire will be smaller and more efficient.

These are only a few of many possible uses of superconductors, research and development of *HTS* may still yield many more uses for materials that can carry electricity without resistances.

11.10 NUMERICAL PROBLEMS BASED ON SOLID STATE PHYSICS

Problem 1. *Consider silver in metallic state with one free electron per atom. Calculate the Fermi energy. Given that density of silver is 10.5 g/cm³ and atomic weight is 108.*

Hint: $$V = \frac{108\text{ g}}{10.5\text{ g}/\text{cm}^3},\ N = 6.02 \times 10^{23},\ \frac{N}{V} = 5.85 \times 10^{28}\ /\text{m}^3$$

$$E_F = \frac{h^2}{8\pi^2 m}\left[3\pi^2 \frac{N}{V}\right]^{2/3} = \frac{(6.62 \times 10^{-34})^2}{8 \times (3.14)^2 \times 9.1 \times 10^{-31}}\ [3 \times (3.14)^2 \times 5.85 \times 10^{28}]^{2/3}$$

$$= 8.799 \times 10^{-19}\text{ J}\quad \boxed{= 5.5\text{ eV.}}$$

Problem 2. *There are 2.5 × 10²⁸ free electrons per cubic meter of sodium. Calculate the Fermi energy and Fermi velocity.*

Hint: $$\frac{N}{V} = 2.5 \times 10^{28},\ E_F = \frac{h^2}{8\pi^2 m}\left[3\pi^2 \frac{N}{V}\right]^{2/3} = 5.0 \times 10^{-19}\text{ J} = 3.12\text{ eV}$$

$$E_F = \frac{1}{2}\text{ m } v_F^2 = V_F = \frac{\hbar}{m}\left(3\pi^2 \frac{N}{V}\right)^{1/3} = 1.05 \times 10^6\text{ m/s.}$$

Problem 3. *Determine the temperature at which there are one percent probability that a state with an energy 0.25 eV above the Fermi energy will be occupied by an electron?*

Hint: $$f(E) = \frac{1}{(e^{E-E_F/k_BT}+1)} \Rightarrow \frac{1}{100} = \frac{1}{\left(e^{\frac{0.25\times1.6\times10^{-19}}{1.38\times10^{-23}T}}+1\right)}$$

$$\Rightarrow \quad T = \boxed{631 \text{ K.}}$$

Problem 4. *The Fermi level in silver is 5.5 eV at 0K. Calculate the number of free electrons per unit volume in silver at the same temperature.*

Hint: $$E_{F_0} = \left(\frac{h^2}{8m}\right)\left(\frac{3}{\pi}\right)^{2/3} n^{2/3}$$

$$\Rightarrow \quad 5.5 \times 1.6 \times 10^{-19} = \frac{(6.63\times10^{-34})^2}{(8\times9.1\times10^{-31})}\left(\frac{3}{\pi}\right)^{2/3} \times n^{2/3}$$

$$n^{2/3} = \frac{8.811\times10^{-19}}{5.85\times10^{-38}} = 1.506 \times 10^{+19}$$

$$n = (1.506 \times 10^{19})^{3/2} = \boxed{5.84 \times 10^{28}/\text{m}^3.}$$

Problem 5. *An intrinsic semiconductor has an energy gap of 0.4 eV. Calculate the probability of occupation of the lowest level in conduction band at 0° C, 50° C and 100° C.*

Hint: $$E_g = 0.4 \text{ eV} = 0.4 \times 1.602 \times 10^{-19} \text{ J} = 6.408 \times 10^{-20} \text{ J}$$

$$E_F = \frac{1}{2} \cdot E_g \text{ (as intrinsic semiconductor)}$$

$$f(E) = \frac{1}{e^{(E-E_F)/k_BT}+1} \Rightarrow f(E) = \frac{1}{\exp\left(\frac{E_g}{2k_BT}\right)+1}$$

At 0°C = 273 K $$f(E) = \frac{1}{\exp\left(\frac{6.408\times10^{-20}}{2\times1.38\times10^{-23}\times273}\right)+1} = 2.025 \times 10^{-4}$$

At 50°C = 323 K $$f(E) = \frac{1}{\exp\left(\frac{6.408\times10^{-20}}{2\times1.38\times10^{-23}\times323}\right)+1} = 7.55 \times 10^{-4}$$

At 100°C = 373 K $$f(E) = \frac{1}{\exp\left(\frac{6.408\times10^{-20}}{2\times1.38\times10^{-23}\times373}\right)+1} = 19.76 \times 10^{-4}.$$

Problem 6. *Calculate the intrinsic concentration of charge carriers at 300 K. Given that $m_e^* = 0.12\ m_0$, $m_h^* = 0.28\ m_0$ and the band gap for Ge = 0.67 eV.*

Hint:
$$m_e^* = 0.12 \times 9.1 \times 10^{-31} \text{ kg},\ m_h^* = 0.28 \times 9.1 \times 10^{-31} \text{ kg}$$

$$n_i = 2\left(\frac{2\pi k_B T}{h^2}\right)^{3/2} (m_e^* m_h^*)^{3/4} \exp.\left(\frac{-E_g}{2k_B T}\right)$$

$$= 2\left(\frac{2\pi \times 1.38 \times 10^{-23} \times 300}{(6.623 \times 10^{-34})^2}\right) \left[0.12 \times 0.28 \times (9.1 \times 10^{-3})^2\right]^{3/4} \times$$

$$\exp\left[-\frac{0.67 \times 1.6 \times 10^{-19}}{2 \times 1.38 \times 10^{-23} \times 300}\right]$$

$$= \boxed{4.7 \times 10^{18}/\text{m}^3.}$$

Problem 7. *For pure Ge, the forbidden band gap energy $E_g = 0.7$ eV. Determine the position of Fermi level at room temperature if $m_h^* = 6\ m_e^*$. Also calculate the density of holes and electrons at room temperature. What would happen if the band gap is changed to 5 eV?*

Hint: $E_F = \dfrac{E_C + E_V}{2} + \dfrac{k_B T}{2} \log_e\left(\dfrac{m_h^*}{m_e^*}\right)^{3/2}$; $n_i = 2\left[\dfrac{2\pi k_B T}{h^2}\right]^{3/2} (m_e^* m_h^*)^{3/4} \exp.\left(\dfrac{-E_g}{2k_B T}\right)$.

Ans.
$E_F = 0.385$ eV above valence band
$n_i = 3.5 \times 10^{19}/\text{m}^3$ at 300 K
$n_i = 8.6 \times 10^{+34}/\text{m}^3$ if $E_g = 5$eV.

Problem 8. *Determine the wavelength of radiation emitted by a LED made up of GaAs. Given that the band gap energy is 1.43 eV.*

Hint:
$$\lambda_g = \frac{hc}{E_g} = \frac{6.626 \times 10^{-34} \times 3 \times 10^8}{1.43 \times 1.6 \times 10^{-19}} = 0.87\ \mu\text{m}.$$

Problem 9. *A superconducting tin has a critical temperature of 3.7 K at zero magnetic field and a critical magnetic field 0.0306 Testa at 0K. Find the critical magnetic field at 2K.*

Hint:
$$H_c = H_{c(0)}\left[1 - \left(\frac{T}{T_c}\right)^2\right] = 0.0306\left[1 - \left(\frac{2}{3.7}\right)^2\right] = 0.0306\ [1 - 0.2922]$$

$$= 0.0306 \times 0.7078 = 0.0217 \text{ T}.$$

APPENDICES

Problem 10. *The critical temperature T_c for H_g with isotopic mass 199.5 is 4.185 K. What will be its critical temperature when its isotopic mass is increased to 203.4.*

Hint: $T_c M^{1/2} = \text{constant} \Rightarrow T_{c_1} M_1^{1/2} = T_{c_2} M_2^{1/2} \Rightarrow T_{c_2} = 4.15 \text{ K}.$

Problem 11. *Calculate the critical current and critical current density for a sample of lead of diameter 1 mm for 4.2 K. The critical temperature for lead is 7.18 K and critical magnetic field at 0K is 6.5×10^4 A/m.*

Hint: (*i*) $H_c = H_{c(o)}\left[1-\left(\frac{T}{T_c}\right)^2\right] \Rightarrow H_c = 4.276 \text{ A/m}^{-1}.$

(*ii*) Critical current $I_c = 2\pi r\, H_c = 134.3$ A

(*iii*) Critical current density $J_c = \frac{I_c}{A} = \frac{I_c}{\pi r^2} = 1.71 \times 10^8 \text{ A/m}^2$

Problem 12. *The critical temperature T_c for H_g with isotope 199.5 a.m.u. is 4.185 K. Calculate the critical temperature when isotopic mass changes to 203.40 a.m.u.*

Hint: $T_c M^{1/2} = \text{constant} \Rightarrow T_{c_1} M_1^{1/2} = T_{c_2} M_2^{1/2}$

$$T_{c_2} = \left(\frac{M_1}{M_2}\right)^{1/2} T_{c_1} = \left(\frac{199.5}{203.4}\right)^{1/2} \times 4.185 \text{ K}$$

$$= 0.9904 \times 4.185 \text{ K} = 4.145 \text{ K}.$$

EXERCISES

1. Elucidate the difference between the classical free electron theory and quantum free electron theory.
2. State the main assumptions of Sommerfeld's free electron theory of metals.
3. Define Fermi energy and Fermi factor. Discuss about the variation of Fermi energy with temperature and energy.
4. Give an account of Fermi factor and define the same.
5. What is Fermi factor? Give the theory of calculation of Fermi energy at 0K.
6. What is physical basis for concept of Fermi energy? What is the assumptions made for Fermi energy at $T > 0$°K? Justify the same.
7. Discuss the dependence of Fermi factor on temperature.
8. Discuss the probability of occupation of various energy states by electron at $T = 0$K and $T > 0$, on the basis of Fermi factor.
9. Give the concept of discrete energy levels for the atomic state and account for the formation of energy bands in the solid state.
10. Explain band theory of solids.

11. Describe the concept of energy bands in solids.
12. What are the essential features of Band theory of solids?
13. Describe the periodic potential of solid for which Bloch's theorem is applied.
14. Explain the motion of free electrons in periodic lattice.
15. Explain the classification of solids on the basis of energy band theory.
16. Discuss the motion of electrons in a region of periodic potential. How can one understand electrical conductivity in a solid?
17. Discuss the essential features of electron energy band structures of solid on the basis of Kronig-Penney model.
18. Discuss Kronig-Penney model. Using the model show the energy spectrum of electron consisting of a number of allowed energy bands separated by forbidden bands.
19. Distinguish between conductor, semiconductor and insulator on the basis of band theory of solids.
20. Obtain an expression for carrier concentration of holes in intrinsic semiconductor.
21. Obtain an expression for carrier concentration in an *N*-type semiconductor.
22. Obtain an expression for carrier concentration of charges in *P*-type semiconductor.
23. What are traps? Discuss simple model to show the effect of traps on the photoconductivity. Calculate the gain factor.
24. What is photoconductivity? State the principle of photoconductive cell. Describe construction, working and use. Show the illumination characteristic and spectral response.
25. What is photovoltaics? Discuss the energy conversion process in a photovoltaic cell/solar cell.
26. What is superconductivity? What are the basic aspects of it?
27. Describe Meissner's effect. Distinguish between Type I and Type II superconductors. Write various applications of superconductors.
28. Explain in detail the *BCS* theory of superconductivity.
29. Describe and explain following:

    (*a*) Meissner's effect　　(*b*) Isotope effect

    (*c*) *BCS* theory.
30. Prove that Meissner effect contradicts the Maxwell's equation.

SUGGESTION FOR FURTHER READING

1. A. J. Dekker, Solid State Physics, Macmillan India Ltd., New Delhi, 2003.
2. Donald R. Askeland, P. P. Fulay and D. K. Bhattacharya, Materials Science and Engineering, Cengage Learning, New Delhi, 2010.
3. Devraj Singh, Fundamentals of Engineering Physics, Vol. 1 and Vol. 2, Dhanpat Rai & Co. (P) Ltd., New Delhi, 2009.

A. K. Saxena, Solid State Physics, Macmillan India Ltd., New Delhi, 2006.

5. M. A. Wahab, Solid State Physics, Narosa Publishing House, New Delhi, 2008.
6. H. Ibach and H. Liith, Solid State Physics, Springer (India) Pvt. Ltd., New Delhi, 2004.
7. G. D. Raj, Solid State Physics, Anmol Publications Pvt. Ltd., New Delhi, 2002.
8. Devraj Singh, Engineering Physics, Vol. 2, Dhanpat Rai & Co., New Delhi, 2010.
9. Charles Kittel, Introduction to Solid State Physics, Wiley India Pvt. Ltd., New Delhi, 2004.
10. S.L. Kakani and C. Hemrajani, Solid State Physics, Sultan Chand & Sons, New Delhi, 2005.
11. T. V. Ramakrishnan and C. N. R. Rao, Superconductivity Today, Universities Press (India) Ltd., Hyderabad, 1999.
12. P. M. Chaikin and T. C. Lubensky, Principle of Condensed Matter Physics, Cambridge University Press, New Delhi, 2007.
13. D. K. Roy, Physics of Semiconductor Devices, Universities Press, Hyderabad, 2004.
14. K. C. Jain, Electronics and Instrumentation, Jain Brothers, New Delhi, 2008.
15. I. P. Singh, S. Chander and R. K. Prasad, Materials Science and Engineering Jain Brothers, New Delhi, 2009.
16. Navneet Gupta, Electrical and Electronics Engineering Materials, Dhanpat Rai & Co., New Delhi, 2007.

APPENDICES

Appendix I

CONVERSION FACTORS

1 Å	10^{-10} m = 0.1 nm
1 atm	760 mm Hg = 10^5 Pascal
1 bar	10^{-1} Mpa
1 Calorie	4.18 J
1 Coulomb	1 A-s
1 Debye	3.33×10^{-30} cm
1 Dyne/cm	10^{-3} N/m
1 dyne/cm^2	0.1 N/m^2
1 erg	10^{-7} J
1 erg/cm	10^{-5} J/m
1 eV	1.602×10^{-9} J
1 gram-calorie	4.185 J
1 Hertz	1 cycle/s
1 HP	0.7457 kW
1 Newton	1 kg.m/s^2
T°C	$(T + 273.15)$ K

T°F	$5/9(T + 459.67)$ K
1 Watt	1 Joule/sec
1 Joule	10^7 ergs
1 K-cal	4.18 kJ
1 degree	1.745×10^{-2} rad (360 degree = 2π rad)
1 fermi	10^{-15} m
1 light year	9.46×10^{15} m
1 Inch	2.54 cm
1 eV	1.6022×10^{-19} J
1 kWh	3.6×10^6 J

Appendix II

Prefix Names, Symbols and Multiplication Factors

Constant	Symbol	Value
1 Deci	d	10^{-1}
1 exa	E	10^{18}
1 Giga	G	10^{9}
1 Kilo	k	10^{3}
1 Mega	M	10^{6}
1 micro	μ	10^{-6}
1 milli	m	10^{-3}
1 Nano m	nm	10^{-9} m
1 peta	p	10^{15}
1 tera	T	10^{12}

Appendix III

Accepted Values of Some Common Physical Constants

Physical Constant	Value
Velocity of light in vacuum	$c = 3 \times 10^{8}\ ms^{-1}$
Planck's constant	$h = 6.625 \times 10^{-34}\ Js$
Boltzmann constant	$k_B = 1.381 \times 10^{-23}\ JK^{-1}$
	$= 8.62 \times 10^{-5}\ eVK^{-1}$
Universal gas constant	$R = 8.317\ J\ mol^{-1}\ K^{-1}$
Avogadro's number	$N = 6.023 \times 10^{23}\ mol^{-1}$
Gravitational constant	$G = 6.67 \times 10^{-11}\ Nm^{2}\ kg^{-2}$
Atomic mass unit	$1\ amu = 1.66 \times 10^{-27}\ kg$
Acceleration due to gravity (at sea level 45° altitude)	$g = 9.806\ ms^{-2}$
Proton rest mass	$m_p = 1.672 \times 10^{-27}\ kg$
Electron charge	$e = 1.602 \times 10^{-19}\ C$
Electron rest mass	$m_e = 9.108 \times 10^{-31}\ kg$
Radius of electron	$r_e = 2.82 \times 10^{-15}\ m$

Permittivity of free space	$\varepsilon_0 = 8.854 \times 10^{-12}\ C^2/N\text{-}m^2$
Permeability of free space	$\mu_0 = 4\pi \times 10^{-7}$ (H/m) $= 1.26 \times 10^{-6}\ N/A^2$
Intrinsic impedance of free space	$\eta_0 = 120\ \pi\Omega$ or $377\ \Omega$
Electron rest energy	$m_e c^2 = 5.11$ MeV
Proton rest energy	$m_p c^2 = 938.28$ MeV
Neutron rest energy	$m_n c^2 = 939.28$ MeV
Proton-electron mass ratio	$\frac{m_p}{m_e} = 1836.15$
Rydberg constant	$R_d = 1.0974 \times 10^7/m$

PAPER

B.TECH (CSE, IT, ECE, MAE, CE, AE & B. TECH NANOTECH, INT. M. TECH NANOTECH.)

First Semester End Term Examination : November, 2009

Applied Physics–I (Fields and Waves)

Time : 3 Hrs. *Maximum Marks : 70*

Note: *Values of Physical constants are given at the end of paper.*

SECTION–A **(30 Marks)**

Attempt any **5** *questions.*
Each question carries **6** *marks.*

1. (*a*) What is effect of damping on the natural frequency of an oscillator? (3)

(*b*) Prove that when damping forces are present, the frequency of an oscillator is reduced by $12.5/Q^2$, where Q is quality factor. (3)

2. (*a*) If φ is a scalar. Prove that curl grad $\varphi = 0$. (3)

(*b*) Prove that div curl of a vector is equal to zero. (3)

3. (*a*) What is poynting vector? Explain. (2)

(*b*) Write down Maxwell's equations explaining each term. (4)

4. (*a*) Derive differential equation of wave motion. (3)

(*b*) Compare the characteristics of the progressive and stationary waves. (3)

5. (*a*) Describe a method to determine the wavelength and velocity of ultrasonic waves. (4)

(*b*) State the Gauss divergence theorem for electromagnetism. (2)

6. (*a*) Define specific rotation. (2)

(*b*) A sugar solution in a tube of length 20 cm produces optical rotation of 13°. The solution is than diluted to one third of its previous concentration. Find the optical rotation produced by 30 cm long tube containing the diluted solution. (4)

SECTION–B (20 Marks)

Attempt any **2** *questions.*

Each question carries **10** *marks.*

7. (*a*) Derive an expression for the intensity distribution due to Fraunhofer diffration at a single slit and show that the intensity of first order maxima is about 4.5% of that of the principle maxima. (6)

(*b*) In Newton's rings arrangement, the diameter of 10th dark ring changes from 1.5 cm to 1.27 cm when a liquid is introduced between the lens and glass plate. Calculate refractive index of liquid. (4)

8. (*a*) What do you understand by double refraction? Explain how would you use the phenomena to produce plane polarized light and circularly polarized light? (6)

(*b*) How many orders will be visible if the wavelength of the incident radiation is 5000 Å and the numbers of lines on the grating is 2620 in one inch. (4)

9. (*a*) Explain the concept of Maxwell's displacement current and show how it led to the modification of Ampere's law. (6)

(*b*) If the magnitude of magnetic field in a plane wave is 1 A/m, find the magnitude of electric field for plane wave in free space. (4)

SECTION–C (20 Marks)

(Compulsory)

10. (*a*) Show that at the frequency of velocity resonance the absorption in a driven harmonic oscillator is maximum. (5)

(*b*) For a uniformly charged solid sphere of charge density ρ, calculate the electric field intensity at a point (*a*) outside the surface, (*b*) on the surface of the sphere. (5)

(*c*) In biprism experiment, the distance between the slit and screen is 160 cm. The biprism is 40 cm away from the slit and its refractive index is 1.52. When a source of wavelength 5893 Å is used the fringes width is found to be 0.01 cm. Find the angle of prism. (5)

(*d*) Fringes are produced with monochromatic light of wavelength 5800 Å. A thin plate of glass of refractive index 1.5 is kept in the path of one of the interfering beams and the central bright band from centre. Calculate the thickness of glass plate. (5)

$$\mu_0 = 4\pi \times 10^{-7} \text{ Wb/A-m}$$

$$\varepsilon_0 = 8.85 \times 10^{-12} \text{ C/Nm}^2$$

INDEX

A

B

C

D

E

H

I

J

K

L

M

Q

R

W

X

Y

Z